Pattern Recognition

Ideas in Practice

Pattern Recognition
Ideas in Practice

Ideas in Practice

Edited by
Bruce G. Batchelor
University of Southampton, England

Plenum Press · New York and London

Library of Congress Cataloging in Publication Data

Main entry under title:

Pattern recognition.

  Includes bibliographies and index.
  1. Pattern perception. I. Batchelor, Bruce G.
[DNLM: 1. Biomedical engineering. 2. Pattern recognition. QT34 P315]
Q327.P37   001.5'34   77-12488
ISBN 0-306-31020-1

© 1978 Plenum Press, New York
A Division of Plenum Publishing Corporation
227 West 17th Street, New York, N.Y. 10011

Printed in the United States of America

Contributors

W. A. Ainsworth   Department of Communication, University of Keele, Keele, Staffordshire, England

I. Aleksander   Department of Electronic and Electrical Engineering, Brunel University, Uxbridge, Middlesex, England

L. A. Amos   M.R.C. Laboratory of Molecular Biology, Hills Road, Cambridge, England

B. G. Batchelor   Department of Electronics, University of Southampton, Southampton, England

Donald A. Bell   Computer Science Division, National Physical Laboratory, Teddington, Middlesex, England

C. D. Binnie   Instituut fur Epilepsiebestrijding, Heemestede, Netherlands. Formerly at Department of Clinical Neurophysiology, St. Bartholomew's Hospital, 38 Little Britain, London, England

C. M. Brown   University of Rochester, Rochester, New York 14627; Formerly at Department of Artificial Intelligence, University of Edinburgh, Forest Hill, Edinburgh, Scotland

M. J. B. Duff   Department of Physics and Astronomy, University College London, England

A. S. J. Farrow   Medical Research Council, Clinical and Population Cytogenetics Unit, Western General Hospital, Crewe Road, Edinburgh, Scotland

D. K. Green   Medical Research Council, Clinical and Population Cytogenetics Unit, Western General Hospital, Crewe Road, Edinburgh, Scotland

P. D. Green   Department of Computing, North Staffordshire Polytechnic, Blackheath Lane, Stafford, England

J. A. G. Halé   Mullard Research Laboratories, Cross Oak Lane, Redhill, Surrey, England

D. C. Mason   Medical Research Council, Clinical and Population Cytogenetics Unit, Western General Hospital, Crewe Road, Edinburgh, Scotland

E. A. Newman   Computer Science Division, National Physical Laboratory, Teddington, Middlesex, England

R. J. Popplestone   Department of Artificial Intelligence, University of Edinburgh, Forest Hill, Edinburgh, Scotland

J. R. Parks   Computer System and Electronics Division, Department of Industry, Dean Bradley House, Horseferry Road, Westminster, London, England

D. Rutovitz   Medical Research Council, Clinical and Population Cytogenetics Unit, Western General Hospital, Crewe Road, Edinburgh, Scotland

P. Saraga   Mullard Research Laboratories, Cross Oak Lane, Redhill, Surrey, England

G. F. Smith   Neuropsychology Unit, Department of Applied Psychology, University of Aston, Birmingham, England. Formerly at Department of Electronics, University of Southampton, Southampton, England

D. W. Thomas   Department of Electronics, University of Southampton, Southampton, England

J. R. Ullmann   Department of Applied Mathematics and Computing Science, University of Sheffield, Sheffield, England

Preface

Pattern recognition is a child of modern technology; electronics and computers in particular have inspired research and made it possible to develop the subject in a way which would have been impossible otherwise. It is a rapidly growing research field which began to flourish in the 1960s and which is beginning to produce commercial devices. Significant developments have been made, both in the theory and practical engineering of the subject, but there is evidence of a schism developing between these two approaches. Practical machines have usually been designed on an *ad hoc* basis, with little use being made of advanced theory. It is difficult to provide a rigorous mathematical treatment of many problems pertinent to a practical situation. This is due, in part at least, to a conceptual rift between theory and practice. The mathematics of *optimal* systems is well developed, whereas pragmatists are more concerned with vaguer ideas of *reasonable* and *sufficient*. In some situations, the quest for optimality can constrain research and retard practical progress. This can occur, for example, if too narrow a view is taken of "optimal": the accuracy of a system may be optimal whereas its speed, cost, or physical size may be grossly suboptimal.

The objective of this book is to present a glimpse of the pragmatic approach to pattern recognition; there already exist a number of excellent texts describing theoretical developments. Ultimately, pattern recognition will be judged by its contribution to human society and there can be little confidence in its future unless we continually satisfy our paymasters that the research is bearing fruit. This book is offered as an "interim report," showing the breadth and direction of current research. The full potential of pattern recognition has yet to be realized, but the present indications are that very substantial benefits are to be expected — as we hope to show.

It would be nice to begin by formally defining our subject, but so far no pattern recognition worker has been able to do this succinctly and to the satisfaction of the majority of his colleagues. Pattern recognition is inherently a science of ill-defined concepts. It is part of the broader field of information processing and is devoted primarily to (selectively) reducing information content. Information selection tasks occur, for example, in medical diagnosis, biological classification, television bandwidth compression, picture processing and enhancement, automatic inspection, optical character recognition, speech recognition, and a host of other applications.

The physical origin of the information is unimportant to the theoretician but not to the engineer concerned with converting the data from one medium to another. The physical properties of the source of the data may be well understood and this knowledge may provide important clues as to how the information may best be coded. For example, a motor-vehicle sound may be expected to contain components at a frequency equal to the crankshaft rotation rate. The question of what we measure is often dictated by economic considerations; there is then little or no opportunity to *optimize* the measurements. In medicine, for example, pathology laboratories have invested large amounts of money in auto-analysis equipment, which would not be discarded or modified significantly merely at the whim of a pattern recognition worker (who is often regarded with suspicion, as an interloper, by his medical colleagues). In some applications, the primary information rate is very high: 10^8 bits/s in satellite photography is perfectly feasible. A special-purpose machine is then essential and this may severely constrain the algorithms which can be employed. Algorithms must often give way to heuristics or unproved procedures whose mathematical analyses have not yet been completed. There may be many reasons for this, the most common being:

1. Certain applications-oriented studies generate the need for new techniques which must be developed quickly (by experimentation), without the luxury of detailed mathematical analysis.
2. Factors other than system performance may be critical. These may place severe demands upon the complexity of equipment and the operations which it can perform. The size, weight, speed, power consumption, or cost of a machine may be just as important as its accuracy.

We hope to convince the reader that pattern recognition is primarily an engineering discipline, drawing upon a broad spectrum of skills. Indeed, one of the major attractions for many of its practitioners is that it is a truly interdisciplinary subject, calling upon such diverse areas as electronics, optics, statistics, psychology/ergonomics, programming, and systems engineering. Pattern recognition research is often conducted under a different guise, or in laboratories specializing in some applications area, e.g., medicine, factory automation. This is evident here from the list of contributors. Many research workers "drift" into pattern recognition from other areas, seeing it as a source of new techniques for application to their particular problems. Many of my coauthors feel, as I do, that the pattern recognition community has done little to court these researchers and entice them further into our subject. This book is written with the express intention of doing this. However, we believe that we also have a point to make to our theoretically inclined colleagues specializing in pattern recognition; there are many procedures whose genesis is attributable to pragmatism, intuition, and applications-oriented research, and which require formalization by mathematical analysis. We hope to stimulate our colleagues into devoting their attentions to analyzing "practical" procedures.

In view of the interdisciplinary nature of our subject, we have deliberately tried to keep our discourse at a level which is comprehensible to a new graduate in science of engineering.

Numerous "unseen" people contribute to a book such as this. Many people assisted in the formulation of ideas or encouraged our endeavors with their interest. To my mentors, may I say "Thank you." I should like to express my gratitude to my coauthors who have tolerated my whims and fancies with remarkable patience. It is a pleasure to acknowledge the willing help of my secretary, Miss Julia Lawrence. Finally, I should like to thank my wife, Eleanor, for her unending encouragement and interest.

B. G. B.

Southampton

Contents

METHODS AND MACHINES

1 Setting the Scene *1*

   B. G. Batchelor

  1.1.   Sets 1
  1.2.   Motivation 2
  1.3.   Component Problems of Pattern Recognition 4
  1.4.   Relation between Pattern Recognition and Other Subjects 8
  1.5.   Lessons of This Book 13

2 A Review of Optical Pattern Recognition Techniques *17*

   J. R. Ullmann

  2.1.   Introduction 17
  2.2.   Nonholographic Optical Correlation 18
  2.3.   Holographic Cross-Correlation 19
  2.4.   Speech Recognition by Optical Correlation 21
  2.5.   Automatic Inspection 22
  2.6.   Normalized Cross-Correlation 23
  2.7.   Optoelectronic Transduction 25
  2.8.   Linear Discriminant Implementation 26
  2.9.   Numerical Discriminants 27
  2.10.   Transformations 28
  2.11.   Normalization and Segmentation 31
  2.12.   Feature Detection 32
  2.13.   Sequential Recognition Techniques 33
  2.14.   Interactive Pattern Recognition 34
  2.15.   Concluding Comment 34
       References 35

3   *Pattern Recognition with Networks of Memory Elements*   43
   I. Aleksander
   3.1.   Introduction...   43
   3.2.   The Random-Access Memory as a Learning Element   44
   3.3.   Combinational Networks   46
   3.4.   Sequential Learning Automata...........................   53
   3.5.   The Computation of Some Global Properties   53
   3.6.   Other Forms of Feedback...............................   58
            References..   64

4   *Classification and Data Analysis in Vector Spaces*   67
   B. G. Batchelor
   4.1.   Statement of the Problem..............................   67
   4.2.   Classification Methods   75
   4.3.   Additional Data Analysis Techniques   80
   4.4.   Implementation in Hardware............................   92
   4.5.   Current Research...................................... 101
   4.6.   Appendix: Classifier Design Methods................... 103
            References.. 114

5   *Decision Trees, Tables, and Lattices*   119
   Donald A. Bell
   5.1.   Introduction.. 119
   5.2.   Feature Design for Decision Trees...................... 121
   5.3.   The Design of Decision Trees.......................... 121
   5.4.   Decision Tables....................................... 123
   5.5.   Table Conversion Methods 125
   5.6.   Table-Splitting Methods 126
   5.7.   The Common Subtree Problem 131
   5.8.   Equivalence of Subtables 134
   5.9.   Table to Lattice Conversion........................... 136
   5.10. The Development of Decision Rules 137
   5.11. Choosing the Set of Rules............................. 139
            References.. 140

6   *Parallel Processing Techniques*   145
   M. J. B. Duff
   6.1.   Introduction.. 145
   6.2.   Brief Survey of Proposed Parallel Processors............ 150
   6.3.   The CLIP Processors 152
   6.4.   Programming a CLIP Array 159
   6.5.   Processing Larger Image Areas 166

6.6. Future Trends. 173
   References 174

7   Digital Image Processing   177
J. A. G. Halé and P. Saraga
7.1. Introduction. 177
7.2. Motivations for Image Processing 178
7.3. Image Processing Disciplines 179
7.4. Image Processing Equipment 194
7.5. Conclusions 199
   References 200

8   Cases in Scene Analysis   205
C. M. Brown and R. J. Popplestone
8.1. Introduction. 205
8.2. Scope of Scene Analysis 206
8.3. Line Finding. 208
8.4. Polyhedral Vision 212
8.5. Region Finding. 224
   References 227

APPLICATIONS

9   The Control of a Printed Circuit Board Drilling Machine by   231
   Visual Feedback
J. A. G. Halé and P. Saraga
9.1. Introduction. 231
9.2. The Structure of a Practical Visually Controlled Machine 233
9.3. The Application of Visual Control to Drilling Printed Circuit
   Boards. 234
9.4. Picture Processing Operators 246
9.5. Performance and Conclusions 249
   References 250

10   Industrial Sensory Devices   253
J. R. Parks
10.1. Introduction. 253
10.2. Potential Areas of Application 254
10.3. Sensory Modalities. 256
10.4. Applications. 282
10.5. Constraints. 284
   References 285

11   Image Analysis of Macromolecular Structures   289
       L. A. Amos
11.1.   Introduction. 289
11.2.   Two-Dimensional Image Filtering. 291
11.3.   Rotational Image Filtering 292
11.4.   Image Simulation of Three-Dimensional Structures 295
11.5.   Three-Dimensional Image Reconstruction 296
        References. 299

12   Computer-Assisted Measurement in the Cytogenetic
      Laboratory   303
       D. Rutovitz, D. K. Green, A. S. J. Farrow, and D. C. Mason
12.1.   Introduction. 303
12.2.   The Laboratory Workload. 307
12.3.   Automatic Karyotyping 309
12.4.   Classification 319
12.5.   Counting and Finding Cells 324
12.6.   Accurate Measurement 325
        References. 328

13   Vehicle Sounds and Recognition   333
       D. W. Thomas
13.1.   Introduction. 333
13.2.   Planning and Preprocessing 334
13.3.   Feature Extraction 337
13.4.   Moment Feature Space. 345
13.5.   Nonsinusoidal Transforms. 349
13.6.   Homomorphic Filtering 353
13.7.   Conclusions 360
        References. 361

14   Current Problems in Automatic Speech Recognition   365
       W. A. Ainsworth and P. D. Green
14.1.   Preliminaries. 366
14.2.   Isolated Word Recognizers 374
14.3.   Machine Perception of Continuous Speech. 381
14.4.   Conclusions 393
        References. 394

15   *Pattern Recognition in Electroencephalography*   *399*
C. D. Binnie, G. F. Smith, and B. G. Batchelor

15.1.   Introduction and General Description of the Human Electroencephalogram 399
15.2.   Clinical and Research Applications 403
15.3.   The Motive for Using Pattern Recognition in Electroencephalography 407
15.4.   Feature Extraction 407
15.5.   Stepwise Discriminant Analysis 410
15.6.   Pattern Recognition Applications 410
15.7.   Current Research 418
       References 424

16   *Scene Analysis: Some Basics*   *429*
E. A. Newman

16.1.   What Is Scene Analysis? — Robotics Problem 430
16.2.   Views of Pattern Recognition — Summary 430
16.3.   Why Scene Analysis Is Not Trivial 435
16.4.   Examples of Work in the Area of Scene Analysis 444
16.5.   Some Conclusions 459
       References 462

17   *Social Aspects of Pattern Recognition*   *463*
B. G. Batchelor

17.1.   Introduction 463
17.2.   The Case Against Pattern Recognition 464
17.3.   In Defense of Pattern Recognition 468
       References 471

Index   *473*

Setting the Scene

B. G. Batchelor

The comments here are my own and do not coincide in all respects with those of my colleagues. I have tried to present a point of view that is close to the consensus among the authorship, but I realize that it is impossible to represent another person's opinion with exactly the same emphasis and nuance.

1.1. SETS

Some sets of objects can be adequately represented by a few *archetypes* (variously called *paradigms*, *exemplars*, *representatives*, or *typical members*), while others can only be specified by listing every member. Some sets can be defined by a suitable description, but other sets cannot. These two sentences restate the same basic truth in different words. The significance of this will be seen throughout this book, since *pattern recognition* (PR) is concerned with sets.

Consider the sets of animals that children call "cows" (male, female, and eunuchoid). It is difficult to define the limits of such a group, although its members all have certain essential (or nearly essential) properties. For example, cows are, for a child, animals with four legs, a tail, possibly horns. They also eat grass and go "moo." While this is not a scientific definition of domestic cattle, it is sufficient for many purposes, including ours. Despite the inadequacies of this or any other definition, I can recognize cows with great ease and accuracy. By going on many car rides, my children have learned to recognize cows just as well as I can. The only effort needed of me was to show my children enough

B. G. Batchelor • Department of Electronics, University of Southampton, Southampton, England

cows, each time muttering the appropriate word, "cow." My children did the rest. Suppose we could do this with a machine: teach it to recognize not cows but faulty components for a car engine. Such a machine would only need to be shown enough samples of the appropriate part, and it would then learn to recognize "good" (i.e., not faulty) components without any further action from me. (Note that I did not say "build a machine to recognize good components.") This is one aspect of PR. Another important facet is *classification* (or *discrimination*). Even without specifying the limits of the four classes — cows, pigs, horses, sheep — we can, with no special training, classify (almost all) the four-legged animals (in the fields of England). This is the basis of that subdivision of PR known as *pattern classification*. Another topic is called *clustering*. To illustrate this, let us consider subsets of cows. There are many distinct *breeds*, for example, Jersey, Hereford, Ayrshire, which can be identified by the observant person, even without *training*. Of course, a person would not be able to place the correct "label" on a subset (breed) of cows without training, but it is certainly feasible for the untrained observer to identify the characteristics of a group. Could a machine learn to do so? Each type or breed of cow has a distinct archetype. The set "cows" has many archetypes, whereas "Ayrshires" has only one (per sex). Note that the word "learn" has, so far, been used twice. Strictly speaking it should also have been used to emphasize the learning aspects of classification. *Learning is central to all aspects of pattern recognition.*

Let us turn our attention now to a different kind of set, and in particular the set of PR problems. This set cannot be specified as easily as "cows." At the beginning of this chapter two types of sets were mentioned. "Cows" very definitely lies in the first type and can be represented adequately by few archetypes. I am tempted to place "PR problems" in the second type. However, to do so would be to admit defeat; as the editor of a book on PR, I should be able to indicate the common areas of PR problems. This is extremely difficult. Neither I nor any of my colleagues have found a succinct definition of PR which does justice to our personal understanding. PR means different things to each of us. That is why this book contains such diverse contributions, but I hope that the reader will find far fewer than 15 archetypes. (My own feeling is that there are three major forms of *pattern description*, each supporting a group of decision-making techniques. See Section 1.3.3.)

1.2. MOTIVATION

Pattern recognition problems are too diffuse in nature to allow us to define the problem in formal terms. There is, however a deep underlying unity of purpose in PR research: to build machines that will perform those tasks that we have previously regarded as essentially "human." A few examples are given

Table 1.1. Illustrating the Wide Diversity of Pattern Recognition Applications and Input Media[a]

Problem	Application	Medium of input
Reading alphanumeric, typed/handwritten characters (optical character recognition)	Reading invoices, checks, etc.	Visual
Recognizing spoken words (speech recognition)	Speech input to computers Speech control of machines Phonetic typewriter	Acoustic
Distinguishing patients with biochemical imbalance from those who are "normal"	Medical diagnosis	Chemical
Inspecting machined components	Automatic assembly/ inspection	Tactile/visual
Detecting possible people "at risk" for certain diseases	Sociology/psychology/ epidemiology	Questionnaires
Analyzing neurophysiological signals	Electroencephalography, electrocardiology, electromyography, etc.	Electrical

[a]Many more examples are described later in this book.

in Table 1.1, and many more will be encountered in later chapters. Many lay people, when hearing the term PR, automatically turn their attention to visual images. This is far too restrictive, as Table 1.1 indicates. The basic event being observed, the pattern may occur in a visual, acoustic, electrical, chemical, tactile, or more abstract "world." PR has often been regarded as a specialized branch of statistics, which we all accept as a legitimate field of study without specifying where the input data arise. Once the transducing problem has been overcome, there is often a strong similarity in the nature of PR problems arising from any of the "worlds" exemplified in Table 1.1. This is not universally true, and there are two major subdivisions of PR techniques as a result. However, the distinction between these is ill-defined and some problems have characteristics of both, while others do not fit conveniently into either. In fact, when we attempt to apply PR *techniques* recursively to PR *problems*, we fail to gain any clearer insight into the basic nature of the subject: *meta-pattern recognition* is still embryonic. We must resort then to human values and human judgment.

PR is motivated partly by our desire to free human beings from tedious, boring, repetitive, and dangerous work. Why should a person become a "vegetable" through working in a factory, or risk contracting cancer through inspecting or repairing the inside of an operational nuclear reactor? Copytyping is a boring job, especially when the copytypist does not fully understand the text he/she is duplicating. In our attempts to control certain diseases, we use mass-screening techniques, which place boring, mundane "inspection" tasks upon technicians or doctors. There are strong economic motivations for PR as well. It is too expensive to inspect every component of a (mass-produced) motor car manually, but the result is a product which is annoying to the customer and costly to repair under warranty. Copytyping is expensive as well as tedious. The same is true of computer-card punching. Finally, there are certain tasks that people cannot do well under any circumstances. If a person is presented with a table of numbers listing, say, the results of a 10-measurement biochemical assay on each of 10,000 patients, little more than bewilderment will occur. There is little chance that he/she would be able to find a basis for discriminating between "normal" people and those with an "abnormal" profile of measurements. We can list the three major incentives for PR research as:

1. *Social* (i.e., eliminating boring, repetitive, or dangerous work)
2. *Economic*
3. *Improving performance*

1.3. COMPONENT PROBLEMS OF PATTERN RECOGNITION

The factors noted above as motivating PR are at the heart of the subject. They provide the most effective "binding" for our discussion. This must be realized before we begin our discourse in earnest, because the variety of techniques that we shall meet may produce a feeling that the subject is a hodgepodge. However, there are certain identifiable areas of PR which we shall discuss now.

1.3.1. Transducing

First of all, the pattern must be translated into a form suitable for manipulation in a computer. Electronic technology, especially digital computers, provides the processing flexibility and speed that we shall need. At the moment, the only serious contender to electronics is optics, but this is still in a rudimentary state of development, compared to the extraordinary variety of analogue and digital integrated circuits that are now available. Once in electronic form, the data may still be in an inconvenient format. For example, unary- to binary-coded-decimal conversion may be required before the data can be used by our computer. This type of format conversion is a mere technicality and will therefore be dismissed here.

1.3.2. Preprocessing

It is axiomatic that certain features of a pattern are irrelevant. For example, the symbol

$$A$$

would be regarded as a letter, wherever it occurred on an otherwise blank page; position is irrelevant to the correct identification of isolated letters. (This should not be taken to imply that context is unimportant when poor-quality letters are placed into groups.) Further examples of irrelevant variables occur in

1. Speech: loudness
2. Visual scenes: mean brightness is redundant, contrast is not
3. Radio transmissions: carrier frequency

Size and orientation are often regarded as irrelevent in optical character recognition (OCR), although there are dangers. For example, ∃ and ∀ are symbols from mathematical logic and not the letters E and A inverted.

Removal of irrelevent characteristics is termed *preprocessing* or *normalization*. The important point to note is that the redundant feature is obviously so, but we must avoid being too rash in our enthusiasm to discard information.

1.3.3. Pattern Description

We must distinguish between a *pattern* (e.g., a visual image) and its *representation* within a computer. In describing a medical patient, we almost invariably oversimplify his situation. *All pattern recognition problems require information destruction.* Pattern description destroys information — always. We can do little more than hope that most of the useful information is retained and the irrelevent information is discarded. Choosing an appropriate description is still an art, not yet formalized into a science. Much of this book is concerned with this topic. As is only to be expected when rigor fails, intuition comes to the fore and much of the success of PR to date is due to the hunches and cunning of its practitioners. It is a regrettable, but undeniable, fact that there is little theory to help us here. We commented earlier that meta-PR does not exist in any real sense. Pattern representation techniques can, at the moment, only be judged by their performance within a complete PR system. We cannot analyze our problem in advance and predict which method of pattern description will work best. This would result from a meta-PR. This lack of a PR metatheory means that practical progress is slow, because the only method of properly assessing the performance of a pattern representation system is by designing the complete system and then evaluating it experimentally. Even then it is difficult to distinguish the contributions of the pattern description and later units in the PR system. Few workers seem to have realized the full implications of these ideas. One notable exception is evident here in the work of Duff (Chapter 6). Other contributions touch on this: Rutovitz *et al.* (Chapter 12), Halé and Saraga (Chapters 7 and 9) and Batchelor (Chapter 4).

Many different forms of pattern description have been proposed. A few are outlined below.

1.3.3.1. Vectors

A particularly popular form of pattern description has been the *vector*, representing the results of a number of measurements taken in *parallel*. Vectors are a natural form of pattern description when a number of separate measurement units operate in parallel. In one study on breast cancer, I was given the following data on each of 200 women: Age, height, weight, volume of urine excreted/day, number of pregnancies, number of months the patient had spent breast feeding (all children), hormone concentrations in the blood.

Most of these *descriptors*, as they are sometimes called, are analogue quantities but others are integers (e.g., number of pregnancies). Sometimes, descriptors have only two permissible values, and some workers insist that all descriptors are binary; Bell (Chapter 5) and Aleksander (Chapter 3) consider only two-state variables in any detail. My attention has recently turned to another variation: vectors with some *unknown values*. Such vectors have omissions in known elements. For example, the measurement vector:

$$(age, height, weight)$$

might appear for one individual as

$$(33, 180, ?)$$

and for another as

$$(?, 190, 80)$$

Another type of description, related to this, may occur in data collected by questionnaire. Descriptors are sometimes expressed by an inequality. For example, "age" is a particularly common question resulting in an answer of the form >21. It is not difficult to see that in electronic equipment, where descriptors are expressed in parallel digital form, the failure of a gate may result in a measurement appearing like this:

$$110?1 \equiv 25_{10} \text{ or } 27_{10}$$

$$101??? \equiv 40_{10} \text{ to } 47_{10}$$

Yet another variation allows *nonnumeric* descriptors, e.g.,

$$(age, height, color of hair, favorite daily newspaper)$$

1.3.3.2. Feature Lists

To illustrate the idea, let me describe my wife in a feature list:

Age:	>21
Height:	170 cm
Weight:	60 kg
Color of eyes:	brown

Descriptors may be precise or vague numerics or nonnumerics. The list could be shuffled without loss of information, since each descriptor carries a tag. The number of descriptors is variable, unlike a vector where it is fixed and the order predefined. Another man may describe his wife's attributes in an entirely different way, using features that I think are unimportant. For example,

Hobby:	painting
Career:	teacher
Size of shoes:	6

This emphasizes one of the essential points about the feature-list representation: the qualities specified in it vary (and can be in any order). A fully numeric feature list could be forced into a fixed format, like that of a long vector which has nearly all of its descriptors with unknown values. However, this is rather unnatural and does not accommodate nonnumeric attributes. Bell (Chapter 5) discusses decision-making techniques which employ feature lists, but his decision-maker effectively orders the evaluation of the attributes.

1.3.3.3. Graphs

A road system, electrical network, or PERT diagram cannot easily be represented by either a vector or feature list. We could, of course, do so using an adjacency matrix or an array of nodes and pointers. However, neither of these is particularly convenient, and it is better to use a more "natural" form of pattern description. Graphs arise "naturally" in pictorial pattern recognition, as we shall outline. Consider the letter A. This can be represented in the form shown in Fig. 1.1.

This form of description is robust: sloppily written letters often have the same code. In addition, rules can be formulated easily for converting a description from one form to another. For example,

$$\text{node } 1 \xleftarrow{\text{(right)}} \text{node } 2$$

is equivalent to

$$\text{node } 1 \xrightarrow{\text{(left)}} \text{node } 2$$

Techniques for matching graphs have been developed. This problem is not quite

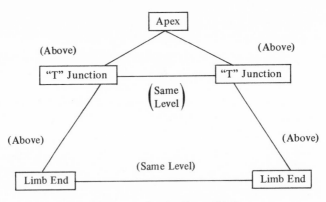

Fig. 1.1. Graph of letter "A."

identical to that of *graph isomorphism*, because two graphs may not fit exactly. One graph may be a subgraph (or nearly so) of the other.

Graphs provide the basic form of pattern description for what is called *syntactic pattern recognition*, vectors for *statistical* PR, and feature lists for *sequential* PR (since decisions are calculated using a tree like strategy).

1.3.4. Making Decisions

Once the pattern description has been fixed, there still remains the problem of how we use it in order to make a decision. The *decision-maker* (or *classifier*) can accept the description given to it without protest, or request new information, before making some statement about the pattern. Aleksander (Chapter 3) and Batchelor (Chapter 4) both base their work on the former policy, while Bell (Chapter 5), Ainsworth and Green (Chapter 14), Newman (Chapter 16), and Brown and Popplestone (Chapter 8) all consider the situation where there is a "dialogue" between the description module and decision-maker. Sometimes, there is no clear distinction between the two units (e.g., see Chapters 10 and 12).

1.4. RELATION BETWEEN PATTERN RECOGNITION AND OTHER SUBJECTS

Cybernetics, Psychology, and Pattern Recognition. PR is used by psychologists in relation to animal and human perception (primarily acoustical and visual). For a time, during the 1960's, there was a lively dialogue between workers in the biological sciences and PR, but this has waned recently. PR workers no longer feel, as they once did, that successful PR machines will emulate biological organisms. Cybernetics still looks to PR for inspiration regarding possible "brain models."

Artificial Intelligence (AI). This subject and PR are intrinsically interwoven. AI has tended to concentrate on theoretical developments even more strongly than has PR. I find it difficult to believe that an "intelligent" robot could fail to possess a PR faculty; I have several times been impressed by seemingly intelligent behavior by PR programs, which have the unpleasant tendency to give the impression that they are working perfectly satisfactorily, when in fact they are not.

Logic Design. The logic design problem may be viewed as partitioning a multidimensional space. This is the same problem as that tackled by Aleksander (Chapter 3) and Batchelor (Chapter 4), although the latter allows continuous variates as inputs. Logic design uses binary input variables and, despite the apparent simplicity of the problem, heuristic techniques still offer the only practical solution when there are a large number of input variables.

Numerical Taxonomy/Cluster Analysis. This area has largely been developed by biologists and psychologists and is very closely allied to PR. I have used some of these techniques but found them to be too slow for running with the large data sets that PR tends to generate.

Finite State Machines. A finite state machine is said to "recognize patterns." However, the classes of input sequence that such a device can detect are rather limited and of little practical value for PR.

Compilers. A compiler is said to "recognize" a syntactically correct segment of the input data stream. As a programmer, I feel that a good compiler should tolerate small (e.g., spelling) errors, although this has not been accepted by the industry. A compiler should be able to recover after detecting a minor (syntax) error, so that the remainder of the program being compiled can be checked for syntax errors. This is tantamount to PR, although the present development of error toleration and recovery owes little to our subject.

Graph Theory. Graph isomorphism is a problem that might be regarded as falling under the aegis of PR. The task of proving that a graph, A, is a subgraph of another graph, B (A $\subseteq$ B), is also a PR problem. PR is also concerned with the much more difficult problem of proving "approximate subgraphs." Furthermore, the arcs and nodes of the graphs used in PR are frequently *labeled*; the (approximate) equality of such graphs is very much more difficult to prove.

Signal Detection and Processing. The communication engineer is concerned with deciding whether a (binary) signal is present, against a high-noise background. Signal detection is more readily amenable to exact mathematical analysis than is PR, since the former is a one-dimensional problem. (That is, deciding where best to place a threshold on a continuum.)

Signal processing and PR have a common interest in certain techniques, especially Fourier, Walsh, and Haar transforms. In PR these are frequently used to describe:

1. One-dimensional patterns (e.g., acoustic waveforms; see Chapters 13, 14, and 15),
2. Two-dimensional patterns (see Chapters 2, 7, and 11),
3. Three-dimensional patterns (see Chapter 11).

Control Engineering/Feedback Systems. Feedback/control systems and PR have much in common. The state-space concept of control engineering is equivalent to that of a measurement space in PR. Control systems are frequently adaptive, like PR learning procedures; both employ feedback to improve performance. An on–off servo operating within a state-space is controlled by a strategy which effectively partitions the space. [A pattern classifier also partitions its measurement space. (See Chapter 4.)] More complex control may be achieved by encouraging the system operating point to follow a certain trajectory in state-space. PR has no obvious parallel for such "dynamic" problems.

Data Processing. This provides one of the strongest "customer areas" for PR. Commercial computing would greatly benefit from some method of bypassing the punch operator. Checks, invoices, etc. might be read by a machine, thereby greatly reducing the cost of providing data for the computer accounting programs.

Management Information Systems. Information retrieval may be regarded as an area of PR research, but it does seem to be developing along distinctive lines, with little direct contact with the remainder of the subject. The most difficult problem in forcing information retrieval into a PR mold seems to lie in deciding how best to describe documents, etc.

Management decisions are often made using techniques reminiscent of those found in PR. Perhaps Bell (Chapter 5) presents the closest point of contact between our subject and management-system design. Dynamic programming has often been used by PR workers, and I commented earlier on the fact that a PERT network could be regarded as a graph describing a pattern and could therefore be associated with syntactic PR.

Information Theory. Regarding this as an area of probability theory, it seems appropriate to defer this discussion to Section 1.4.15 on statistics.

Interactive Computing and Graphics. This is another "customer area." Imagine being able to draw a pencil sketch of a circuit diagram, or of a car, and requesting the computer to draw a good copy! Another area where PR may contribute is in the provision of speech recognition as an alternative to a keyboard input. (See Chapter 14 by Ainsworth and Green.)

Computer-Architecture Studies. The ability to process data at high speed is an essential prerequisite for our subject's development. Even the very fastest general-purpose computers are far too slow for some PR procedures that we should like to use. PR is one of the areas forcing new developments. (See

Chapters 2, 3, 4, and 6.) I also believe that PR should be developing new languages as well (see Chapter 4). In particular, attention should be paid to the practical problems of applying a large number of PR procedures (especially data-description methods) in quick succession, preferably in an interactive mode (see Chapter 6).

Statistics and Probability Theory. In some ways statistics is the savior of PR and, in others, the *bête noire.* My personal view is that statistics is overused to the point of distracting us from more important questions. Statistics seems to offer much more than it achieves. In many theoretical papers one reads a sentence like this:

The variance of the chosen measure of error tends to zero as the number of training samples is increased to infinity.

What happens in practice when the number of training samples is finite and small? Another criticism that one can level against mathematical statistics is that it is often unable to handle anything other than the simplest algorithms. To illustrate the point, let us consider a familiar recurrence relation for finding $\sqrt{A}$, given an initial guess x_0. (PR learning procedures often use recurrence relations.) A new and better approximation x_1 can be found using the equation

$$x_1 = \tfrac{1}{2}[A/(x_0 + x_1)]$$

To find an even better approximation, we can reapply this formula substituting x_1 for x_0. It is known that this procedure converges for all $A > 0$, and all x_0 ($-\infty < x_0 < +\infty$). What happens if during the division we round A/x_0 to the nearest integer multiple of some "quantum," δ? Will the procedure still converge for all x_0? (This particular modification to the original procedure is important, because it models fixed-point division in a digital processor. If this complication is insufficient to inhibit the analyst, then I am sure that I can find some other one that does.) Much the same kind of problem occurs in PR, where many of the procedures are recurrent in operation.

Another common form of PR learning procedure increases the quantity of storage needed without limit. What happens in practice when the store size is limited? Mathematical statistics has little to say about this type of question.

Yet another problem confronts the analytical approach. Suppose that a procedure has been proposed which demands the evaluation of a function $f(x)$. How much does it cost to implement $f(x)$? Would an approximation

$$f'(x) \text{ where } (|f'(x) - f(x)| \leqslant f(x))$$

be "good enough"? [Here, $f'(x)$ is a more easily implemented alternative to $f(x)$.] There is no mathematically computable cost function $G(f)$ which allows us to "balance" the performance improvement of f' over f, with the increased "cost" of implementation. ("Cost" here is a vector quantity measuring money-cost, speed, reliability, weight of equipment, power consumption, etc.) At the moment, the only really effective way of implementing $f(x)$ is to consult an expert systems-design engineer. Perhaps we could list a number of "easy" functions or write software for designing PR hardware, but these would be aids for the mathematician rather than ways of obtaining the function $G(f)$.

My real criticism of the zealous use of mathematics is that it distracts us too easily. We have no direct interest in whether a procedure converges eventually. The really important question is whether it will achieve an *acceptable* level of performance in a *reasonable* time. These words in italics are indefinable, and that is why I believe PR will progress best by pioneering developments using *experimentation*, guided by intuition and experience. Later, if we are fortunate, mathematical analysis will catch up and lay a rigorous framework to guide further work.

The analysts reading this have probably felt that I have presented one side of an argument without doing justice to the other. Mathematical analysis, if it does succeed in making positive statements about asymptotic convergence, can be an encouragement for further work. If the analysis results in a proof that "convergence" does not occur, even in infinite time, there is some discouragement. (It may be that we have not defined "convergence" properly; a technique may be practically useful if it reaches a state where the parameters have a small, but finite, random oscillation about the ideal values.) How can we be sure, if we rely on experimentation, that our solution is (nearly) optimal? To counter this, I should point out that even though the results we obtain may not be optimal, they may be "good enough now." The difference, I believe, between the practical and mathematical approaches is that mathematical statistics offers "cake tomorrow," but for "bread today" we need to resort to experimentation and measurement, guided by anything that can help us.

Statistics does provide a very convenient framework for thinking about problems. For example, the classification of a pattern described by a vector, x, may be performed optimally as follows:

Attribute x to class i iff $P(\text{class } i/x) > P(\text{class } j/x)$, $i \neq j$, where $P(\text{class } i/x)$ is the *conditional probability* that the pattern (represented by x) is in class i, given the measurement vector, x. This conditional probability is equal to

$$P(\text{class } i/x) = \frac{P(x, \text{class } i)}{P(x)} = \frac{P(x/\text{class } i)\, P(\text{class } i)}{P(x)}$$

(Bayes' theorem) where $P(x, \text{class } i)$ is the *joint probability* and $P(x/\text{class } i)$ is the *conditional probability* of x, given that the pattern is in class i. Hence we may attribute the pattern to class i iff the ith term

$$P(x/\text{class } i) \cdot P(\text{class } i)$$

is the largest. While $P(\text{class } i)$ can be estimated easily, $P(x/\text{class } i)$ cannot, because, in general, x is a multidimensional vector. This particular form of analysis is developed in great depth in other texts, but as I have indicated, our approach is concerned with presenting the practical aspects of our subject.

1.5. LESSONS OF THIS BOOK

1. PR has as its broad objective the design and construction of computer-like machines that can perform complex perceptive functions such as reading, visual inspection, and recognizing speech. Numerous specific applications are mentioned and many of these are described in detail.

2. Existing PR methods are still relatively weak, compared to human abilities, and machines are frequently incapable of coping with the most general (difficult) problem. Nevertheless, satisfactory performance has often been obtained using non-PR "tricks," for example, colored or patterned light to illuminate a scene or object (being observed visually by machine). On other occasions, a PR machine has specifically been designed with the feature that it will notify a human operator when it is incapable of making a reliable decision. Limited objectives are, of course, essential to good engineering.

3. Many authors in this book clearly regard PR as an engineering discipline. There are already several excellent books covering the theoretical aspects of our subject. It has been our specific intention to avoid overlapping our discussions with theirs and we have therefore tended to concentrate on *practical* techniques and applications.

4. PR often demands instrumentation techniques that are close to the limits of available technology.

5. It is apparent that PR places very severe demands upon computers and, in many cases, special-purpose machines have been proposed.

6. PR techniques vary from naive to complex. The simpler techniques are frequently "better," since they are more easily understood by the systems designer. The practitioner of this subject must be prepared to do "a bit of everything." Even if the intending researcher has a specific application in mind, other fields may possess important lessons for him. For this reason, PR research is enormously fascinating.

7. Although PR techniques may be built from trivially simple units, the complete procedure may be very difficult to understand and often displays "intelligent" behavior.

8. PR research is slow and expensive. (Personally, I believe that this is in part our fault for not developing good PR software and computing architecture.)

9. PR learning procedures are difficult to program and debug effectively. For 18 months, I ran a program with a significant fault in it, despite careful and thorough testing! PR programs are intelligent and have an unpleasant habit of *seeming* to work, when in fact they are only doing so suboptimally.

10. The bounds of PR are fuzzy and there is no clear borderline between this and numerous other areas of information processing.

11. PR makes widespread use of *self-adaptive* or *learning* procedures for designing its component modules.

12. PR is primarily concerned with large data sets or making decisions at high speed.

2

Editorial Introduction

J. R. Ullmann discusses a host of ideas which are considered elsewhere in this book, for example:

1. Correlation methods (Chapter 11)
2. Template matching (Chapter 10)
3. Moments (Chapter 13)
4. Fourier and Walsh transforms (Chapters 7, 11, 13, 14, 15)
5. Linear discriminant methods (Chapters 3, 4)
6. "Distance" (Chapter 4)
7. Nearest-neighbor classification methods (Chapter 4)
8. Interactive pattern recognition (Chapters 6, 12)
9. Face recognition (Chapter 16)
10. Karyotyping (Chapter 12)
11. Automatic inspection (Chapter 10)
12. Speech recognition (Chapter 14)

However, his main contribution lies in the consideration he gives to implementation. Aleksander (Chapter 3), Batchelor (Chapter 4), and Duff (Chapter 6) consider advanced electronic techniques of implementing pattern recognition procedures. The advances in optoelectronic transducers, especially light-emitting diodes and large photosensitive diode arrays, make it likely that more consideration will be given to optical or hybrid optical—electronic realizations of pattern recognition procedures in the future. The greatest problem with optical methods lies in the frequent need for transparent masks. However, ingenuity is circumventing this difficulty in many cases, and opaque reference patterns can now be used. Ullmann presents many fascinating instances like this and his contribution provides a valuable review and introduction for the remainder of the book.

2

*A Review of Optical Pattern Recognition Techniques**

J. R. Ullmann

Automatic pattern recognition is the subject of a large literature, much of it theoretical, dealing primarily with optical pattern recognition techniques, and concentrating on the principles of recognition rather than the detailed physics and technology of optoelectronic transduction. Mathematical, sequential, and interactive techniques are mentioned only briefly because they have already been well reviewed by various authors.

2.1. INTRODUCTION

Automatic recognition of patterns may be worthwhile in practice if the number of patterns to be recognized at the required rate of recognition is too great for economically available human effort.

   Recognition is the assignment of a pattern, or at least part of a pattern, to a class. Such a class is called a *recognition class*. For instance, in character recognition, the recognition class "5" generally contains a very large number of somewhat different patterns that we would call 5's. In the screening of cervical smears, "malignant" is the name of a recognition class of cells that may differ from each other to a limited extent in shape, color, and texture. In bubble

---

*This chapter was originally published in *Opto-Electronics*, Vol. 6, 1974, pp. 319–332. Permission to republish it was kindly supplied by Chapman & Hall Ltd.

J. R. Ullmann · Department of Applied Mathematics and Computing Science, University of Sheffield, Sheffield S10 2TN, England

chamber work in nuclear physics, events must be correctly classified despite differences in the detailed accidents of bubble formation and the presence of irrelevant crossing tracks.

There may be a mathematical definition of a recognition class of circles, but practical recognition classes of characters, blood cells, fingerprints, fine particles, bone fractures in radiological photographs, and so on, have no known analytical definition. Furthermore, it is generally impractical to define a recognition class by storing all the constituent patterns, because there are far too many. What we actually do is to hypothesize that a particular definition, or definitory system, will yield tolerably low recognition error rates in practice. If the error rates turn out experimentally to be too high, we try a somewhat different hypothesis, and so on.

A popular hypothesis has been that for each recognition class we can choose one specimen pattern that serves as a reference pattern: an unknown pattern can be cross-correlated with all of the reference patterns and assigned to the class of the reference pattern that yields the highest correlation score. This idea has been popular because it is practicable and, with proper normalization, it is plausible in at least one application, namely, in the recognition of good quality print in a single font.

2.2 NONHOLOGRAPHIC OPTICAL CORRELATION

.   Stevens (1961) has provided a historic bibliography of template matching techniques. A more recent technique of this sort has been described by Howard (1966) who focused positive images of a printed character onto an array of negative transparencies or templates. The total light transmitted through each template was measured by a separate photomultiplier. Horizontal movement of the document swept the images horizontally across the masks, and a rotating mirror swept the images at a much higher speed vertically across the masks, so that at some instant the images and masks would be optimally aligned. At that instant the printed character was assigned to the class of the template that transmitted least light.

Lange (1967) has described Meyer-Eppler's technique for obtaining a visual display of the cross-correlation function of two patterns. One version of this technique uses an extended noncoherent source in the source plane in Fig. 2.1. T_1 and T_2 are transparencies of the two patterns $f_1(x, y)$ and $f_2(x, y)$ and the cross-correlation function

$$\int\int f_1(x,y) f_2(x + \alpha, y + \beta)\, dx\, dy$$

is displayed in the output plane, although the region near the optic axis may appear unduly bright (Rosenfeld, 1972). Alternatively we can dispense with lens

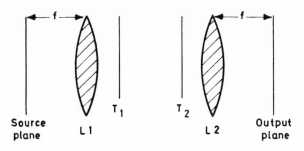

Fig. 2.1. A setup for nonholographic optical correlation.

L1 and replace the transparency T_1 by a strongly illuminated opaque pattern from which light is reflected onto T_2. Crane and Steele (1968) described experiments in which light from a document bearing a line of characters was reflected onto T_2, and the output plane contained a column of photomultipliers in the y direction. The document moved in the x direction parallel to the line of characters and T_2 was a reference character transparency. At some instant the point of maximum cross-correlation passed close to one of the photomultipliers. There was in effect a similar setup for each reference character and the input character was assigned to the class of the reference character that had the greatest single photomultiplier output.

For Lahart (1970), T_1 was a stationary transparency of a target that was being sought in a stationary transparency photograph T_2. The extended source in Fig. 2.1 was replaced by the surface of a flying spot CRT. Movement of the spot brought the image of T_1 into different alignments with T_2, and a photomultiplier measured the total light transmitted through lens L2, thus measuring the cross-correlation score at each instant. The magnification of the target with respect to the photograph T_2 could of course be varied by adjusting the distance between lens L1 and the face of the CRT.

2.3. HOLOGRAPHIC CROSS-CORRELATION

It is well known that a Fraunhofer hologram is a photograph of the interference pattern between a spatial Fourier transform and reference irradiation. For pattern recognition Vander Lugt *et al.* (1965) used a single beam of parallel light as the reference irradiation, whereas Gabor (1965, 1969, 1971) used a set of beams differing slightly from each other in direction, and derived from an array of holes in an otherwise opaque masking plate.

Figure 2.2 shows a simple setup for Vander Lugt's holographic cross-correlation. A transparency of a function $f_2(x,y)$ is illuminated with collimated

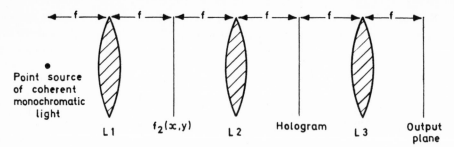

Fig. 2.2. A setup for holographic correlation.

coherent monochromatic light, and the Fourier transform of $f_2(x, y)$ is formed in the focal plane of lens L2. This plane contains the hologram of a function $f_1(x, y)$ and the cross-correlation function of $f_1(x, y)$ and $f_2(x, y)$ is obtained in the output focal plane of lens L3. If $f_1(x, y)$ is an array of different reference characters and $f_2(x, y)$ is a single unknown character, then the output plane contains a relatively bright spot in a position corresponding to any reference character that the input character resembles. If the input character is moved in its own plane, the bright spots move correspondingly. For purposes of recognition, alignment must be dealt with by using an array of photodetectors in the output plane, or by mechanical shifting of the input transparency, or an image of it, with respect to the optic axis, as in nonholographic optical correlation.

If $f_1(x, y)$ is a single character and $f_2(x, y)$ is a page of text, then the output plane ideally contains bright spots to indicate the positions of occurrence of $f_1(x, y)$ in the text. This arrangement would be more useful for detecting targets in photographs than for character recognition, except that it allows very little tolerance for variation in perspective of the target. Detailed implementations of Vander Lugt's technique have been described for recognition of characters (Dickinson and Watraseiwicz, 1968; Dye, 1969; Phillips, 1974; Hard and Feuk, 1973), fingerprints (KMS Industries, 1973; Ackroyd, 1973; Eleccion, 1973), star patterns (Gorstein *et al.*, 1970), and abnormal cells in rat liver (Bond *et al.*, 1973). An advantage of holographic vs nonholographic cross-correlation is that in the holographic case we can carry out spatial frequency filtering (Binns *et al.*, 1968; Lohmann and Werlich, 1971).

Nakajima *et al.*, (1972) have experimented with a character recognition technique similar to that of Gabor (1965, 1969, 1971). For each different reference character, a hologram is obtained with reference irradiation obtained from a different set of holes in a 9-hole shutter plate. In recognition using the Fig. 2.2 setup, if the input character matches the holographic reference character, and if the hologram has been properly corrected (Lohmann, 1970; Dickinson and Watraseiwicz, 1968), then the output plane ideally contains a pattern of bright points corresponding to the pattern of holes used in forming the hologram.

This pattern can be read out by means of an array of nine photomultipliers, and redundancy in the 9-element output code may perhaps yield some immunity to noise. Dickinson *et al.* (1968) were experimentally disenchanted with a technique of this sort.

Most of the holographic techniques that we have so far mentioned require the input pattern in the form of a transparency. Jacobson *et al.* (1973) and Nisenson *et al.* (1972) have described devices that very rapidly produce a transparency of an incoherently illuminated pattern. This transparency can be used as in Fig. 2.2 with a laser source for holographic cross-correlation. The same material of the transparency can be repeatedly used with different patterns.

In character recognition there are an increasing number of applications where the number of documents that must be archivally stored is so large that it is worthwhile to store them on microfilm. Small characters on microfilm have Fourier transforms that are relatively large.

For recognizing embossed patterns such as credit cards or fingerprints, the embossed pattern can be pressed against a glass surface. At the region where the pattern is pressed against the glass, a collimated laser beam can be internally reflected, having been modulated by the pattern. This modulation occurs because of the refractive difference where the skin does or does not touch the glass, and the internally reflected laser beam can be used for holographic cross-correlation (Ackroyd, 1973; Eleccion, 1973).

Coherent light and nondiffuse objects are not indispensable for holographic correlation (Bieringer, 1973; Lohmann, 1970; Lohmann and Werlich, 1971). To introduce a simple plausibility argument, suppose that in Fig. 2.2 the hologram is an ordinary Vander Lugt type hologram of $f_1(x, y)$ and that $f_2(x, y)$ is incoherently illuminated. Let us consider any bright point (x'', y'') in $f_2(x, y)$ and regard this as a coherent point source. This coherent source produces, inter alia, an inverted image $f_1(x'' + \gamma - x, y'' + \delta - y)$ in the output plane, and this image includes a bright point at (γ, δ) if $f_1(x'', y'')$ is bright. Since this holds for each bright point (x'', y'') in $f_2(x, y)$, the total intensity at (γ, δ) in the output plane is a correlation score for $f_1(x, y)$ and $f_2(x, y)$. It is important that in noncoherent holographic correlation (Lohmann and Werlich, 1971) the lateral position of the hologram is not critical, whereas this position is critical in the coherent case.

2.4. SPEECH RECOGNITION BY OPTICAL CORRELATION

A spectrum analyzer is a device that takes a speech pattern from the amplitude/time to the frequency domain. By means of a two-dimensional light-modulator, Preston (1972a, b, c) proposes to convert a frequency-domain speech pattern into a coherent optical pattern that can be recognized by holographic cross-correlation with reference patterns.

The use of resonating optic fibres may provide cheaper speech recognition techniques. Cutler *et al.* (1970) attach an electromechanical transducer to a transparent plate that is mounted vertically. Optic fibers are fixed at one end to the plate and protrude horizontally from the plate. The fibers are of different lengths and different mechanical resonant frequencies. A masking plate is mounted parallel to the fiber-bearing plate, and this masking plate is transparent except for one opaque spot corresponding to each fiber. The spots are positioned so that when a laser beam is shone through the fiber-bearing plate and thence through the fibers, the spots prevent onward transmission of light from all fibers that are not vibrating. Speech is used to drive the electromechanical transducer and thus to cause vibration of the resonant optic fibers. The optical pattern emerging from the spotted plate is recognized by holographic cross-correlation.

Tscheschner's (1970) method is to allow light passing through resonating optic fibers to impinge on a screen. The pattern on the screen is recognized by nonholographic template matching.

2.5. AUTOMATIC INSPECTION

It may eventually be possible to distinguish authorized people from imposters by means of automatic speech (e.g., Lummis, 1973; Wolfe, 1972), or signature (e.g., Nemcek and Lin, 1974; Nagel and Rosenfeld, 1973), verification. It may not be quite so difficult to compare a newly manufactured integrated circuit mask or currency note with a perfect standard reference item in order to determine faults in manufacture. Automatic inspection differs from a problem such as character recognition in at least two respects. First, a character recognition system only has a limited number of recognition classes, and it is not required to classify patterns that do not belong to any of these classes. For instance, a machine for recognizing Roman characters is not required to make correct classifications of Greek characters: Greek characters are assumed not to turn up. In automatic inspection, however, it may be desirable to distinguish bona fide patterns from all other patterns whatsoever, which is a more difficult problem. On the other hand, in automatic inspection there may only be two recognition classes, "accept" and "reject," and it may be economically viable to use an implementationally costly recognition process that we could not afford to use if it had to be repeated for many different recognition classes.

To introduce an optical interference method for automatic inspection of, for instance, integrated circuit masks, let us recall that in Young's double-slit experiment, a black cancellation fringe is obtained where light from corresponding points in the two slits is in antiphase. If the slits are replaced by two patterns that are not quite identical, then for a point in one pattern there may be no corresponding point in the other, so that there is no cancellation. Clear black fringes are observed only if the two patterns are almost identical (Aagard, 1971; Weinberger and Almi, 1971). When used for comparing a test pattern with a standard, this method has the advantage of positional tolerance. There is also positional tolerance in Watkins' (1973) technique that works by superimposing antiphase images of the test and reference patterns.

Positional alignment is a major problem in techniques that scan corresponding points in two patterns synchronously to check whether corresponding points have the same blackness or color. Hillyer (1954) patented a system in which two patterns were synchronously raster-scanned by two flying spot scanners, and a total mismatch score was computed. The patterns remained stationary, and the complete raster scans were repeated with the scan of one pattern slightly displaced. The scans were repeated for a succession of different displacements so that a function of total mismatch score vs displacement could be computed, on the assumption that there was no distortion of one raster with respect to the other. This problem of distortion was mitigated in a system of Price *et al.* (1960), which focused separate images of the same flying spot onto the two patterns that were to be compared. A recent IBM patent (Druschel, 1972) describes synchronous mechanical scanning of two patterns such as printed circuit masks.

Accurate comparison of aerial photographs requires sophisticated means for determining which point in one photograph corresponds to which point in the other (Nagy, 1972; Webber, 1973; Yao, 1973). For instance, it may be necessary to compensate for differences in the point of view of the camera and difference in atmospheric refraction. We can tackle this by cross-correlating a small part of one pattern with the other pattern and finding the position of the best match. This can be repeated with further small parts of the first pattern, and the correct mapping of the two patterns can be interpolated from the relative positions of optimal correlation of the small parts (Barnea and Silverman, 1972; Webber, 1973; Yao 1973).

Instead of making a point-by-point comparison of two patterns such as currency notes, a Mitsubishi system (1970) moves each pattern past a slit whose length is transverse to the direction of motion, and a photodetector is arranged to measure the total light from a slit-shaped area of the pattern. The photodetector-output vs time waveforms for the two patterns are cross-correlated.

2.6. NORMALIZED CROSS-CORRELATION

One of the well-known problems of simple correlation is that if, for instance, the character "O" is identical to "Q" except for the tail of the "Q," or, in other words, if the "O" is included in the "Q," then the matching score for "O" with "Q" may be the same as for "O" with "O." Similarly, the matching score for "F" and "E" may be the same as for "F" and "F." A simple but not always applicable method for dealing with this can be introduced in terms of Euclidean distance.

Hitherto we have regarded a pattern $f(x, y)$ as a function of x and y. It is now convenient to sample this function at N discrete points and to denote the values of $f(x, y)$ at these N points by $x_1, x_2, \ldots, x_N$, respectively. The vector defined by $X = \{x_1, x_2, \ldots, x_i, \ldots, x_N\}$ can be regarded as the position vector of a point in N-dimensional space. In the literature, the vector X is usually called a *pattern*.

The Euclidean distance d between an unknown pattern X and a reference pattern $X' = \{x_1', x_2', \ldots, x_i', \ldots, x_N'\}$ is given simply by the N-dimensional version of Pythagoras' theorem. Thus

$$d^2 = \sum_{i=1}^{i=N} (x_i - x_i')^2 = \sum_{i=1}^{i=N} (x_i^2 - 2x_i x_i' + x_i'^2)$$

The reference pattern X' for which d is minimal is the reference pattern for which

$$\sum_{i=1}^{i=N} x_i x_i' - \tfrac{1}{2} \sum_{i=1}^{i=N} x_i'^2 \qquad (2.1)$$

is maximal, because d^2 is a monotonically increasing function of d and $\Sigma_{i=1}^{i=N} x_i^2$ is independent of X'. Thus the minimum Euclidean distance rule "assign X to the class of X' such that d is minimal" is equivalent to the rule "assign X to the class of X' such that equation (2.1) is maximal." The second term in equation (2.1) ensures that if X is included in X', (e.g., "F" in "E") then the value of equation (2.1) is less than when $X = X'$. The first term in equation (2.1) is essentially a spatially sampled version of

$$\iint f_1(x, y) f_2(x + \alpha, y + \beta)\, dx\, dy \qquad (2.2)$$

at $\alpha = 0$, $\beta = 0$, where $f_2(x, y)$ is the nonsampled original of X'. Assigning X to the class of X' such that the matching score [equation (2.2)] is maximal is roughly equivalent to using equation (2.1) without the second term, so that we have no protection against the "F," "E" inclusion problem. To remedy this we can use equation (2.1), or a nondiscrete counterpart of it, instead of equation (2.2). Although from the point of view of "F," "E" inclusion, equation (2.1) is better than equation (2.2), we have no *a priori* guarantee that the use of equation (2.1) will yield tolerably low recognition error rates in a given specific practical application. Indeed the hypothesis that maximization of equation (2.1) will yield tolerably low error rates is often not even plausible. For instance, we may find in practice that every pattern $kX = \{kx_1, \ldots, kx_N\}$ belongs to the recognition class of X, for all positive real values of k. In particular, different values of k may correspond to different levels of illumination. In this case the matching score should be independent of k, and equation (2.1) is inappropriate because it depends strongly on k.

The angle between the vectors X' and kX is independent of k. Instead of using equation (2.1) we can gain invariance to k by assigning X to the class of X' such that the cosine of the angle between X and X' is maximal (Lange, 1967). This cosine is given by

$$\frac{\sum_{i=1}^{i=N} x_i x_i'}{[\sum_{i=1}^{i=N} x_i^2]^{\frac{1}{2}} [\sum_{i=1}^{i=N} x_i'^2]^{\frac{1}{2}}}$$

Since $\sum_{i=1}^{i=N} x_i^2$ is independent of X', we can use the recognition rule: assign X to the class of the reference pattern X' such that

$$\sum_{i=1}^{i=N} x_i w_i \tag{2.3}$$

is maximal, where

$$w_i = \frac{x_i'}{[\sum_{i=1}^{i=N} x_i'^2]^{\frac{1}{2}}}$$

The matching score [equation (2.3)] differs from the discrete version of equation (2.2) in that the components of X' are divided by the square root of the sum of their squares. The values of $w_1, \ldots, w_N$ can be computed prior to recognition, and the division does not have to be done again for every unknown pattern (Gorbatenko, 1970).

The use of equation (2.3) to some extent deals with the "F," "E" inclusion problem, but a practical improvement can be obtained by means of Highleyman's (1961) penalty weighting. Other ideas for dealing with "F," "E" inclusion are discussed by Caulfield and Maloney (1969), Maloney (1971), and Pursey (1973).

2.7. OPTOELECTRONIC TRANSDUCTION

The simple correlation techniques that we have so far considered are not always adequate for practical purposes. More advanced techniques are usually implemented electronically rather than optically, and many different techniques of optoelectronic transduction used. We will not attempt to review these techniques here. Hawkins (1970) and Palmieri (1971) have provided reviews from the viewpoint of pattern recognition, and a more general source is Bieberman and Nudelman (1971).

CRT and laser flying spot scanners, and television pickup tubes such as vidicons, are used in commercial character recognition (British Computer Society, 1971). Mechanical scanners and photodetector arrays are also used. An array may be either one- or two-dimensional; pattern alignment being dealt with by document or film motion and/or by moving mirrors. The readout of an integrated circuit photodiode array may either be serial as (Styan and Mason, 1974) or random access (Nippon Electric Co., 1972). CCI have patented a lenseless scanner in which the photodiodes in an array are very close to the document that is to be scanned, and the document is illuminated from the remote side of the photodiodes, which are semitransparent (CCI Aerospace Corp., 1972). The IBM 3886, which is the latest IBM character recognition system, uses a light-emitting diode array scanner, in which successive diodes illuminate successive small areas of a document with infrared light.

For purposes of recognition we are usually more interested in *contrast*, which is conveniently defined to be a *ratio* of video amplitudes, than in *differences* between video amplitudes. In order to realize ratios rather than differences, we can logarithmically scale the video scanner output by using the nonlinear characteristic of a diode (Hanson, 1971; Wilson, 1971).

In character recognition we are primarily interested in the shape of the distribution of black on white, and except for purposes of determining this shape, the precise variation of whiteness of white and blackness of black may be of little practical interest. One of the first stages of video processing is usually to quantize the grey scale into just two levels, called *black* and *white*, so that the detailed variations of whiteness and blackness are eliminated from the later stages of processing. This two-level quantization is known as *binarization*; a bibliography of binarization techniques is given by Ullmann (1974).

2.8. LINEAR DISCRIMINANT IMPLEMENTATION

Equations (2.1) and (2.3) have the general form

$$w_0 + \sum_{i=1}^{i=N} w_i x_i \tag{2.4}$$

We will now briefly mention hardware techniques that can be used for evaluating equation (2.4) for any given vector $\{w_0, w_1, \ldots, w_i, \ldots, w_N\}$, even when some of the components of this vector are negative.

There is a resistor mask technique in which we employ a separate resistor for each of $w_0, w_1, \ldots, w_N$. For $i = 0, 1, \ldots, N$, the resistance value of the resistor corresponding to w_i is $\alpha/|w_i|$, where α is a constant. The $N + 1$ resistors are all connected together at one end, and thence to earth via a device that measures the total current through all the resistors. The other end of the resistor for w_i is connected to a source of $+\beta x_i V$ if w_i is positive or $-\beta x_i V$ if w_i is negative, where β is a constant. This holds for all $i = 0, 1, \ldots, N$, taking $x_0 = 1$, and the total current through the $N + 1$ resistors is therefore proportional to equation (2.4). Further resistor techniques are mentioned by Ullmann (1973).

Instead of using resistors we can use an ordinary digital computer or a stochastic computer (Goke and Doty, 1972) to evaluate equation (2.4). For use with binary patterns, we can employ a reactive rather than a resistive network (Gorbatenko, 1970). Jackson (1969) has patented a further idea in which the transmission coefficient of part of a photochromic plate is made proportional to positive w_i, and equation (2.4) is evaluated by simple noncoherent optical summation. A further suggestion is to do holographic correlation with a computer-generated (Lohmann and Paris, 1971; Lee, 1973) hologram of $\{w_1, \ldots, w_N\}$.

2.9. NUMERICAL DISCRIMINANTS

Given a set of patterns that belong to the same recognition class we can obtain a mean vector $M = \{m_1 m_2, \ldots, m_N\}$, where for $i = 1, \ldots, N$, m_i is the average of the ith components of all patterns in the set. We can then use M instead of X' in equation (2.1) or (2.3), and this may yield better experimental results (Highleyman, 1961). Indeed, when recognition classes are normally distributed with equal scalar covariance matrices, a statistically optimal decision rule uses M in equation (2.1) (Duda and Hart, 1973; Fukunaga, 1972; Patrick, 1972; Andrews, 1972).

A set of specimen patterns used, for instance, in determining M, is known as a *training set*. There are procedures (Duda and Hart, 1973; Nagaraja and Krishna, 1974) which, given training sets from all classes to be recognized, automatically and *optimally* determine one vector $\{w_0, w_1, \ldots, w_N\}$ per class. These vectors are optimal in the sense that they yield some sort of minimum-error recognition of the training set patterns when we use the rule: assign X to the class of W such that equation (2.4) is maximal. Though minimal with respect to all possible differently chosen W's, the error rates may still be intolerably high. We can tackle this by using more than one W per class and assigning X to the class of the W such that equation (2.4) is maximal. Methods are available for automatically determining W's for this purpose (Firschein and Fischler, 1963; Chang 1973).

Instead of computing M's or nontrivial W's, we can use nearest-neighbor techniques. The simplest nearest-neighbor technique stores many reference patterns per class and assigns X to the class of the nearest reference pattern. The "nearest" reference pattern may, for instance, be that to which the Euclidean distance is least, so that we can use equation (2.1), possibly with an optical-correlation implementation. In practice it may be neither necessary nor desirable to use all of the patterns in a given training set as the reference patterns in a nearest-neighbor system. Instead, a subset can be automatically selected for use as reference data (Ullmann, 1974a,b). Patrick has provided a detailed review of nearest-neighbor techniques and their statistical theory (Patrick, 1972).

A further approach is to employ a separate vector W for each pair of classes. The vector associated with any two given classes A and B is chosen so that ideally equation (2.4) is negative for all patterns in A and positive for all patterns in B. An unknown pattern is recognized by determining for which W's equation (2.4) is or is not positive. This method generally gives lower error rates than one-W-per-class methods, but it uses more W's. It has been used experimentally in, for instance, the recognition of mass spectra (Bender and Kowalski; 1973), a lung disease (Hall et al., 1973) and handprinted characters (Sammon, 1970).

Yet another approach is to use a polynomial function instead of the linear function [equation (2.4)], and several theoretical approaches yield methods for automatically determining the coefficients of the product terms (Duda and Hart, 1973; Meisel, 1972; Ullmann, 1973).

2.10. TRANSFORMATIONS

The mathematical techniques that we have just briefly mentioned are very general. They might be applicable, for instance, in automatic medical diagnosis when $x_1, \ldots, x_N$ are the results of N clinical tests on a patient (Patrick et al., 1974). They might also be applicable for recognizing the color at a given point in a visual pattern where x_1 is the ratio of the red filter output to total output, and x_2 is the ratio of the green filter output to total output (Yachida and Tsuji, 1971; Young, 1970). They might also be applicable for recognizing blood cells where, for example, x_1 might be the area of the nucleus, x_2 might be the (perimeter)^2/area of the nucleus, x_3 might be the difference in brightness between the nucleus and background, and so on (Bacus and Gose, 1972). Indeed the mathematical techniques may be applicable when $x_1, \ldots, x_N$ are the results of any N arbitrary tests applied to a pattern.

When $x_1, \ldots, x_N$ are simply digitized video samples, X must be properly aligned or normalized in position, size, orientation, etc., before the mathematical techniques can be applied, since these techniques presuppose that x_i corresponds to w_i for $i = 1, \ldots, N$.

To mitigate this problem of alignment, a common idea is to transform a visual pattern into a new pattern that is invariant to shift and possibly to stretch and orientation. Shelton (1962) focused the Fraunhofer diffraction of a character onto a frosted glass plate, and recognized the position-invariant diffraction pattern by simple optical template matching. To gain limited invariance to size,

Shelton used a rotating Fabry—Perot interferometer as the light source, so that the wavelength repeatedly swept through a range of values. A disadvantage of this method is that the Fraunhofer diffraction patterns for the characters "L" and "T" are far more similar than the original patterns, as can be seen in the illustration of Armitage and Lohmann (1965). This is because of limited insensitivity to changes of relative position of parts of a pattern (Holeman, 1968).

To determine spatial frequencies in a given direction, we can move a transverse slit in this direction, and measure the total light obtained from a slit-shaped area of the pattern at each instant. This total vs time can be Fourier-transformed by a filter bank or FFT, and the result is invariant to shift of the original pattern (Nakano *et al.*, 1973; cf. Declaris and Greeson, 1968, 1969). To improve the detection of a given spatial frequency we can use not just one slit but a set of parallel slits disposed at that frequency, although this requires a separate grating, or separate optical scaling of a grating (Rogers *et al.*, 1971) for each frequency.

Spatial Fourier coefficients provide global information about an optical pattern, and it may be sufficient to use only a few coefficients as $x_1, x_2, \ldots, x_N$ for recognition purposes. Thus the transformation may achieve information reduction. As an alternative to the use of Fourier transformation for this purpose, Walsh transformation has been considered because of its implementational simplicity (Andrews, 1972; Carl and Hall, 1972; Inokuchi *et al.*, 1972; Nemeck and Lin, 1974), although it does not provide shift invariance. A further alternative is the Karhunen—Loeve transformation (Kittler and Young, 1973; Han *et al.*, 1973).

For use in classifying texture in aerial photographs, Lendaris and Stanley (1970) integrated the Fourier power spectrum over each of a set of concentric annuli centered on the optic axis, and the resulting set of values was invariant to orientation as well as position of the original visual pattern. Kruger *et al.* (1974) have used a similar technique in the classification of texture in X-ray photographs of lungs, for health screening purposes. Hawkins (1970) has provided a general survey of texture recognition techniques.

To gain invariance to size, we can integrate the Fourier power spectrum over strips bounded by radii extending outward from the optic axis in the Fourier plane, and divide the total energy per strip by the overall total energy. A further idea is to place in the Fourier plane an opaque rotating mask containing a single long slit passing through the optic axis. If the axis of rotation is coincident with the optic axis, and the total light transmitted at any instant through the mask is measured by a photomultiplier, then the photomultiplier output is a periodic function of time. When normalized with respect to total energy, the Fourier power spectrum of this function is invariant to size, position, and orientation of the original visual pattern. This technique has been considered for the recognition of fingerprints (Reid and Redman, 1969) and fine particles (Kaye and Naylor, 1972). Tanaka and Ozawa (1972) have described an approximate implementation of this technique that uses static rather than rotating masks.

To recognize a cell in a cervical smear, McCarthy (1971) finds the center of gravity of the cell, regarding blackness as analogous to mass. He measures the distance from the center of gravity to the periphery of the cell, and applies recognition techniques to the autocorrelation function of the radius/angle function, since this is independent of rotation and location of the cell.

Using optic fibers, a Bendix system (Bendix Corp., 1970) sums and squares the light over each of a set of parallel strips of a pattern. The squares are summed and the total is stored. The pattern is then rotated by means of, for instance, a Dove prism, and a new total is obtained and stored, and so on up to $180°$ rotation. The stored totals constitute a vector that is independent of position of the original pattern, and cyclic permutation of the vector components can be made to yield invariance to orientation.

Greanias (1966) servo-controls a flying spot to trace round the periphery of an object in a pattern, and measures the time taken for the spot to return to its original position. This measurement is used to adjust the scanning speed so that the spot takes exactly a prescribed time T to trace round the periphery a second time. At times T/N, $2T/N$, $3T/N$, ... the system digitizes and stores the direction of motion of the spot, and the set of N differences between successive directions is recognized by correlation with reference vectors. Cyclic permutation of the vector components yields invariance to orientation, as well as to size and position of the original pattern. If the second trace is repeated many times, the angular direction/time function of the trace is a periodic function. The autocorrelation function or the Fourier power spectrum of this periodic function is invariant to orientation, position, and size of the original pattern (Zahn and Roskies, 1972), and noise can be ignored by ignoring high-frequency Fourier components.

The autocorrelation function of a pattern is position-invariant (Horwitz and Shelton, 1961; Doyle, 1962), and we can gain size invariance by computing the autocorrelation function of the logarithm to the autocorrelation function of a pattern (Beurle, 1971, 1972). An autocorrelation function can be realized optically as the Fourier transform of a Fraunhofer intensity pattern (Horwitz and Shelton, 1961; Rose *et al.*, 1971). Alternatively an autocorrelation function can be obtained by transmitting light from an extended noncoherent source through a transparency of a pattern. The transmitted light is then transmitted back through the transparency to yield the autocorrelation function (Kovasznay and Arman, 1957; Holden, 1972).

The higher moments of a pattern afford a global parameterization (Hu, 1962). The values of higher moments can be normalized to yield invariance to shift, size, and shear (Alt, 1962; Casey, 1970). To evaluate the higher moment

$$M_{pq} = \iint x^p y^q f(x, y) \, dx \, dy$$

Giuliano *et al.* (1961) have suggested integrating light that has passed through a transparency of $f(x, y)$ and also through a transparency whose transmission coefficient is proportional to $x^p y^q$.

2.11. NORMALIZATION AND SEGMENTATION

To recognize a pattern regardless of position and size, we can apply some sort of transformation or set of tests whose result is invariant to position and size. Alternatively we can adjust the position and size of the original pattern to standard values, and apply recognition technique to the adjusted original pattern. This process of adjustment is called *normalization*. There are two main categories of normalization methods. The first is normalization by measurement. In this, we measure, for instance, the height and position of an input pattern, and then rescan the pattern (Philco, 1966) or copy it into a further storage area (Andrews and Kimmel, 1973) so that its height and position now have prescribed values. The other category consists of methods of normalization by recognition. Here, normalization is not carried out prior to recognition, but occurs contemporaneously with recognition. For instance in cross-correlation we usually try all possible alignments and use the correlation score itself to determine the optimal alignment. In practice this idea is commonly implemented by means of the important shift-register alignment technique that is explained by Ullmann (1973). A more advanced normalization-by-recognition technique is exemplified by Velichko and Zagoruyko (1970) and Nakano *et al.* (1973), where the normalization problem is to determine which element of an unknown pattern corresponds to which element of a known pattern; and the solution to this problem need not be distinct from the recognition decision-making process.

Segmentation is the demarcation of a specific object in a scene that contains many objects. For instance, segmentation of a line of text usually means splitting the line into constituent characters. This is easy if there are gaps between successive characters. In automatic karyotyping of chromosomes, segmentation of the spread into individual chromosomes is easy so long as they do not touch or overlap (Ledley, 1973).

Certain bacteria can be identified by recognition of the shapes of their approximately spherical colonies (Glaser and Ward, 1972). When two colonies partially overlap there are telltale cusps where the peripheries of their approximately circular images join. The positions of these cusps can be used, in effect, to segment the colonies (Glaser and Ward, 1972). In the recognition of touching characters whose widths are approximately known in advance, Bond and

Shatford (1973) look for certain telltale shapes in the region where the characters are known to meet. The positions of these shapes determine where the characters should be segmented for recognition purposes. In a line drawing of a collection of polyhedra, the detailed shapes of the vertices provide evidence as to which face belongs to which polyhedron, and this can be used to segment the drawing into individual polyhedra, even if these polyhedra overlap each other drastically (Eleccion, 1973; Winston, 1972).

In these examples, segmentation takes place prior to recognition. There is, however, a further class of segmentation techniques where segmentation is indistinguishable from recognition. A simple example is in character recognition, where a character-sized window is scanned along a line of text. Any bona fide character appearing in the window is recognized, and the window blocks out touching characters (e.g., Recognition Equipment Inc., 1968, 1970; cf. Lahart, 1970). It is the output of the recognition system, not prior segmentation, that determines the proper position of the window for recognition. In Fischler and Elschlager's (1973) more advanced illustration of this principle, there is neither segmentation nor normalization prior to the nontrivial matching process.

2.12. FEATURE DETECTION

Let us consider two handprinted 5's that have been normalized in position, height, and width, so as just to fit into a standard rectangular box. These normalized patterns will generally be related to each other by complicated distortions that defy simple analytical description. A handprint recognition system should only be invariant to some such distortions, and we call these *admissible* distortions. Fourier, Walsh, and higher moment expansions do not necessarily yield invariance to all admissible distortions, and this is one reason why such techniques are not much used commercially in character recognition.

A better technique can be introduced as follows. Let us take a size and position normalized reference character and select some of its limb ends, bends, and junctions to serve as features for use in recognition. Let us assign a size and position normalized unknown character to the class of this reference character only if the unknown character contains an admissibly distorted version of each of the selected features. This idea ensures invariance to all admissible distortions, but Fourier, Walsh, and other such expansion techniques generally do not.

Admissibly distorted versions of a feature of a size and position normalized character generally lie within a limited zone. This is why commercial systems such as those of Patterson (1973) and Scan-Data Corp. (1972) and the system discussed by Rohland *et al.* (1968) detect features in zones. A feature can be detected by means of a switching circuit that is designed to tolerate admissible variations in feature shape (Andrews *et al.*, 1968; Scan-Data Corp.,

1972). Although feature detection is usually done electronically, a nonholographic optical feature detection system is described by Hitachi (1968). Marko (1974) has considered holographic correlation for feature detection. In Hitachi's (1969) work a pattern to be recognized is focused onto a collection of extended photosensitive surfaces (e.g., solar batteries), the shapes and positions of these surfaces being specially designed to facilitate feature detection. A further patent (Freedman, 1972) describes an image dissector that focuses a small part of a pattern onto an array of (e.g., nine) channel multipliers, this being sufficient for detection of localized features. Elaborate computer programs have also been used for feature detection, particularly for edge detection (Davis, 1973).

Besides yielding invariance to various distortions, the use of features enables us to capitalize the fact that, for instance, the top part of a "3" may be the same as the top part of a "2." We may be able to economize by using the same detector for this top part, regardless of the character that it occurs in. This aspect of feature detection is clearly illustrated by Thomson-CSF (1973), where there are optical templates for parts of characters, and a character is recognized according to which parts it is or is not found to match.

A further aspect of feature detection is that it may achieve a reduction in dimensionality, and thus possibly a reduction in the cost of the final recognition process.

There is a considerable literature (e.g., Fukunaga, 1972; Andrews, 1972; Kruger *et al.*, 1974; Wolf, 1972; Hall *et al.*, 1973) on techniques for automatically selecting a good set of features from a candidate set of features for use in recognition. These techniques can be applied to any set of candidate tests, not only to geometrical features. Although it is often possible to achieve automatic selection that yields better experimental results than random selection, this is still acknowledged to be an open area for research (Andrews, 1972).

2.13. SEQUENTIAL RECOGNITION TECHNIQUES

To recognize a given unknown pattern it may not be necessary to carry out all of a fixed set of feature-detection (or other) tests. Instead the tests can be carried out in a fixed sequence and after each test we can decide whether the results obtained so far will yield a tolerably confident recognition decision so that no further tests need be carried out, and we therefore can avoid the cost of carrying them out (Fu, 1968; Andrews, 1972; Smith and Yau, 1972). Instead of using a fixed sequence of tests we can, after carrying out a test, statistically compute which of all the remaining tests it would be best to carry out next, in the light of the results so far obtained (Slagle and Lee, 1971). We may, however, prefer to avoid statistical computation during recognition, but instead do the statistical computation in advance, thereby automatically constructing a decision tree for subsequent use in recognition (Banks, 1970; Slagle and Lee, 1971). It has been more usual to design decision trees by intuitive trial and error (e.g., Naylor, 1973; Spanjersberg, 1974; Reddy, 1967).

Modern linguistics and pattern recognition meet at two distinct points. One of these is at a high level where, for instance, a computer is required to converse in natural language concerning a visual scene (Winograd, 1973). The other is at a low level where we merely use the formal machinery of a syntactic acceptor for recognizing streams of sequential data obtained from original patterns, and practical examples of this are given by Ullmann (1973).

2.14. INTERACTIVE PATTERN RECOGNITION

Kanal (1972) has surveyed man–machine interactive techniques of pattern recognition. A human operator and a computer may work cooperatively in designing a recognition system which will subsequently recognize patterns without human assistance (e.g., Naylor, 1973; Sammon, 1970). It is more usual to find that recognition itself is a man–machine cooperative effort. For instance, a human operator may pick out objects in a complicated scene, so that the computer has no segmentation problem. If the automatic recognition process does not arrive at a confident decision, it can request help from a human operator who is equipped with a visual display (e.g., Ledley, 1973; Royston, 1972).

Goldstein *et al.* (1972) have described an experimental face-recognition system in which a human looks for specific features in the visual pattern and gives the result to a computer which then works out statistically what features the human should look for next. This illustrates the prevalent idea that a computer should be used for tedious repetitive computational tasks, and a human operator should do that part of the job which we do not know how, or cannot afford, to computerize.

2.15. CONCLUDING COMMENT

Commercial character recognition is usually accomplished in special-purpose electronic equipment although there may be an associated minicomputer for use in format control and the control of manual correction of reject errors. In research on automatic processing of satellite photographs, radiological photographs, and microscope slides, it has been customary to rely heavily on general purpose computers; but the highly sequential nature of an ordinary digital computer may make it intolerably slow for processing visual patterns. For instance, Olsztyn *et al.* (1973) performed a feasibility experiment in which a robot manipulator automatically mounted car wheels onto hubs under the control of a computer. A vidicon provided the computer with video data that was used, for example, to locate the stud holes. One of the findings of this experiment was that the computerized process would have to be speeded up by a factor of about 10 in order to become useful for production-line purposes.

To speed up the automatic processing of visual patterns it is possible to use special purpose computers that have a considerable degree of parallelism (Gray, 1971; Preston and Norgren, 1972; Allen *et al.*, 1973). For attaining high speed at relatively low cost, optical techniques may perhaps deserve consideration.

REFERENCES

Aagard, R. L., 1971, US Pat. 3,624,605, Nov. 1971.

Ackroyd, J., 1970, 1973, UK Pats. 1,185,910, Mar. 1970; 1,188,302, April 1970; 1,305,248, Jan. 1973.

Allen, G. R, Bonrud, L. O., Cosgrove, J. J., and Stone, R. M., (1973) *Machine Processing of Remotely Sensed Data*, Conference Proceedings, Purdue University, Oct. 1973, IEEE Catalogue No. 73 CHO 834-2GE, pp. 1A-25 to 1A-42.

Alt, F. L, 1962, In: *Optical Character Recognition*, (G. L. Fischer, ed.), Spartan Books, New York, pp. 153–179.

Andrews, D. R., Atrubin, A. J, and Hu, K. C., 1968, *IBM J. of R&D* 12:364–371.

Andrews, D. R., and Kimmel, M. J., 1973, US Pat. 3,710,323, Jan. 1973.

Andrews, H. C., 1972, *Introduction to Mathematical Techniques in Pattern Recognition*, Wiley-Interscience, New York.

Armitage, J. D., and Lohmann, A. W., 1965, *Appl. Optics* 4:461–467.

Bacus, J. W., and Gose, E. E., 1972, *IEEE Trans. on Systems, Man, Cybernetics* SMC-2: 513–526.

Banks, E. R., 1970 *MIT Project MAC Artificial Intelligence Memo No. 189.*

Barnea, D. I., and Silverman, H. F., *IEEE Trans. Computers* C-21:179–186.

Bender, C. F., and Kowalski, B. R., 1973, *Anal. Chem.* 45:590–592.

Bendix Corporation, 1970, UK Pat. 1,216,458, Dec. 1970.

Beurle, R. L., 1971, 1972, UK Pats. 1,237,534, June 1971, 1,294,891, Nov. 1972, 1,300,224, Dec. 1972.

Biberman, L. M., and Nudelman, S., (eds.), 1971, *Photoelectric Imaging Devices*, Plenum Press, New York and London.

Bieringer, R. J., 1973, *Appl. Optics* 12:249–254.

Binns, R. A., Dickinson, A., and Watraseiwitz, B. M., 1968, *Appl. Optics* 7:1047–1051.

Bond, M. F., and Shatford, J. F., 1973, UK Pat. 1,304,429, Jan. 1973.

Bond, R. L., Mazunder, M. K., Testerman, M. K., and Hsieh, D., 1973, *Science* 179:571–573.

British Computer Society, 1971, *Character Recognition 1971*, a handbook, London.

Carl, J. W., and Hall, C. F., *IEEE Trans. Computers*, 1972, C-21:785–790.

Casey, R. G., 1970, *IBM J. of R&D* 14:548–557.

Caulfield, H. J., and Maloney, W. T., 1969, *Appl. Optics* 8:2354–2356. US Pat. 3,622,988, Nov. 1971.

CCI Aerospace Corporation, 1972, UK Pat. 1,271,394, April 1972.

Chang, C-L., 1973, *IEEE Trans. Computers* C-22:859–862.

Crane, H. D., and Steele, C. M., 1965, *Pattern Recognition* 1:129–145.

Cutler, C. C., Julesz, B., and Pennington, K. S., 1970, UK Pat. 1,191,004, May 1970.

Davis, L. S., 1973, *Computer Science Center U. of Maryland., Tech. Rept. TR 273.*

Declaris, N., and Greeson, J. C., 1968, 1969, UK Pat. 1,121,954, July 1968; US Pat. 3,482,211, Dec. 1969.

Dickinson, A., and Watraseiwicz, B. M., 1968, IEE/NPL Conference on Pattern Recognition, *IEE Conf. Publ.* 42:207−213.

Doyle, W., 1962, *J. Assoc. Computing Machinery* 9:259−267.

Druschel, W. O., 1972, US Pat. 3,645,626, Feb. 1972.

Duda, R. O., and Hart, P. E., 1973, *Pattern Classification and Scene Analysis*, Wiley-Interscience, New York.

Dye, M. S., 1969, *Marconi Rev.* 32:111−128.

Eleccion, M., 1973, *IEEE Spectrum* 10:36−45.

Firschein, O., and Fischler, M., 1963, *IEEE Trans. Electronic Computers* EC-12:137−141.

Fischler, M. A., and Elschlager, R. A., 1973, *IEEE Trans. Computers* C-22:67−92.

Freedman, M. D., 1972, US Pat. 3,694,806, Sept. 1972.

Fu, K. S., 1968, *Sequential Methods in Pattern Recognition and Machine Learning*, Academic Press, London and New York.

Fukunaga, K., 1972, *Introduction to Statistical Pattern Recognition*, Academic Press, New York and London.

Gabor, D., 1965, *Nature* 208:422−423; UK Pat. 1,143,086, Feb. 1969; US Pat. 3,600,054, Aug. 1971.

Giuliano, V. E., Jones, P. E., Kimball, G. E., Meyer, R. F., and Stein, B. A., 1961, *Inform. Control* 4:332−345.

Glaser, D. A., and Ward, C. B., 1972, In: *Frontiers of Pattern Recognition* (S. Watanabe, ed.), Academic Press, London and New York, pp. 139−162.

Goke, L. R., and Doty, K. L., 1972, *IEEE Trans. Computers* C-21: 1347−1354.

Goldstein, A. J., Harmon, L. D., and Lesk, A. B., 1972, *Bell System Tech. J.* 51:399−427.

Gorbatenko, G. G., 1970, UK Pat. 1,179,839, Feb. 1970.

Gorstein, M., Hallock, J. N., and Valge, J., 1970, *Appl. Optics* 9:351−358.

Gray, S. B., 1971, *IEEE Trans. Computers* C-20: 551−561.

Grenias, E. C., 1971, 1966, US Pat. 3,273,124, Sept. 1966.

Hall, E. L., Crawford, W. O., Preston, K., and Roberts, F. E., 1973, *Proc. First Int. Joint Conf. on Pattern Recognition*, Washington, D.C., IEEE Publication 73 CHO 821−890, 77−87.

Han, K. S., McLaren, R. W., and Lodwick, G. S., 1973, *IEEE Trans. Systems, Man, Cybernetics* SMC-3: 410−415.

Hanson, C. C., 1971, UK Pat. 1,251,117, Oct. 1971.

Hard, S., and Feuk, T., 1973, *Pattern Recognition* 5:75−82.

Hawkins, J. H., 1970, In: *Advances in Information Systems Science*, Vol. 3 (M. T. Tou, ed.), Plenum Press, New York and London, pp. 113−214.

Hawkins, J. H., 1970, In: *Picture Processing and Psychopictorics*, (B. Lipkin and A. Rosenfeld, eds.) Academic Press, New York and London, pp. 347−370.

Highleyman, W. H., 1961, *IRE Trans. Electronic Computers* EC-10:501−512. US Pat. 2,978,675, April 1961.

Hillyer, C., 1954, US Pat. 2,679,636, May 1954.

Hitachi, 1968, UK Pat. 1,123,564, Aug. 1968.

Hitachi, 1968, UK Pat. 1,175,517, Dec. 1969.

Holden, S., 1970, 1972, UK Pats. 1,207,679, Oct. 1970, 1,300,050, Dec. 1972.

Holeman, S., In: *Pattern Recognition*, (L. N. Kanal, ed.), Thompson Book, Washington, D.C., pp. 63−78; UK Pat. 1,151,190, May 1969.

Horwitz, L. P., and Shelton, G. L., 1961, *Proc. IRE* 49:175−185.

Howard, P. H., 1960, US Pat. 3,267,430, Aug. 1966.

Hu, M. K., *IRE Trans. Information Theory* IT-8:179–187.
Inokuchi, S., Morita, Y., and Sakurai, Y., 1972, *Appl. Optics* 11:2223–2227.
Jackson, M. C., 1969, UK Pat. 1,166,693, Oct. 1969.
Jacobson, A. D., Beard, T. D., Bleha, W. P., Margerum, J. D., and Wong, S.-Y., 1973, *Pattern Recognition* 5:13–19.
Kanal, L. N., 1972, *Proc. IEEE* 60:1200–1216.
Kaye, B. H., and Naylor, A. G., 1972, *Pattern Recognition* 4:195–199.
Kittler, J., and Young, P. C., 1973, *Pattern Recognition*, 5:335–352.
KMS Industries Inc., 1972, 1973, UK-Pats. 1,292,646, Oct. 1972, 1,338,787, Nov. 1973.
Kovasznay, L. S., and Arman, A., 1957, *Rev. Sci. Instr.* 28:793–797.
Kruger, R. P., Thompson, W. B., and Turner, A. F., 1974, *IEEE Trans. Systems, Man, Cybernetics* SMC-4:40–49.
Lahart, M. J., 1970, *J. Op. Soc. Am.* 60:319–325.
Lange, F. H., 1967, *Correlation Techniques*, Iliffe, London; Van Nostrand, Princeton, N.J.
Ledley, R. S., 1973, *Proc. First Int. Joint. Conf. on Pattern Recognition*, Washington, D.C., Oct. 1973, IEEE Publication 73 CHO 821-9C, 89–112.
Lee, S. H., 1973, *Pattern Recognition* 5:21–35.
Lendaris, G. G. and Stanley, G. L., 1970, Proc. IEEE 58:198–216.
Lohmann, A. W., 1970, US Pat. 3,501,221, March 1970.
Lohmann, A. W., 1970, US Pat. 3,514,177, May 1970.
Lohmann, A. W., and Paris, D. P., 1968, 1969, *Appl. Optics* 7:(1968) 651–655; *IBM J. R&D* 13:160–169.
Lohmann, A. W., and Werlich, H. W., 1971, *Appl. Optics* 10:670–672.
Lummis, R. C., 1973, *IEEE Trans. Audio Electro. Acoust.* AU-21:80–89.
Maloney, W. T., 1971, *Appl. Optics* 10:2127–2131.
Marko, H., 1974, *IEEE Trans. Systems, Man, Cybernetics* SMC-4:34–39.
McCarthy, B. D., 1971, UK Pat. 1,221,071, Feb. 1971.
Meisel, W. S., 1972, *Computer-Oriented Approaches to Pattern Recognition*, Academic Press, New York.
Mitsubishi, 1970, UK Pat. 1,187,451, April 1970.
Nagaraja, G., and Krishna, G., 1974, *IEEE Trans. Computers* C-23:421–427.
Nagel, R. N., and Rosenfeld, A., 1973, *Proc. First Int. Joint Conf. on Pattern Recognition*, Washington, D.C., Oct. 1973, IEEE Publication 73CHO 821-9C, 59–66.
Nagy, G., 1972, *Proc. IEEE* 60:1177–1200.
Nakajima, M., Morikawa, T., and Sakurai, K., 1972, *Appl. Optics* 11:362–371.
Nakano, Y., Nakata, K., Uchikura, Y., and Nakajima, A., 1973, *Proc. First Int. Joint Conf. on Pattern Recognition*, Washington, D.C., Oct. 1973, IEEE Publication 73CHO 821-9C, 172–178.
Naylor, W. C., 1973, UK Pat. 1,327,325, August 1973.
Nemcek, W. F., and Lin, W. C., 1974, *IEEE Trans. Systems, Man, Cybernetics* SMC-4:121–126.
Nippon Electric Company, 1972, UK Pat. 1,287,205, Aug. 1972.
Nisenson, P., and Iwasa, S., *App. Optics* 11:2760–2767.
Olsztyn, J. T., Rossol, L., Dewar, R., and Lewis, N. R., 1973, *Proc. First Int. Joint Conf. on Pattern Recognition*, Washington, D.C., Oct. 1973, IEEE Publication 73 CHO 821-9C, 505–513.
Palmieri, G., 1971, *Pattern Recognition* 3:157–168.
Patrick, E. A., 1972, *Fundamentals of Pattern Recognition*, Prentice-Hall, Englewood Cliffs, N.J.
Patrick, E. A., Stellmack, F. P., and Shen, L. Y. L., 1974, *IEEE Trans. Systems, Man, Cybernetics* SMC-4:1–16.

Patterson, J. V., 1973, UK Pat. 1,320,243, June 1973.

Philco, 1966, UK Pats. 1,033,531 and 1,033,532, June 1966.

Phillips, 1974, UK Pat. 1,354, 182, May 1974.

Preston, K., 1972a, *Coherent Optical Computers*, McGraw-Hill, New York.

Preston, K., 1972b, 1972c, US Pat. 3,636,261, Jan. 1972, UK Pat. 1,289,202, Sept. 1972.

Preston, K., and Norgren, P. E., 1972, *Electronics* 45:89–98.

Price, P. R., Crosfield, J. F., Allen, G. S., and Brimelow, P., 1960, UK Pat. 834,125, May 1960.

Pursey, H., 1973, *Optics Laser Technology* 5:24–27.

Recognition Equipment Incorporated, 1968, 1970, UK Pat. 1,115-909, June 1968, US Pat. 3,509,533, Jan. 1970.

Reddy, D. R., 1967, *J. Acoust. Soc. Am.* 42:329–347.

Reid, C. D., and Redman, J. D., 1969, UK Pat. 1,172,539. Dec. 1969.

Rogers, G. L., Leifer, I., and Stephens, N. W. F., 1971, UK Pat. 1,232,029, May 1971.

Rohland, W. S., Traglia, P. J., and Hurley, P. J., 1968, 1968 Fall Joint Computer Conf. *AFIPS Conf. Proc.* 33, (Pt. 2):115–1162.

Rose, H. W., Williamson, T. L., and Yu, F. T. S., 1971, *Appl. Optics* 10:515–518.

Rosenfeld, A., 1972, *Picture Processing by Computer*, Academic Press, New York and London, Section 5.3.

Royston, R. J., 1972, In: *Machine Perception of Patterns and Pictures, Inst. Physics Conf. Ser.* 13:171–180.

Sammon, J. W., 1970, 1972, *IEEE Trans. Computers* C-19:594–616.

Scan-Data Corporation, 1972, UK Pat. 1,271,705, April 1972.

Shelton, G. L., 1962, US Pat. 3,064,519, Nov. 1962.

Slagle, J. R., and Lee, R. C. T., 1971, *Comm. Assoc. Computing Machinery* 14:103–110.

Smith, S. E., and Yau, S. S., 1972, *IEEE Trans. Inf. Theory* IT-18:673–678.

Spanjersberg, A. A., 1974, UK Pat. 1,345,686, Jan. 1974.

Stevens, M. E., 1961, *Automatic Character Recognition*, Technical Note No. 112, National Bureau of Standards, Washington, D.C., 1961.

Styan, P. O., and Mason, S. J., 1974, UK Pat. 1,350,200, April 1974.

Tanaka, K., and Ozawa, K., 1972, *Pattern Recognition* 4:251–262.

Thomson-CSF, 1973, UK Pat. 1,304,193, Jan. 1973.

Tscheschner, W., 1970, UK Pat. 1,216,756, Dec. 1970.

Ullmann, J. R., 1973, *Pattern Recognition Techniques*, Butterworths, London, and Crane Russak, New York.

Ullmann, J. R., 1974a, *Pattern Recognition*, 6:127–135.

Ullmann, J. R., 1974b, *IEEE Trans. Inform. Theory* IT-20:541–543.

Vander Lugt, A., Rotz, F. B., and Klooster, A., 1965, In: *Optical and Electro-Optical Information Processing* (J. T. Tippet, ed.), MIT Press, Cambridge, Mass. and London, pp. 125–141.

Velichko, V. M., and Zagoruyko, N. G., 1970, *Int. J. Man. Machine Studies* 2:223–234.

Watkins, L. S., 1973, *Appl. Optics* 12:1880–1884.

Webber, W. F., 1973, *Machine Processing of Remotely Sensed Data, Conf. Proc.*, Purdue University, Oct. 1973, IEEE Catalogue No. 73 CHO 834-2GE, 1B-1 to 1B-7.

Weinberger, H., and Almi, U., 1971, *Appl. Optics* 10:2482–2487.

Wilson, M. G., 1971, UK Pat. 1,255,759, Dec. 1971: US Pat. 3,626,092, Dec. 1971.

Winograd, T., 1973, *Computer* 6 25–30.

Winston P. H., 1972, In: *Machine Intelligence 7*, (B. Meltzer and D. Michie, eds.), Edinburgh U. Press, Edinburgh, pp. 431–463.

Wolf, J. J., 1972, *J. Acoust. Soc. Am.* 51:(Pt. 2) 2044–2056.

Yachida, M., and Tsuji, S., 1971, *Pattern Recognition* 3:307–323.

Yao, S. S., 1973, In: *Machine Processing of Remotely Sensed Data, Conf. Proc.*, Purdue
    University, Oct. 1973, IEEE Catalogue No. 73 CHO 834-2GE, pp. 1B-8 to 1B-23.
Young, I. T., 1970, In: *Automated Cell Identification and Cell Sorting*, (G. L. Weid and
    G. F. Bahr, eds.), Academic Press, New York, pp. 187–194.
Zahn, C. T., and Roskies, R. Z., 1972, *IEEE Trans. Computers* **C-21**:269–281.

3

Editorial Introduction

In Chapter 3, I. Aleksander begins with the observation that random-access memory is now relatively cheap and offers enormous flexibility as a means of implementing logic functions. In this respect, his contribution to pattern recognition research is unique, although he is not alone in acknowledging the importance of implementation. His distinctive approach relies upon a fundamental reappraisal of the pattern recognition problem and the realization that it may be better to study ways in which cheap, flexible modules can be interconnected to build algorithms, rather than deferring implementation consideration as a "mere detail" to be considered later. Aleksander relates his work to other well-known proposals such as the Perceptron and linear separator (or classifier) (which is also discussed by Batchelor) and points out that there are many seemingly naive operations which a Perceptron cannot perform, but which feedback networks of RAMs can execute properly.

It is interesting to compare this to Chapter 6 by Duff. There is some evidence of "convergent evolution," although the objectives and the proposed solutions differ. The convergence is manifest by the desire of both authors to use up-to-date electronic technology in a parallel-processing machine. In Chapter 4, Batchelor is doing essentially the same, while Ullmann (Chapter 2) is looking ahead to optical processing. This concentration of attention to implementation is strongly felt in Great Britain (although perhaps less so in other countries). Further evidence of this can be seen in the work of W. K. Taylor at University College, London, who has for many years successfully championed analogue "building bricks," while the editor has an interest in applying *microprocessors* to pattern recognition. Aleksander cannot be said to typify such a diverse group, but he is certainly one of its most eloquent disciples and he gives a clear message here that "clever" things can be done by building with simple "bricks."

3

Pattern Recognition with Networks of Memory Elements

I. Aleksander

3.1. INTRODUCTION

This chapter deals with a development of ideas stemming from the early work of Bledsoe and Browning (1959) in which a binary pattern field was broken up into several randomly selected sets of n binary points called n-tuples. The scheme is adaptive and requires a set of patterns with which the system can be trained: the *training set*.

Bledsoe and Browning (and several others, e.g., Ullmann in Chapter 2) saw their scheme as an algorithm for computer-based pattern recognition. However, with recent advances in the technology of solid state circuits, it is possible to envisage the n-tuple and its implied storage merely as a circuit component: the RAM (random-access memory). The first section of this chapter deals with the way in which a RAM may be used as an adaptive element. There are several important practical implications of this, the main one being that an adaptive pattern recognizer made of RAMs (or the Bledsoe and Browning algorithm) can be implemented directly as a network of ROMs (read-only memories). There are several applications for pattern recognizers, such as in reading aids for the blind, instrumentation data interpreters [e.g., mass spectrometers (see Stonham *et al.*, 1973)] or medical aids (for recognizing heart conditions, etc.), where one

I. Aleksander • Department of Electronic and Electrical Engineering, Brunel University, Uxbridge, Middlesex, England

cannot afford to have a sophisticated computer and its attendant software as part of the recognizing equipment. In such cases a ROM implementation of logic learned by the Bledsoe and Browning algorithm or a RAM machine is invaluable.

The classical ways of organizing an n-tuple system as a combinational network are outlined in Section 3.2. This includes a comparison with other adaptive schemes such as the *linear separator* and the *Perceptron*. By considering some limitations of Perceptrons one can discuss similar limitations in combinational nets of RAMs. These limitations center largely around the lack of being able to recognize certain "global" properties of patterns.

It is argued in Section 3.3 that many of these limitations are due to the combinational nature of the RAM networks. In fact, looking at the n-tuple method as a network of RAMs rather than as a computer algorithm leads one to conceptual developments which go far beyond the Bledsoe and Browning schemes. Naturally, these can be simulated, but it helps to think of these concepts as networks of RAMs rather than ways of accessing the main store in a digital computer. Section 3.3, in fact, initially presents some simple sequential circuits which do recognize global pattern properties. The explanation is kept at the level of one-dimensional patterns to prove the point regarding sequentiality (i.e., the presence of feedback) without considering unduly complicated circuits.

In Section 3.4, one considers two other forms of feedback. The first is intended to show that the state variables (i.e., feedback loops) of the system can be made to model their input environment. This is a "clustered" model which, as a result of the clustering, has advantageous pattern recognition properties. The second form of feedback is intended to improve the reaction of single-layer net by making "strong decisions" even in situations where, informationally, the number of bits indicating the presence of a pattern feature is low but significant. This is the sort of situation one encounters when one is trying to distinguish, say, between an O and a Q.

3.2. THE RANDOM-ACCESS MEMORY AS A LEARNING ELEMENT

3.2.1. Description of the Memory Element (ME)

A random-access memory has the following essential sets of terminals (all binary).

Address terminals ($X = \{x_1, x_2, \ldots, x_n\}$): These select one of 2^n flip-flops situated within the device. There is one flip-flop corresponding to every pattern at the input terminals. In learning tasks this set of terminals is called the *input*.

Data-in terminal (t): This single terminal determines what shall be inserted into the flip-flop selected by the address terminal. In the literature on learning systems this is sometimes referred to as the *teach* terminal, as it effectively "teaches" the device what its response should be for a given input pattern at X.

Data-out terminal (z): This again is a single terminal which (nondestructively) outputs the contents of the selected flip-flop. This is the effective logical *output* of the device.

Write clock terminal (w): This single terminal determines the moment at which the binary value on the t terminal may be transferred to the selected flip-flop.

Several other control terminals may be found in practical versions of the above devices, such as a setting or resetting facility for all the flip-flops; or a "device enable" input. For the purposes of this discussion, however, it is sufficient to consider the interaction between the above terminals. This may be formally set out as follows.

Let each flip-flop having a label ϕ_j be selected by the input iff the binary value of X taken as an ordered binary number is j. For example, assume that

$$X = \{x_1, x_2, x_3\}$$

then ϕ_3 will be addressed if $x_1 = 0$, $x_2 = 1$, and $x_3 = 1$. That is "X taken as an ordered binary number" is (0 1 1), decimal 3. Define a variable $\langle X \rangle_j$ which is 1 iff X has a binary value of j in the above manner. That is $\langle X \rangle_3 = 1$ for $x_1 = 0$, $x_2 = 1$, $x_3 = 1$, but all other values of x_1, x_2, x_3, give $\langle X \rangle_3 = 0$. Then if ϕ_j is the present binary state of the jth flip-flop and ϕ_j' its next state, then ϕ_j' equals ϕ_j iff $\langle X \rangle_j = 0$ or $\omega = 0$, otherwise ϕ_j' becomes 1 iff $\langle X \rangle_j = 1$ and $\omega = 1$ and $t = 1$. When expressed as a Boolean equation this becomes

$$\phi_j' = \phi_j\,(\overline{\langle X \rangle_j + \overline{w}}) + \langle X \rangle_j wt \tag{3.1}$$

The above describes the setting or resetting of flip-flops which is, in fact, the way in which the internal state of the device is changed. Each internal state is one of the truth tables for the combinational Boolean function performed by the device between a pattern input X and the output z. This may be expressed as

$$z = \sum_{j=0}^{2^n - 1} \phi_j \langle X \rangle_j \tag{3.2}$$

For example, if for $n = 3$, $\phi_7 = 1$ and all other ϕ_j's $= 0$, then the device will behave as an AND gate because (3.2) becomes

$$z = \langle X \rangle_7 \qquad\qquad \text{i.e., } z = 1 \text{ iff the input pattern is (1 1 1)}$$

Clearly, there are 2^{2^n} states for an n-input element. This means that an element can be "trained" to perform any one of 2^{2^n} possible logic functions of n inputs.

3.2.2. A Note on "Electronic Intelligence"

One often sees references to "artificial intelligence" or electronic networks that behave "intelligently." The word *intelligence* in a technological context remains undefined. However, we wish to define it here with respect to learning automata so as to state as clearly as possible what is meant by this term within the confines of this chapter.

A learning automaton is said to behave "intelligently" if, on the basis of its "training" data, which is provided within some context together with information regarding the desired action, it takes the correct action on other data within the same context not seen during training. [Batchelor (Chapter 4) refers to this as *generalization*.] For example, the context may be the recognition of handprinted numerals. Examples of numerals are fed into the system together with the desired classification. That is, a shakily written 4 may be presented to a system, together with information that the system must output a logical 1 on the fourth of 10 wires (say). This is repeated for examples of other handprinted numerals. The "intelligence" of the system may then be gauged by the proportion of correct responses to unseen patterns.

On this intelligence scale it is seen below that the learning element described above, when left on its own, scores very poorly. Such a device can classify information into two categories only: 0 and 1. Let us say that the device is initially reset with all ϕ_j's at 0, and that it is then trained to output a 1 for certain input patterns. It is clear that the system can respond with a 1 only to the patterns seen during training and no other, because each input pattern corresponds to one and only one ϕ_j as each input is one and only one $\langle X \rangle_j$. Even though such a system would have perfect memory of the classification of patterns seen during training, any nontrivial number of inputs would require a great deal of storage. A classifier of patterns on a 10×10 matrix would require 2^{100} ϕ_j's, that is 10^{30} bits of memory: a high price to pay for zero in intelligence!

It will be shown in the next section that networks of the elements just described are more intelligent, and do not require exorbitant amounts of storage.

3.3. COMBINATIONAL NETWORKS

A typical combinational network of memory elements is the single-layer arrangement shown in Fig. 3.1. This consists of a data input matrix (binary), the set of memory elements (MEs) and an output decision circuit. There are many ways of organizing such a scheme and two will be discussed here. The first describes a system whose output is required to give merely a yes–no decision (dichotomy), while the second is a scheme for classifying inputs into several classes. This will be followed by a description of devices that are not purely

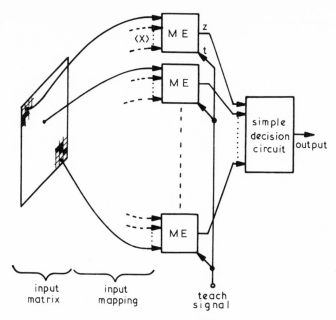

Fig. 3.1. A simple single-layer network.

digital: Perceptrons. A note on the overall limitations of such systems is included in Section 3.3.5.

3.3.1. A Dichotomy System

Assume that the input matrix in Fig. 3.1 has B binary points and that each such point is connected to one and only one randomly chosen ME input terminal. If the MEs have n inputs there are B/n MEs in the system. For the time being assume that the *simple decision circuit* is merely a (B/n)-input AND-gate, and that the t connections are all wired together. Also, assume that all the ϕ_j's in all the MEs are initially set to zero. To understand the "intelligent" properties of the system, assume that the net has been trained to respond with a 1 at the output of each ME (i.e., also with a 1 at the overall output) for two input patterns I_1 and I_2.

Clearly, depending on whether I_1 and I_2 are less or more similar,* less or more MEs will "see" two distinct input $\langle X \rangle_j$'s during training. For example, if I_1 and I_2 differ in only one bit, only one ME can "see" two distinct $\langle X \rangle_j$'s during

---

*Similarity being measured in terms of the number of bits by which the patterns differ (i.e., *Hamming* distance).

training. Now assume a more general situation, where a total of l MEs have "seen" two $\langle X \rangle_j$'s each. Then not only will I_1 and I_2 produce an overall 1, but also precisely $2^l - 2$ other patterns will do the same. This number is given as follows. There are 2^l ways in which one can address precisely those ϕ_j's which have been set to 1 during training, one way for each combination of addresses of the l MEs which have been addressed twice and differently. One subtracts 2 to account for I_1 and I_2.

In general, if I_1 and I_2 are similar in the sense described above, then l will be low, but those patterns provoking a 1-output classification will also be similar to I_1 and I_2. The reader is referred to the literature for a statistical assessment of such systems (Aleksander, 1971).

It is common practice in pattern recognizers not to make the output decision circuit perform an AND function, but rather to let it respond with a 1 iff the number of 1's at its (B/n) inputs exceeds some predetermined threshold. This basically increases the number of patterns to which the system responds positively after training.

It can be seen from the literature quoted above that a strong response (i.e., many MEs responding with 1) is obtained when an unseen pattern is close in Hamming distance (see last footnote for definition) to one seen during training. The general effect of lowering the threshold in the last method mentioned above is to allow patterns with greater Hamming distances from the training set to give a l-response at the overall output and thus be placed in the same class as the training set.

This raises questions of the effect of training and threshold on the apparent "intelligence" of the system. There are, in fact, three parameters involved in this process: number of training patterns T, the value of n, and the threshold θ. For a given problem, all these have optimum values, and generally experimentation is required to find them. For example:

> As T increases, while keeping n and θ constant, the number of $\langle X \rangle_j$'s seen by the MEs increases. Thus the total number of patterns giving an above-threshold response increases. At first, this gives an improvement in "intelligence" (as discussed in Section 3.1.2), but taking T too far will cause erroneous patterns to be drawn into the classification set.

The reader is encouraged to develop his own reasoning to show that the variation of n and θ has a similar effect.

3.3.2. Multicategory Classification

Given several dichotomizing classifiers above, one could imagine that a number C of such devices could classify input data into 2^C categories (with one binary output pattern from the dichotomizers taken as a whole representing each category). Almost by definition this will require some dichotomizers to be trained on widely differing data. For example, in the recognition of alphanumerics (numbers and letters), at least six dichotomizers are required ($2^6 = 64$) each of which would have to be trained on about half of the data set. That is, parts of the system might have to accept that A and Z are in the same class, while there is no obvious similarity between A and Z, and this is bound to give a poor performance.

To overcome this, one generally uses C distinct networks which we call *discriminators*, for a C-class classification. These are organized and trained as follows.

A discriminator is merely a dichotomizer without the output decision circuitry. Thus, the kth discriminator outputs a number r_k, that is, the number of 1's at its output terminals.

The system is trained by "teaching" (after an initial setting to 0) only that discriminator which corresponds to the class of the current input to output all 1's. For example, if handprinted numerals are involved, only the 4th discriminator would be taught when a 4 is input, the 5th for 5, and so on.

After training, one applies the following rule in order to classify an output pattern:

The pattern is classified as belonging to the class whose discriminator has the highest response.

This system has two advantages. First, each discriminator is trained only on patterns that are ostensibly similar, and, second, there is no need to set predetermined thresholds on output decisions, since the output is determined on a comparative basis. The reader is referred to the work of Stonham (1973) for a discussion of a practical application of this work to the automatic classification of mass spectral patterns.

3.3.3. Linear Separators

Although linear separators are discussed in Chapter 4, they are considered here in the context of being possible alternatives to a structure of MEs. A typical device is shown in Fig. 3.2. The weights ω_k in this system are adjustable and training proceeds by applying an input to $i_1 \cdots i_B$ and adjusting the weights (using one of several algorithms) until an input response of 1 is obtained. The threshold θ is also variable, the overall function of the elements being always

$$F = 1 \text{ iff } \sum_B \omega_k i_k > \theta$$

$$F = 0 \text{ otherwise}$$

This type of system is called a *linear* separator owing to the fact that the

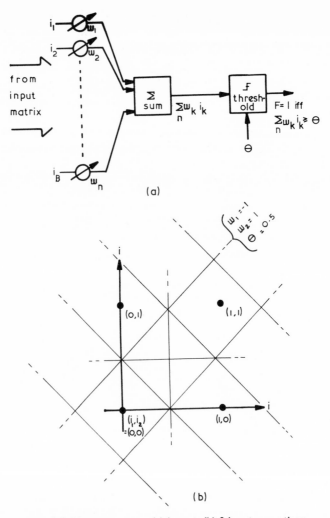

Fig. 3.2. Linear separator: (a) layout; (b) 2-input separations.

above equation is a linear one. Generally the related equation

$$\sum_B \omega_k i_k = \theta$$

is referred to as the "separating function" or "discriminant function." Taking an example of a 2-input device, we see (in Fig. 3.3b) that, due to the binary nature of i_1 and i_2, there are only eight significant ways in which the equation

$$\omega_1 i_1 + \omega_2 i_2 = \theta$$

may be drawn, and these discriminant functions are clearly seen to yield straight lines.

For example, the discriminant function given by

$$-i_1 + i_2 = 0.5 \qquad (\omega_1 = -1, \omega_2 = +1, \theta = 0.5)$$

separates the input pattern $(i_1, i_2) = (0,1)$ from patterns $(0,0)$, $(1,0)$, $(1,1)$. It is left as an exercise for the reader to write down the equations for the other discriminant functions.

One notes in the above example that small variations about values of ω_1, ω_2, and θ do not affect the classification. Also, a function giving the classification effect of, say,

$$-i_1 + i_2 = 0.5$$

may be obtained by training the device to output a 1, say, for pattern $(0,1)$ and a 0 for patterns $(0,0)$ and $(1,1)$. The system may then automatically classify pattern $(1,0)$ as a 0 and this may contribute to its "intelligence." In general, the linearity of linear separators itself is a restricting factor, making them less flexible than ME systems. Also, the need to alter weights and thresholds is more complex in practice than the direct setting of ME flip-flops.

3.3.4. Perceptrons

There is a connection between the Perceptron as proposed by Rosenblatt (1962) and ME systems. In Fig. 3.3 it is seen that a Perceptron contains a linear

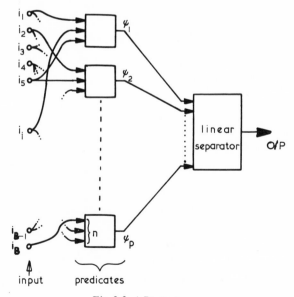

Fig. 3.3. A Perceptron.

separator with the addition of a set of *fixed function* "predicates." The question asked in Perceptron applications is: "Can one find a set of predicates $\psi_1 \cdots \psi_p$ which perform a desired dichotomy on a set of patterns at inputs i_1 to i_B?" Predicate boxes are assumed all to have the same number of inputs n, and n is referred to as the "order" of the Perceptron. The Perceptron, in a sense, includes ME dichotomizers since, if the answer to the above question is "no," then no ME dichotimizer can be found for the dichotomy in question. This is seen from the fact that the predicates are, in fact, "fixed" MEs. The difference between ME dichotomizers and Perceptrons lies in the fact that in the former the "learning" takes place in the predicates themselves, whereas in the latter, the learning takes place in the linear separator.

3.3.5. Limitations of Combinational Learning Nets

The Perceptron concept is more important than appears at first, if looked at solely in the context of learning automata. It is the archetype of a certain class of parallel processing system. The input may be thought of as a mass of data that requires parallel processing, the $\psi_1 \cdots \psi_p$ predicates are independent parallel processes, each of which operates on part of the data. The final output processor gathers the results of the predicates and performs some simple operation such as a linear separation.

However, Minsky and Papert (1969) have shown that Perceptrons, and hence other combinational learning schemes and parallel processes, have certain inherent limitations. These appear in the context of the classification of *global* properties of input data. Typical global properties, for example, are:

Parity:	The detection of all patterns with an even number of 1's
Specificity:	The detection of all patterns with a given specific number of 1's
Uniqueness:	The detection of all patterns with only one 1
Connectedness:	The detection of all patterns with 1 only in mutually adjoining positions.

The thread of Minsky and Papert's argument begins with the thought that the order, n, of the Perceptron must be less than B to avoid triviality. It is easy to show that parity cannot be computed by a Perceptron with, say, ρ, ψ's each with B/ρ inputs: the ψ's sampling disjoint input sets. The best each ψ can do is to compute the parity of its own input. The linear separator would then have to compute parity on the ψ outputs. It is easily shown that parity is not a linearly separable function. Indeed, if it were, there would be no need for the ψ's and the linear separator would suffice.

Similar arguments hold true for the other global properties mentioned above.

It will be seen in the next section that if the ψ's are allowed to communicate with one another, global pattern properties can be computed.

3.4. SEQUENTIAL LEARNING AUTOMATA

In this section we shall first show that automata which allow feedback from the output to the input of Perceptron-like predicates can compute global properties of patterns. This type of consideration does not involve the concept of "training," and it is this concept which will concern us in Section 3.5.

3.5. THE COMPUTATION OF SOME GLOBAL PROPERTIES

To simplify the nature of the problem we assume that the automata have a one-dimensional binary vector at their input. That is, any input i_j has only two neighbors, i_{j-1} and i_{j+1}.

Consider the system shown in Fig. 3.4. The object of the exercise is to show that this system can compute disparity (the inverse of parity), uniqueness, and connectedness with respect to the vector $(i_1 \cdots i_4)$.

First, let us establish that a combinational layer of MEs connected to an AND-gate *cannot* compute these functions with MEs of less than four inputs.

The functions themselves are shown in the form of combinational truth tables below:

Disparity:

		$i_1 i_2$			
F		00	01	11	10
$i_3 i_4$	00	0	1	0	1
	01	1	0	1	0
	11	0	1	0	1
	10	1	0	1	0

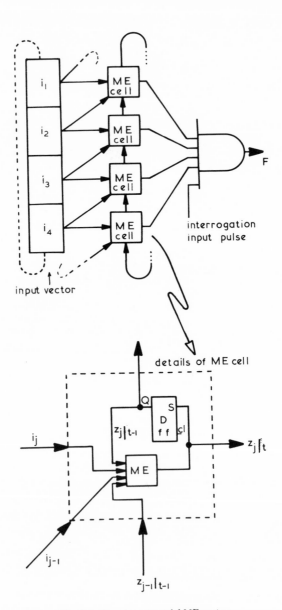

Fig. 3.4. Simple sequential ME system.

Uniqueness:

F	i_1i_2 00	01	11	10
i_3i_4 00	0	1	0	1
01	1	0	0	0
11	0	0	0	0
10	1	0	0	0

Connectedness:

F	i_1i_2 00	01	11	10
i_3i_4 00	0	1	1	1
01	1	0	1	1
11	1	1	1	1
10	1	1	1	0

The output AND-gate in Fig. 3.4 implies that a product-of-sums type of Boolean equation is to be implemented. Indeed, these Boolean equations are:

Disparity:

$$F = (i_1 + i_2 + i_3 + i_4)(\bar{i}_1 + \bar{i}_2 + i_3 + i_4)\ldots$$

(one term for each 0 in the truth table)

Uniqueness:

$$F = (i_1 + i_2 + i_3 + i_4)(\bar{i}_2 + \bar{i}_4)(\bar{i}_1 + \bar{i}_4)(\bar{i}_1 + \bar{i}_3)(\bar{i}_2 + \bar{i}_3)(\bar{i}_1 + \bar{i}_2)(\bar{i}_3 + \bar{i}_4)$$

Connectedness:

$$F = (i_1 + i_2 + i_3 + i_4)(i_1 + \bar{i}_2 + i_3 + \bar{i}_4)(\bar{i}_1 + i_2 + \bar{i}_3 + i_4)$$

It is seen that at least one term of each equation is dependent on all four variables, implying that at least one 4-input ME is required. Indeed, this can perform any function of its four inputs, making other circuitry unnecessary. Hence, the above functions cannot be performed by a single-layer ME net with an AND-gate output.* It is now possible to show that there exists at least one function for the ME of the form

$$|z_j|_t = |f(i_j, i_{j-1}, z_j, z_{j-1})|_{t-1}$$

where $|x|_t$ reads x at time t etc.; for which, assuming that $|z_j|_{t=0} = 0$ for all j, at some time $t = T$ all $|z_j|_T = 1$. In this example we take $T = 4$ as being the crucial time (this generalizes to $T = n$ for an n-element input vector). T is the time at which an interrogation pulse is supplied to the output AND-gate (see Fig. 3.4). It is only at this time that information from any ME output has had the opportunity of propagating to all other MEs. It is noted that due to the property of position invariance of the desired global properties (i.e., 1000 is in the same class as 0010) all the MEs are assumed to perform the same function.

Disparity:

The required function is

$$|z_j|_t = |(i_j \oplus z_{j-1})|_{t-1}$$

where $\oplus$ denotes exclusive-OR. Here $|z_j|_t$ is independent of $|(z_j, i_{j-1})|_{t-1}$. If $i_j = 1$, $|z_j|_t = |\overline{z_{j-1}}|_{t-1}$, whereas if $i_j = 0$, $|z_j|_t = |z_j|_{t-1}$. That is, the system acts as a ring counter with d inversions, where d is the number of 1's in the input vector. In such a system, the state at $T = 4$ (or $T = n$) will be the inversion of the state at $t = 0$ iff d is an odd number, otherwise the state will be the same as the starting state. Given that the starting state is 0000, it is easy to see that the proposed function causes the system to compute disparity.

Uniqueness:

One possible function is

$$|z_j|_t = |(\overline{i_j} \cdot z_{j-1} + i_j(\overline{z_j \cdot z_{j-1}})|_{t-1}$$

This is best expressed as a state transition table for each element, where the feedback loop from z_j back to the input of the element is considered as an internal

---

*The astute reader will have noted that for the connectedness case the function in its sum-of-products form is

$$F = i_1 \overline{i_3} + \overline{i_1} i_3 + i_2 \overline{i_4} + \overline{i_2} i_4$$

This means that the function can be achieved with 2-input MEs and an output OR-gate. In fact, it may be shown that for an n-element vector and an output OR-gate, connectedness may be computed combinationally by means of MEs with $(n - 2)$ inputs.

state of the element itself, with z_{j-1} and i_j as the inputs. The scheme is independent of i_{j-1}.

$\overrightarrow{|(i_j, z_{j-1})|_{t-1}}$		00	01	11	10
$|z_j|_{t-1}$	0	⓪	1 ↑	1 ↑	1 ↑
	1	0	①	0	①

In the above transition table ○ indicates that the feedback variable will not change with time, whereas ↑ and ↓ show the changes which take place when the delay elements are clocked. Again it is assumed that the system may be started with all feedback variables reset to 0. If a 1 is detected at any i_j, the corresponding element is set to 1 as in the 10 column of the transition table. This causes (at the next time instant) the adjacent element to be set to 1 (column 01) and so on. If there is only one i_j at 1 then after n time periods all elements will output a 1. If there are more than one i_j at 1 some elements will be forced to oscillate according to the 11 column before $t = n$. It may be seen that this prevents the AND-gate from firing at $t = n$. Unfortunately, this scheme cannot be used for all values of n, and the reader should consult published work on this subject for details (Aleksander *et al.*, 1974).

Connectedness:

The required function may be derived from that used in the computation of uniqueness and is

$$|z_j|_t = |\overline{(\bar{i}_j \cdot i_{j-1})} \cdot z_{j-1} + (\bar{i}_j \cdot i_{j-1}) \cdot \overline{(z_j \cdot z_{j-1})}|_{t-1}$$

[The variable i_j in the previous equation has now been replaced by the conjunction $(\bar{i}_j \cdot i_{j-1})$.] Instead of detecting the presence of a single $i_j = 1$, as in the computation of uniqueness, the above equation detects the presence of the pair $i_j \cdot i_{j-1}$. If there is one and only one such pair the system goes to the all-1 state at $t = n$. It is noted that this method does not output a 1 for the all-1's input, which it treats as being disconnected. However, this is thought not to be a major disadvantage for large n, since that input pattern is easily detected in other ways. It has thus been shown that by the addition of simple feedback connections many of the Perceptron disadvantages disappear. The reader might not feel that things like disparity and uniqueness are functions one need worry about. That is not the point: they are symptomatic of a large class of problems not available to static Perceptron-like systems, and the point made here is that the solution lies with dynamic networks such as described above.

3.6. OTHER FORMS OF FEEDBACK

3.6.1. OR-ed Feedback

The essence of these systems is that one starts with an inputless automaton consisting of interconnected MEs on which an external input is imposed by OR-ing it with the feedback wires which carry the state variables. In published papers it has been shown that such systems may be *trained* to have several interesting properties (Fairhurst and Aleksander, 1972). These include the ability to "recognize" certain inputs and "remember" the last seen input. The characteristics of such automata will be illustrated here on the basis of the simple structure shown in Fig. 3.5.

Assume that the MEs are initially all reset. Inputs may be obtained from the set

$$I = \{i_0, i_1, \ldots, i_{15}\}$$

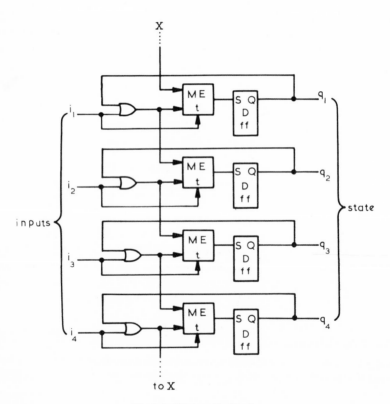

Fig. 3.5. OR-type feedback.

where

$$\mathbf{i_0} \text{ is the } \begin{bmatrix} i_1 \\ i_2 \\ i_3 \\ i_4 \end{bmatrix} \text{ input vector} = \begin{bmatrix} 0 \\ 0 \\ 0 \\ 0 \end{bmatrix}$$

$$\mathbf{i_1} \text{ is the } \begin{bmatrix} i_1 \\ i_2 \\ i_3 \\ i_4 \end{bmatrix} \text{ input vector} = \begin{bmatrix} 0 \\ 0 \\ 0 \\ 1 \end{bmatrix}$$

$$\mathbf{i_2} \text{ is the } \begin{bmatrix} i_1 \\ i_2 \\ i_3 \\ i_4 \end{bmatrix} \text{ input vector} = \begin{bmatrix} 0 \\ 0 \\ 1 \\ 0 \end{bmatrix}$$

and so on in a binary numbering fashion until

$$\mathbf{i_{15}}, \text{ which is the input vector } \begin{bmatrix} 1 \\ 1 \\ 1 \\ 1 \end{bmatrix}$$

The state variables are assumed to be labeled in a similar way for the set of states $Q = \{q_0, q_1, \ldots, q_{15}\}$, where q_0 is the state vector

$$\begin{bmatrix} q_1 \\ q_2 \\ q_3 \\ q_4 \end{bmatrix} = \begin{bmatrix} 0 \\ 0 \\ 0 \\ 0 \end{bmatrix} \qquad \text{etc.}$$

The state itself is considered to be the output of the automaton. Apply the input vectors

$$\mathbf{i_{10}} = \begin{bmatrix} 1 \\ 0 \\ 1 \\ 0 \end{bmatrix} \qquad \text{and} \qquad \mathbf{i_5} = \begin{bmatrix} 0 \\ 1 \\ 0 \\ 1 \end{bmatrix}$$

enabling the memory writing mechanism (t) in each case. Part of the resulting state diagram is shown in Fig. 3.6. Only those inputs which are of interest in this discussion are shown here: the reader is encouraged to investigate what happens with others.

We note that, starting in state q_0, the $\mathbf{i_5}$ and $\mathbf{i_{10}}$ inputs cause an entry into "sympathetic" states q_5 and q_{10}, respectively. This is termed the *perception* or

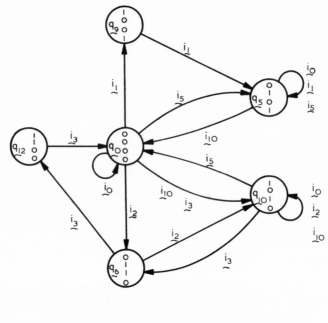

$$i_0 = \begin{bmatrix} 0 \\ 0 \\ 0 \\ 0 \end{bmatrix}; \quad i_1 = \begin{bmatrix} 0 \\ 0 \\ 0 \\ 1 \end{bmatrix}; \quad i_2 = \begin{bmatrix} 0 \\ 0 \\ 1 \\ 0 \end{bmatrix}; \quad i_3 = \begin{bmatrix} 0 \\ 0 \\ 1 \\ 1 \end{bmatrix}; \quad i_5 = \begin{bmatrix} 0 \\ 1 \\ 0 \\ 1 \end{bmatrix}; \quad i_{10} = \begin{bmatrix} 1 \\ 0 \\ 1 \\ 0 \end{bmatrix}$$

Fig. 3.6. Learned state diagram.

recognition of those inputs previously seen during training. There is a very strong element of memory here in the sense that even if the input is removed (i.e., i_0 is applied) the system remains in q_5 or q_{10} having once got there. That is, the system *remembers* the last seen input. The system also displays some "intelligence" in the sense that it classifies i_1 into q_5 and i_2 into q_{10}. Clearly, i_1 differs in only one bit from i_5 (and q_5) as does i_2 from i_{10} (and q_{10}).

Typically, i_3, which is equally dissimilar from i_5 and i_{10}, throws the system into some confusion by causing it to enter a cycle: $q_0 \to q_{10} \to q_6 \to q_{12} \to q_0 \cdots$. Finally, one should note that q_5 and q_{10} are not total "trap" states, as i_{10} can cause an exit from q_5 to q_{10} via q_0 and similarly for i_5 from q_{10} to q_5. In summary, it has been seen that the OR-ed feedback system learns to recognize the "environment" to which it has been exposed (i_5, i_{10}) by the creation of internal states related to this environment (q_5, q_{10}). Environmental events *similar* to i_5 and i_{10} (i.e., i_1 and i_2) are also recognized. The system also exhibits short-term memory of these events. Details of similar behavior in larger networks may be found in the literature mentioned earlier.

3.6.2. Decision Feedback

A second form of feedback is considered here, where the output of the net is fed back to particular inputs in the MEs. For example, a 4-input ME in a matrix can have three inputs connected to the pattern matrix, and its fourth input randomly connected to one of the net outputs. The overall aim here is to make the net aware of its own decisions: that is, *decision feedback*. Broadly, the system is trained by associating the input pattern with a desired decision (on the feedback wires) and then forcing it to make this decision by training. This, under certain conditions, ensures stable areas in the state space of the system which correspond to forceful decisions, as illustrated in the example below.

In a way, this may be related to the need for *action* in animals in response to seen patterns. Take the example of a frog. Assume that it can move in two directions: left by activating one set of muscles L, or right with muscles R. Let us assume that we wish to build a network whose response (in terms of the number of output 1's) determines the amount of effort applied to a muscle. Clearly, we would have two such networks, one for each set of muscles. Say that the frog would be required to move left if it were to see a fly in the left of its field of view, and right if in the right field of view. A single-layer net would not be able to cope with this situation as the effect of a small dot in a large field can at best only produce slightly differing responses in the two nets and later muscles would tend to contract. It will be seen that if the action taken by the net is fed back to the input of the nets, a regeneration action can take place amplifying the contraction of one muscle, while reducing that of the other.

Although the above example is drawn from cybernetics (i.e., a study of living things through the use of formal notions such as automata theory) the processing of speck-like data often occurs in engineering problems. To be assured on this point one has only to think of the need for rapid processing of radar displays, two-dimensional pictures in medicine, or even the identification of small flaws in textiles and metals.

However, we shall proceed with the frog analogy as this gives us a vehicle for a vivid illustration of the problem (Fig. 3.7).

The system is "looking" at a "retina" consisting of B binary points and the "fly" consists of f black binary points among $B - f$ white ones, where $f \ll B$. The net consists of k MEs each with $(n + 1)$ inputs, of which n receive information (through a random connection) from the retina, the additional input receiving information (again through a random connection) from the net outputs.

The net outputs are partitioned into two equal groups, one feeding the left muscle L and the other R. Let l_t and r_t be the number of 1's in the left and right output fields at time t, implying that forces proportional to l_t and r_t are supplied to the respective muscles at that time.

For simplicity, say that the system has been trained on just two patterns,

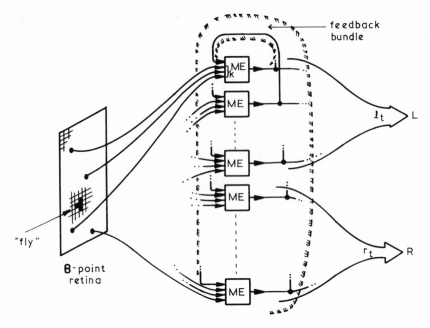

Fig. 3.7. Decision feedback.

T_L and T_R, which represent the fly in the left and the right of the field of view, respectively. Training on T_L, say, consists of applying 1's to the teach terminals of only those MEs associated with muscle L (the L-net) *and* simultaneously setting to 1 all those incoming feedback lines connected to the net output to muscle L. This clearly creates a stable state with $l_t = k/2$ (this is the maximum response for l_t) under the input T_L. One does the same for T_R.

Consider now the sequence of events which takes place if the net starts at $t = 0$ with $l_t = 0$ and $r_t = 0$ and say pattern T_L is present at the retina. One-half of the L-net will have its feedback inputs as they would be during training (at 0) and the other half have 0's, instead of the 1's as during training; thus the response from half of that occurring during training, that is, the L-net will be

$$l_1 = \frac{1}{2} \cdot \frac{k}{2} = \frac{k}{4}$$

The half of the R-net which was trained to respond with 1's on T_R and all 0's at its feedback inputs is likely to respond with $\frac{1}{2}[(k/2) - f]$ logical 1's, since it is likely that f out of the $k/2$ MEs are differently addressed. Thus,

$$r_1 = \frac{k}{4} - \frac{f}{2}$$

We note that the response of two nets, of size $k/4$, one trained on T_L and the other on T_R, would be precisely $k/4$ and $[(k/4) - (f/2)]$ under these conditions. One must now see whether allowing the system to cycle brings about a better separation of these two responses.

It is shown in Aleksander (1974) that, in general,

$$l_{t+1} = \frac{k}{2} \cdot \frac{p}{2}(l_t + r_t + 1)$$

$$r_{t+1} = \frac{k}{2} \cdot \frac{q}{2}(r_t - l_t + 1)$$

where p is the proportion of n-tuples that the test pattern has in common with the totality of the patterns trained to activate L, whereas q is the number of n-tuples it has in common with R. In the above example

$$p = 1 \qquad \text{and} \qquad q = \frac{(k/2 - f)}{(k/2)}$$

It is also shown that the values l_t and r_t converge asymptotically with t to

$$l_\infty = \frac{p(1 - q)}{2 - p - q}$$

where ∞ indicates an asymptotic tendency and

$$r_\infty = \frac{q(1 - p)}{2 - p - q}$$

For the earlier example we see that substituting for p and q

$$l_\infty = 1$$

$$r_\infty = 0$$

which is a perfect response demonstrating the desired improvement.

It is easily seen that this system tends to a perfect response if p or $q = 1$ (but both are *not* 1), which is true for all the patterns in the training set. However, it is left as an exercise for the reader to show that an improved ratio of responses is obtained even when $p \neq 1$ and $q \neq 1$.

Clearly, the above equations and parameters are continuous and therefore only describe the *trends* of the actual system, which is discrete. This means that the learning system will ultimately cycle somewhere near the values given by l_∞ and r_∞.

3.5. SUMMARY

The salient characteristic of the methods presented in this chapter lies in the fact that the proposed systems may be economically constructed using solid-state memory elements. Such elements are readily available, and are benefiting in terms of their cost by the fact that they are being mass-produced in ever-increasing numbers.

The pattern recognition schemes based on such elements are particularly important in areas where a recognition machine is restricted in cost and cannot incorporate a conventional computer with sophisticated software. For example, a blind reading aid which recognizes standard fonts found in books and news-papers and which produces an audible indication of its recognition is being developed in the Electrical Engineering Department at Brunel University. Such a machine when fully implemented with solid-state read-only memories should be economical enough to be made available to large numbers of blind people, possibly under National Health Schemes.

This chapter has also illustrated that dynamic (i.e., sequential) memory element networks succeed in the recognition of properties of patterns where the static recognizer fails. This topic is being taken further in present research, where the sensitivity of dynamic networks to sequences of patterns is being exploited. One hopes to develop schemes for the recognition and correction of languages. At present formal computer languages are being studied, while hopefully this work could extend in the future to natural language.

Most of the structures discussed in this chapter were randomly connected, with the exception of the ring circuit in Section 3.3.1. This work is being actively extended to the study of array-like structures, which hold promise in the area of image processing (see Chapter 6 by Duff). It is felt that the economic implementation and the freedom from predefined programming through the adaptability of the memory elements are likely to provide new avenues of application in this field.

REFERENCES

Aleksander, I., 1971, *Microcircuit Learning Computers*, Mills and Boon, London.

Aleksander, I., 1974, Action-oriented learning networks, *Kybernetes*, 4:39–44.

Aleksander, I., Stonham, T. J., and Wilson, M. J. D., 1974, Adaptive logic for artificially intelligent systems, *Radio Electron. Engr.*, 44:39–44.

Bledsoe, W., and Browning, P., 1959, Pattern recognition and reading by machine, *Proc. Eastern Joint Computer Conf.*, p. 225 American Federation of Information Processing Science, Academic Press, New York.

Fairhurst, M. C., and Aleksander, I., 1972, Dynamics of the perception of patterns in random learning nets. In: *Machine Perception of Pictures and Patterns*, The Physical Society, London.

Minsky, M., and Papert, S., 1969, *Perceptrons*, MIT Press, Cambridge, Mass.

Rosenblatt, F., 1962, *Principles of Neurodynamics*, Spartan Books, New York.

Stonham, T. J., Aleksander, I., Camp, M., Shaw, M. A., and Pike, W. T., 1973, Automatic classification of mass spectra by means of digital learning nets, *Electron Lett.*, 9:No. 17.

4

Editorial Introduction

This chapter is primarily concerned with the problem of making decisions, based upon a set of parallel measurements (descriptors), each of which may assume any value within a finite, continuous range. [This qualification is important since it distinguishes the fields of application of this chapter and those by Aleksander (Chapter 3) and Bell (Chapter 5).] How can decisions be made? How can the decision-making device (the classifier) be designed? How can the classifier be implemented? Subsidiary questions also arise: Does the pattern "resemble" anything that has been seen before now? Is the given pattern "typical" or "atypical" of its class? Can we represent the data structure in a form that is suitable for human visual interpretation? The importance of these questions may be realized from a brief list of application areas:

1. Medical diagnosis, prognosis, choice of therapy
2. Identification of crops from rocket/satellite photographs
3. Automatic quality control

The incentives for this work vary from *hope*, that a machine will be faster, cheaper, and more accurate than a human decision-maker, to the *certain knowledge* that in many applications there is no known solution that could even be called "poor." This chapter is concerned with techniques. Applications are not discussed, since this is done in detail in other chapters. Techniques are described for making complicated decisions in 150 ns (if you have the money to pay for it!) or in 150 ms (if a really cheap solution is required). Methods of making decisions must be selected in the light of the problem requirements; sometimes a fast, crude classifier is needed, while in other situations a precise, slow device is indicated.

4

Classification and Data Analysis in Vector Spaces

B. G. Batchelor

4.1. STATEMENT OF THE PROBLEM

4.1.1. Vectors

Here, as in Chapters 3 and 5, we shall primarily be concerned with methods for making decisions. We shall assume that the primary pattern has already been coded to yield a vector containing numeric descriptors. Such a pattern description is natural in a wide variety of applications, as the following examples show:

1. An autoanalyzer* may be used to define a multielement vector which describes the hormone, protein, salt, and sugar concentrations in human blood.

2. A time-varying signal, such as an EEG or ECG, may be applied to a set of parallel band-pass filters whose outputs are rectified and then integrated. The outputs from the integrators represent the elements of the measurement vector.

3. The color of vegetation, as seen from a satellite, may be used to identify certain crops. A "color" vector might contain three measurements on components from the visible spectrum, as well as ultraviolet or infrared measurements.

---

*A multichannel instrument for performing chemical titrations on a routine basis.

B. G. Batchelor · Department of Electronics, University of Southampton, Southampton, England

4.1.2. Measurement Vectors, Formal Definition

1. A pattern π may be represented by a vector $\mathbf{X}(\pi)$, which contains numeric variables (i.e., binary, ordinal, integer, and real numbers). Each vector $\mathbf{X}(\pi)$ is uniquely specified by π, but the converse is not necessarily true.

2. $\mathbf{X}(\pi)$ contains a fixed number, q, of elements, called descriptors and denoted by $X_1(\pi)$, ..., $X_q(\pi)$. When there is no chance of confusion we shall omit the brackets and index π.

3. The order of the X_i is invariant. Thus (2,3) and (3,2) correspond to different patterns.

4. All X_i have known values. "Don't know" is not allowed.

4.1.3. Spaces

It is convenient to be able to picture a vector as a point in space. This gives us the opportunity of thinking about our subject in visual terms. Of course, we can only do this for two- or three-dimensional spaces ($q = 2,3$), and in a printed book we can only draw two-dimensional spaces. Our subject has largely been developed by intuition, based upon a visual image of a space, and it is a particularly convenient medium for discussing ideas.

The vector $\mathbf{X}(\pi)$ is usually called the *measurement vector* for π, since the $X_i(\pi)$ are usually measurements, or observations, upon π. The corresponding space is called the *measurement space*, $\mathbf{X}$-space, or q-space. The terminology of q-space is identical with that in 3-space, except that the prefix hyper- is added.

4.1.4. Similarity, Compactness, Clusters, and Distance

The following assumption, called the *compactness hypothesis*, is the central axiom of pattern classification research:

> If two patterns π_1 and π_2 are *similar*, then their corresponding measurement vectors $\mathbf{X}(\pi_1)$ and $\mathbf{X}(\pi_2)$ are *almost always close together* in $\mathbf{X}$-space.

The key words are set in italics. The compactness hypothesis is a useful working rule, not a universal truth (Fig. 4.1). We shall endeavor to explain its significance by illustrating the similarity between printed letters (Fig. 4.2). Note the following points:

1. In some cases, an upper case "A" may be more like an "H" than another "A."
2. Font variations lead to large variations of basic shape and the distribution of serifs (Fig. 4.3).
3. Type quality can cause significant variations, even for a fixed font.
4. Upper- and lowercase letters are usually quite dissimilar.

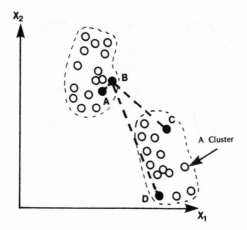

Fig. 4.1. Illustrating the compactness hypothesis and the notion of clustering. Points A and B probably correspond to "similar" patterns, whereas C and D probably represent "dissimilar" patterns. While the line BC is shorter than BD, it should not be inferred that B and C are necessarily more similar than B and D; the compactness hypothesis relates only to highly similar patterns, which are represented by very close points in X-space.

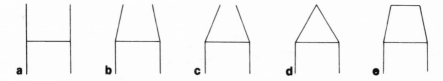

Fig. 4.2. There is no clear distinction between "A" and "H"; there is a continuum of variation between the archetypes, represented in (a) and (d). The character in (e) has a greater resemblance to that in (b), which most people would regard as an "H"; (e) may be regarded as another archetypal "A."

Within a fixed font, we might reasonably expect to find a *cluster* of points in X-space which correspond to printed letters from that font. Another font may yield a cluster which partially overlaps the first. A third font may be separable from the other two (Fig. 4.3b). Another letter may possess a font-cluster which partially overlaps one of the clusters produced by the first letter. We may regard the distributions for a single letter as being composed of a number of *archetypes* which have been "fuzzed" by imperfections in print quality.

In biology, clusters are identified as species, genera, families, and so on. In medicine, clusters of symptoms are called "diseases." Of course, there is a large

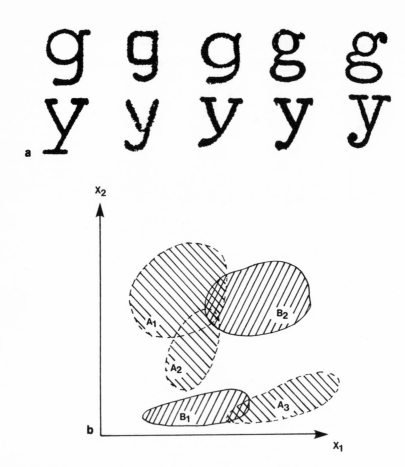

Fig. 4.3. Type font variations suggest the existence of multicluster classes. (a) Type fonts display enormous variety in the basic shape of the letter and in the distribution of serifs. (From Batchelor, 1974a.) (b) Showing a hypothetical situation which might arise in a multifont character recognition problem. Class A contains three clusters A1–A3, while class B contains two clusters B1–B2. The classes are nonseparable in the space shown here.

variety of symptoms within any one disease category, but clustering does seem to occur in many situations (see Fig. 4.10 in Batchelor, 1974a).

What are the implications, if the compactness hypothesis is valid? Suppose that a finite sample of the population contains a cluster which is both small in diameter and far removed from all other clusters. Instead of storing each member of this small, isolated cluster, we need only store its mean vector and diameter. In this idealized case, the compactness leads to a substantial saving in storage (and, as we shall realize later, processing work). It is compactness which allows us to build a machine which can *generalize* to new data, which it has not encountered during its design. Compactness allows a *finite* (and small) machine to be built which can handle data derived from infinitely large sets.

How can we measure distance in q-space? There have been numerous suggestions (Rosenfeld and Pfaltz, 1968; Batchelor, 1971, 1974a). Most of these can be obtained as special cases of the Minkowski r-distance. This is defined as $D_r(\mathbf{U},\mathbf{V})$, where

$$D_r(\mathbf{U},\mathbf{V}) = [\sum_{i=1}^{q} |U_i - V_i|^r]^{1/r} \qquad (4.1)$$

and $\mathbf{U} = (U_1, \ldots, U_q)$, $\mathbf{V} = (V_1, \ldots, V_q)$ are the two points (vectors) whose separation we wish to measure.

When $r = 2$, equation (4.1) defines the *Euclidean* distance. Other distances are frequently used in pattern recognition (see Fig. 4.4):

1. $D_1(\mathbf{U},\mathbf{V})$ is the *city block* (sometimes called *Manhattan*) distance;
2. $D_\infty(\mathbf{U},\mathbf{V})$ is the *square* distance;
3. $[D_1(\mathbf{U},\mathbf{V}) + \sqrt{q}D_\infty(\mathbf{U},\mathbf{V})]$ is the *octagonal* distance.

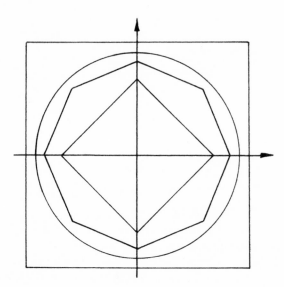

Fig. 4.4. Comparing various distance measures. A figure can be defined for each distance in the same way as the circle is derived from the Euclidean distance. The four circle-analogs are drawn concentrically for ease of comparison. Moving out from the center of the diagram the distances used are: city block, octagonal, Euclidean, square.

Which distance is best? Intuitively the Euclidean distance is most satisfactory, since it is the most familiar in everyday life. However, it is relatively expensive to calculate, compared to the city block or square distance, which do not require the squaring operation. (In many situations the square-root operation of the Euclidean distance can be avoided.) The octagonal distance is attractive, because it is easily computed, and it is a much better approximation to the Euclidean distance than are either the city block or square distance. The generalized Minkowski r-distance is of little more than academic interest, since it is expensive to calculate.

There are some severe disadvantages in using the city block, square, octagonal, or indeed any non-Euclidean distance. The problems are discussed in detail elsewhere (Batchelor, 1973b, 1974a). The discovery of these difficulties led us to reappraise the Euclidean distance, which we had previously regarded as too expensive in computing time/hardware. It is possible now to build a device, using parallel digital techniques, for computing the *squared* Euclidean distance between two given vectors (Fig. 4.19). The cost of such a device is about $35q$ where, as usual, q is the dimensionality. It is fast; using series 74 TTL and a mask-programmed MOS ROM acting as a squarer, the propagation delay is $\sim 1.5\ \mu s$. Using Schottky TTL and a bipolar ROM would increase the cost by a factor of about 5, but the speed would be improved by a factor of $10-15$. Let us briefly consider the situation in software. Many large modern computers possess a hardware multiplier, which allows the Euclidean distance to be calculated almost as rapidly as the city block distance. The CDC 7600 computer can find the squared Euclidean distance between two 64-dimensional vectors in about $100\ \mu s$.

Modern electronic hardware and computers can both be used to calculate the squared Euclidean distance at high speed and at reasonable cost. Hence, there is little need to consider the non-Euclidean distances further, since they are known to produce undesirable effects in many situations.

4.1.5. Teaching Data, Formal Definition

We are given a set of P measurement vectors, $\mathbf{X}(\pi_1), \ldots, \mathbf{X}(\pi_p)$. The only other information available to us during the classifier training phase is a set of decisions $T(\pi_1), \ldots, T(\pi_p)$ which give the *correct classifications* of $\pi_1, \ldots, \pi_p$.[*] Note that T is a function of π, not $\mathbf{X}(\pi)$. This means that $\mathbf{X}(\pi)$ may not contain enough information for us to predict $T(\pi)$ with perfect accuracy. For simplicity, we restrict our attention to 2-class problems; T may take the values ± 1. We shall

---

[*] T is sometimes called the *teacher* decision. The teacher is a conceptual aid and may have no physical existence. The teacher is that (device, observation, experiment, or person) which defines T.

say that π [or $\mathbf{X}(\pi)$] is in class 1, if $T = +1$ and in class 2, if $T = -1$. It is not unduly restrictive to allow only two classes, since any multiclass problem can be replaced by a number of 2-class subproblems.

4.1.6. The Classification Problem, Formal Definition

A *classifier* is a "black box" with input $\mathbf{X}$ and output $M(\mathbf{X})$, $(M = \pm 1)$. In order to design the classifier we are given the information

$$\left.\begin{array}{c} \mathbf{X}(\pi_i) \\ T(\pi_i) \end{array}\right\} \qquad i = 1, \ldots, P \tag{4.2}$$

We are required to design the classifier so that M is a good predictor of T. That is, we wish to design a classifier so that the quantity

$$R \stackrel{\Delta}{=} \sum_{i=1}^{P} \frac{M[\mathbf{X}(\pi_i)] \, T(\pi_i)}{2} \tag{4.3}$$

is a maximum. The quantity $(P - R)/P$ lies in the range $[0,1]$ and is called the *error rate*.

Newcomers to this subject often question why we need to design a classifier which calculates M when we already have the teacher. There are a number of possible reasons:

1. T might be an *a posteriori* statement, while M is a *prediction*.
2. It might be too expensive to continue using the signal T.
3. It might be impossible to obtain the decision T at sufficiently high speed.
4. A machine is deterministic, whereas a human teacher is liable to change as a result of fatigue, mood changes, or recent experience.

4.1.7. A Trivial Example

Given the measurement vector (height, weight) it is possible to predict the sex of an individual with reasonable accuracy. In Fig. 4.5, we plot a scattergram which demonstrates this. By choosing some suitable boundary, it is possible to partition this space into two regions and thereby separate the sexes. This boundary is called a *decision surface*. The problem of pattern classification may be regarded as that of optimizing the position of a decision surface. We use a decision surface because it is more *economical* to store a (small) number of parameters which fix its position, rather than retaining a large number of high-dimensional vectors. This economy is a direct result of the compactness hypothesis. Note that in Fig. 4.5 we obtained two clusters, corresponding to the two

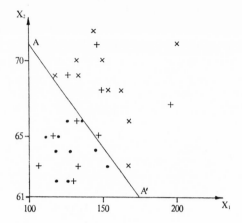

Fig. 4.5. A trivial classification problem, requiring the sexing of 10 people (+) given their measurement vectors (weight, height). (Weight, X_1, in pounds; height, X_2, in inches.) Data are provided on 10 women (•) and 10 men (×) upon which the classifier is designed. No surface can perfectly separate the sexes, but a straight line (AA') provides a nearly optimal solution. Points below this line are classified as women; above the line as men. (From Batchelor, 1974a.)

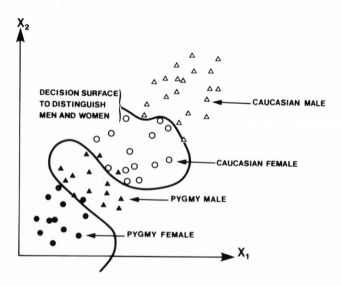

Fig. 4.6. The data collected from Fig. 4.5 were obtained from adult, white Caucasians. If we included Pygmy races, the situation would be more complicated. Pygmies are smaller and lighter than Europeans. (These data are hypothetical, not based on direct measurement.)

sexes. This indicates that men and women have different archetypes (sexual dimorphism). Note also that these clusters overlap, showing that it is impossible to distinguish men and women perfectly on the basis of these two measurements. Numerous people have informed me that there are better ways of distinguishing men and women! In a more realistic pattern classification problem, it is not quite so obvious which measurements most effectively separate the classes.

Let us take this problem a little further. Figure 4.5 shows data which was collected from adult Europeans. If we had included other human races the picture would be very different (Fig. 4.6). Note that in Fig. 4.6 it would be necessary to use a complicated decision surface, in order to distinguish men and women.

4.1.8. A Medical Example

We were presented with medical records on 200 women. These records consisted of the following measurements and observations: age; height; weight; volume of urine excreted/day; concentration of hormones in the blood (two quantities were given); number of pregnancies resulting in live births; number of pregnancies resulting in stillbirths; patient's estimate of the total time she has spent breast feeding (all children). Each woman had been diagnosed either as having cancer of the breast or as having no sign of cancer. We were asked to design a classifier to perform this diagnosis. The results are discussed by Batchelor (1974a).

4.2. CLASSIFICATION METHODS

4.2.1. Introduction

Many classification schemes have been discussed in the literature, although in this chapter we can describe only three. Until recently, the author, like most other workers, had one favorite classification method, which he tried to apply to all problems. This approach is not ideal; pattern classification problems are too variable for any single technique to be optimal in all situations.

4.2.2. Linear Classifier (LC)

Let W_i ($i = 1, \ldots, q + 1$) be parameters, called *weights*, and let X_{q+1} be equal to +1 for all patterns. A linear classifier* computes a binary decision M,

---

*The definition given in equation (4.4) has the merit of being homogeneous in X_i and W_i.

where

$$M = \text{sgn} \left\{ \sum_{i=1}^{q+1} W_i X_i \right\} \tag{4.4}$$

and

$$\text{sgn} (\alpha) = \begin{cases} +1, & \alpha \geqslant 0 \\ -1, & \alpha < 0 \end{cases} \tag{4.5}$$

The decision surface defined by (4.4) is a *straight line* in (X_1, X_2)-space (Fig. 4.7); a *plane* in (X_1, X_2, X_3)-space; and a *hyperplane* in $(X_1, \ldots, X_q)$-space. The hyperplane does not intercept the origin, unless $W_{q+1} = 0$. The weight vector $(W_1, \ldots, W_q)$ controls the orientation of the hyperplane. It is clear that this vector and W_{q+1} must be optimized for each new problem. One method for performing this optimization is given in the appendix (Section 4.6) to this chapter. Further methods are discussed by Nilsson (1965) and Batchelor (1968).

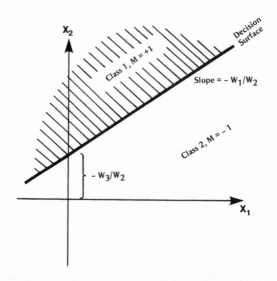

Fig. 4.7. Showing the decision surface of the linear classifier $M = \text{sgn}[\Sigma_{i=1}^{3} (X_i W_i)]$, where $X_3 \equiv 1$. The homogeneous equation is preferable to the alternative form $M = \text{sgn}[\Sigma_{i=1}^{2} (X_i W_i) + W_3]$. The advantage of the homogeneous definition is felt when we consider the optimization of the weights $W_1 - W_3$.

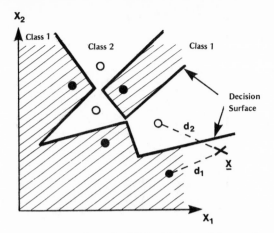

Fig. 4.8. Showing the decision surface of a NNC which uses 4 locates from class 1 (●) and 3 locates from class 2 (○). To classify the point X, the NNC finds the closest locate and attributes X to the same class. In this example, X is classified as belonging to class 1 since $d_1 < d_2$.

4.2.3. Nearest-Neighbor Classifier (NNC)

The NNC requires that we store

N_+ vectors which represent class 1. Denote these by $\mathbf{B}_i^+$.
N_- vectors which represent class 2. Denote these by $\mathbf{B}_i^-$.

The decision of the NNC is given by M, where*

$$M = \text{sgn}\{ \min_{i=1,\ldots,N_-} ([D_2(\mathbf{B}_i^-,\mathbf{X})]^2) - \min_{i=1,\ldots,N_+} ([D_2(\mathbf{B}_i^+,\mathbf{X})]^2)\} \quad (4.6)$$

As before, $D_2(\cdot)$ is the Euclidean distance.

The NNC is easily understood with reference to Fig. 4.8. It is important to note that its decision surface is *piecewise* linear. In general, the complexity of the decision surface is an increasing function of N_+ and N_-. In a practical problem, we must calculate optimal positions for the $\mathbf{B}_i^+$ and $\mathbf{B}_i^-$, which are usually called *locates*. Moreover, we must decide upon suitable values for N_+ and N_-, remembering that the total quantity of data stored by the NNC is equal to $(N_+ + N_-)q$ real, scalar variables. Two techniques for finding suitable values for these parameters are given in the appendix (Section 4.6) to this chapter. Other methods are described by Ford *et al.* (1970) and Batchelor (1974a).

---

*Equation (4.6) is written in this way to emphasize that we need not compute the Euclidean distance; the squared Euclidean distance is perfectly satisfactory.

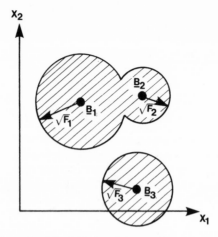

Fig. 4.9. Decision surface of a CC which uses 3 locates $B_1 - B_3$. The locate B_i determines the center of the circle and its radius is given by F_i. The CC produces a decision $M = +1$ for all points within the cross-hatched region; $M = -1$ elsewhere.

4.2.4. Compound Classifier (CC)

The CC requires that we store the following parameters:

N *size parameters*, denoted by $F_1, \ldots, F_N$
N locates, denoted by $B_1, \ldots, B_N$

The decision of the CC is given by M, where

$$M = \operatorname{sgn} \left\{ \max_{i=1, \ldots, N} [F_i - D_2^2(B_i, X)] \right\} \tag{4.7}$$

Again, $D_2(\cdot)$ is the Euclidean distance.* Equation (4.7) is just one of many equivalent forms. The following form is probably more convenient in practice:

$$M = \begin{cases} +1, & \text{if } D_2^2(B_i, X) \geqslant F_i \quad \text{for any } i = 1, \ldots, N \\ -1, & \text{if } D_2^2(B_i, X) < F_i \quad \text{for all } \ i = 1, \ldots, N \end{cases} \tag{4.8}$$

In 2-space, the decision surface produced by a CC consists of segments of circles (Fig. 4.9). The flexibility of the CC can be seen by dropping a handful of coins onto a table. (First remove all noncircular coins and those with holes in them.) The "shadow" cast on the table is associated with the decision $M = +1$. All regions of the tabletop which are not covered by one, or more, coins are associated with the decision $M = -1$. Notice that circles may, but need not, overlap. In multidimensional space, the decision surface of a CC consists of segments of hyperspheres.

---

*Equation (4.7) is written in this way to avoid the square root operation.

The locates, B_i determine the centers of the circles. The radii of the circles are given by F_i. The number of circles, N, their positions, B_i ($i = 1, \ldots, N$), and size parameters, F_i ($i = 1, \ldots, N$), must be optimized for each new problem. A procedure to do this is presented in the appendix (Section 4.6) and is discussed in detail by Batchelor (1974a).

4.2.5. Development of Ideas on Classification

Many pattern recognition problems have been investigated using a LC, and often it has proved to be perfectly adequate. We shall see later that the LC is easily implemented in hardware or software. However, there are certain situations in which the LC is unable to provide a low error rate. For example, problems which require that we discriminate between "normal" and "abnormal" are ill-suited to the LC. In this type of problem the "normal" class is usually compact and embedded within a shell containing the class "abnormal" (Fig. 4.10). Problems of this kind are, of course, ideally suited to the CC, less well to the NNC.

We should distinguish between small-store and large-store classifiers. A small-store device is one in which the number of locates is small, say, 5–20, while a large-store device might hold hundreds, or even thousands, of locates. (Between these two extremes there is, of course, an indeterminate area.) There is a further distinction between large-store and small-store classifiers.

Small-store classifiers are normally designed by sophisticated procedures, which "condense" the essential information out of a large data set. It is, of course, important to minimize the size of the classifier if it is to be used on large data sets; a large-store classifier would be too slow to allow it to be used on large data sets. A small-store classifier can be compared to a polynomial interpolation function. Once such a function has been derived, its design data may be discarded. The polynomial function requires few parameters to be stored and indicates the general trend of the given data. In the same way, a small-store classifier attempts to approximate the general trend of the (Bayes') optimal decision surface. By doing this, we may effect a significant reduction in the storage requirements of the classifier without seriously degrading its performance.

Large-store classifiers are usually designed using fairly simple procedures. The result is that greater processing work will be needed by a large-store device during its classify phase. A large-store classifier has the potential to be more accurate than a small-store system but, as we have already explained, where speed is at a premium, the small-store classifier is preferred.

The CC was intended by its inventor to be a small-store classifier (Batchelor, 1968, 1973a, 1974a). Most authors regard the NNC as a large-store classifier, although there is considerable interest in using it with reduced data storage (Hart, 1968). Occasionally, attempts have been made to use the NNC as a small-store classifier (Ford *et al.*, 1970), but the CC is better suited to this role (Batchelor, 1974a). The LC is, of course, a small-store classifier.

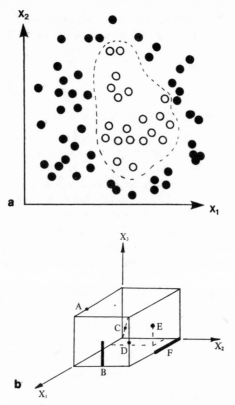

Fig. 4.10. Discriminating between "normal" and "abnormal" classes. (a) Hypothetical distribution in 2-space. Open circles represent "normal" patterns and solid circles "abnormal."
(b) Distribution of the vector (X_1, X_2, X_3) in various kidney diseases, where X_1 = protein
concentration in urine; X_2 = daily urinary output volume; X_3 = specific gravity of urine.
Disease categories are: A, acute nephritis; B, nephrotic syndrome; C, normal; D, acute renal
infection; E, essential hypertension; F, chronic renal failure. (Fig. (b) taken from Batchelor,
1974a.)

4.3. ADDITIONAL DATA ANALYSIS TECHNIQUES

4.3.1. Introduction

Unlike classification, the procedures which form the subject of this section
do not have easily defined objectives, except that they are intended to support
and complement the pattern classification techniques discussed above. These
additional procedures are intended to provide a greater understanding of the
"structure" of the data. The following are perhaps the most important questions
that require consideration:

1. What is the best error rate that can be obtained with the given data?
2. Which descriptors provide the best basis for discriminating between the given groups?
3. Which patterns are "typical"/"atypical" of their class?
4. Which patterns lie in the overlap region (are likely to be misclassified)? (These patterns are not necessarily the same as the "atypical" patterns identified in question 3.)
5. How many clusters are present in each class? How can the coclassed clusters be isolated?
6. Where are the clusters situated and how large are they?
7. Which clusters are responsible for the overlap between classes? (This may be the same as question 4.) Should those clusters which lie within the overlap region be attributed to a new class?
8. How can we display the intraclass structure and interclass relationships to a human observer? Can he be provided with a means for searching for data structure, while interacting with the computer?

A large amount of research has been dedicated to answering these deceivingly simple questions but much more work is still needed. Many procedures have been developed for each of these questions, but in the short space available here, we can introduce only a few of the most significant techniques.

Question 2 is regarded by many people as the central problem of pattern recognition. Despite this, very little progress has been achieved. Difficulties arise because descriptors cannot be assessed in isolation. A complete vector can be evaluated in terms of the discriminating power it provides, but its component descriptors cannot be scored individually. Mucciardi and Gose (1971) compare seven methods for selecting a good subset of the given descriptors. Several of these adopt the heuristic that descriptors which alone provide good discriminating power will, when taken together, provide a "good" vector. Other workers have attempted to use an evolutionary approach of modifying a vector at random and investigating whether the changes achieve an improved ability to separate the classes. This is successful where the pattern classes are nearly linearly separable and are known to be so (Wilkins and Batchelor, 1970). To solve the more general problem requires a procedure for estimating the best attainable error rate. However, the existing methods for this are very slow and could not be used inside an iterative loop, such as that required by an evolutionary procedure. Evolution without some form of bias in the "birth" and "death" processes can be extremely slow. We do not, at the moment, know how to introduce bias effectively. It seems, at the moment, that human judgment combined with heuristic procedures, such as those described by Mucciardi and Gose, provide the best basis for feature selection. Progress with the other questions listed above could well change this situation.

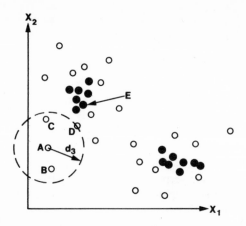

Fig. 4.11. Illustrating the k-NN method of estimating probability density functions. The PDF at point A is to be estimated using the 3-NN method. Point D is the 3-NN to A; C is the 2-NN and B the 1-NN. The PDF at A can be calculated, using the formula given in (4.9). Notice that the 3-NN to point D is *not* A. (In fact, it is point E.) Open circles will yield low values of estimated PDF; solid circles, high values.

Many of the questions listed above can be reduced to the same basic problem, namely, estimating a probability density function (PDF) from a finite (and probably small) set of vectors.

4.3.2. Estimating a Probability Density Function

For the moment we shall consider only one class, which is represented by a finite sample (S) of vectors which have been selected at random. Let $f(\mathbf{X})$ denote the PDF at the point $\mathbf{X}$ and $\hat{f}_k(\mathbf{X})$ be an estimate of $f(\mathbf{X})$. (The parameter k is explained later.) Let $d_k(\mathbf{X})$ be the Euclidean distance from $\mathbf{X}$ to its kth nearest neighbor (k-NN) among the vectors contained in S. An estimate of $f(\mathbf{X})$, called the *k-NN estimate*, may be calculated as follows (Fig. 4.11):

$$\hat{f}_k(\mathbf{X}) = \frac{A(q, k, S)}{[d_k(\mathbf{X})]^q} \tag{4.9}$$

where $A(q, k, S)$ is a constant for varying $\mathbf{X}$. In fact $A(q, k, S)$ need not normally be calculated, since *relative* values of $\hat{f}_k(\mathbf{X})$ are required for different $\mathbf{X}$. It should also be noticed that $A(q, k, S)$ depends only on the *number* of vectors in S, so that interclass comparisons of estimated PDF may be made.* The theoreti-

---

*$A(q, k, S)$ is given by

$$\frac{q\Gamma(q/2) \cdot (k - 1)}{2\pi^{(q/2)} P(S)}$$

where $\Gamma(\cdot)$ is the gamma function and $P(S)$ is the number of vectors in S.

cal analysis of the k-NN PDF estimate is well developed. For example, in the large sample situation (S infinite) and subject to certain constraints on k (Fukunaga, 1972), the estimate $\hat{f}_k(\mathbf{X}) \to f(\mathbf{X})$. The choice of the parameter k is important. If too large a value of k is selected, the estimate is inaccurate, because the true PDF, $f(\mathbf{X})$, varies within that neighborhood of $\mathbf{X}$ which is being used to calculate the value of $\hat{f}_k(\mathbf{X})$. [This neighborhood is a hypersphere, centered on $\mathbf{X}$ and having a radius $d_k(\mathbf{X})$.] If k is too small, inaccuracies occur which are apparent as "noise" on $\hat{f}_k(\mathbf{X})$. Fortunately, only crude estimates are needed in many of the applications for PDF estimators. The major disadvantage of the k-NN method lies in the long computation time required to obtain the estimate $\hat{f}_k(\mathbf{X})$. Suppose that we wished to find the PDFs at every point in a set S containing $P(S)$ points, using a "leave one out" principle. Then there would be a total of

$$\tfrac{1}{2}P(S) \cdot [P(S) - 1] \doteq \tfrac{1}{2}[P(S)]^2 \tag{4.10}$$

distance calculations required.

Other forms of PDF estimator have been developed. For example, the Parzen estimate $\hat{f}(\mathbf{X})$ is calculated in the following way:

$$\hat{f}(\mathbf{X}) = \frac{h}{P(S)} \sum_{\mathbf{X}_i \, \epsilon \, S} \phi \left\{ \frac{\mathbf{X} - \mathbf{X}_i}{h} \right\} \tag{4.11}$$

where $\mathbf{X}_i$ are the vectors from the sample S, h is a positive constant, and $\phi(\mathbf{y})$ is any one of a large class of multivariate functions, such as

$$\left. \begin{array}{c} \exp\,(-|\mathbf{y}|) \\[2mm] \dfrac{1}{1 + \mathbf{y} \cdot \mathbf{y}} \end{array} \right\} \tag{4.12}$$

(see Fukunaga, 1972). The Parzen PDF estimate is also well understood, following detailed mathematical analysis. It is difficult to use in practice, because the parameter h must be chosen carefully. If h is too small, the estimate $\hat{f}(\mathbf{X})$ is "noisy," and if h is too large, the estimator is insensitive to the variations in $f(\mathbf{X})$. It is not unusual to find that the parameter h is optimal in one part of the space but grossly suboptimal in another. The choice of h is perhaps a little more difficult than that of k in the k-NN method.

Other PDF estimators should be mentioned, notably *orthogonal function expansions* (Fukunaga, 1972) and *adaptive sample set construction* (Sebestyen, 1962; Sebestyen and Edie, 1966; Mucciardi and Gose, 1971). Both of these estimators attempt to model the PDF, $f(\mathbf{X})$, using a small-store procedure. Orthogonal series expansion bears a close resemblance to Fourier series expansion; $f(\mathbf{X})$ is approximated by a series, such as (4.13):

$$\sum_{\mathbf{T}} A_{\mathbf{T}} \exp\,(j\mathbf{T} \cdot \mathbf{X}) \tag{4.13}$$

where $\mathbf{T}$ is a q-dimensional vector containing integer elements; $j = \sqrt{(-1)}$; and $A_{\mathbf{T}}$ are complex coefficients. Provided the PDF, $f(\mathbf{X})$, is *smooth*, containing few peaks, a series expansion can be an economical method of approximating it (Anderson, 1969).

Adaptive sample set construction typifies the *ad hoc* approach to PDF estimation. (Of course, this does not find favor with purists who desire to have a pedigree of mathematical theory for their procedures. On the other hand, pragmatists ignore such niceties and concentrate immediately on the important questions of computational feasibility and whether the procedure is effective and useful in a practical situation.) Adaptive sample set construction is a fast procedure, develops a small-store estimator and, according to Mucciardi and Gose (1972) is useful. (This last statement is based on experimentation.) Adaptive sample set construction develops an estimate which is a mixture of normal distributions. Such a mixture has the form:

$$\sum_i A_i \exp\left[-(\mathbf{X} - \mathbf{X}_i)^T \, \mathbf{B}_i(\mathbf{X} - \mathbf{X}_i)\right] \tag{4.14}$$

where A_i are scalars, controlling the mixture proportions; $\mathbf{B}_i$ are diagonal matrices, determining the lengths of the semiaxes of the elliptical contours of the ith "hill" in the mixture; and $\mathbf{X}_i$ are vectors determining the positions of the "hills" in the mixture. (See Fig. 4.12.) These parameters and the number of terms in the mixture are determined in an iterative procedure, which closely resembles the training technique for the CC.

We cannot leave this topic without mentioning the parametric methods, such as that due to Wolfe (1967, 1970). He models $f(\mathbf{X})$ by a mixture of normal distributions, but he allows the matrices $\mathbf{B}_i$ to be nondiagonal. This has the effect of allowing the elliptical contours of the hills to rotate to any arbitrary

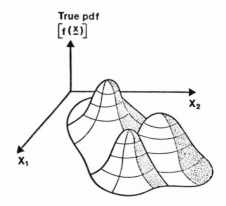

Fig. 4.12. A mixture of normal distributions.

orientation. Wolfe calculates the parameters A_i, X_i, B_i in a modified Newton–Raphson procedure for solving the maximum-likelihood equations.

4.3.3. Using Probability Density Function Estimates

4.3.3.1. Estimating the Best Attainable Error Rate

Let $f_1(X)$ be the PDF for class 1 at the point X; similarly for $f_2(X)$; $\hat{f}_1(X)$ be an estimate of $f_1(X)$; $\hat{f}_2(X)$ is an estimate of $f_2(X)$; θ_1 be the probability of selecting a point from class 1; similarly for θ_2; and $\hat{\theta}_1$ be an estimate of θ_1. $\hat{\theta}_2$ is an estimate of θ_2. Then the best attainable classifier is the Bayes' classifier which calculates a decision B, where

$$B = \text{sgn}\left[\theta_1 f_1(X) - \theta_2 f_2(X)\right] \tag{4.15}$$

The probability that B is in error is (assuming $B = +1$)

$$\frac{\theta_1 f_1(X)}{\theta_1 f_1(X) + \theta_2 f_2(X)} \tag{4.16}$$

In general, the probability that B is in error, at a given point X is $E(X)$, where

$$E(X) = \frac{\min\left[\theta_1 f_1(X), \theta_2 f_2(X)\right]}{\theta_1 f_1(X) + \theta_2 f_2(X)} \tag{4.17}$$

To find the error probability (β) over the whole space, we must integrate $E(X)$:

$$\beta = \int_{-\infty}^{+\infty} \cdots \int_{-\infty}^{+\infty} E(X) \cdot dX_1 \cdots dX_q \tag{4.18}$$

This quantity, β, is important because it measures the quality of the data. Of course, the form of the PDFs, $f_1(X)$ and $f_2(X)$, is not known precisely. While we have assumed estimates $\hat{f}_1(X)$ and $\hat{f}_2(X)$ exist, we do not propose using these in a multidimensional integration. It is easier to estimate β by counting the mis-classifications produced by that classifier which uses equation (4.15) to calculate its decisions but with the *estimates* $\hat{\theta}_1$, $\hat{\theta}_2$, $\hat{f}_1(X)$, and $\hat{f}_2(X)$ replacing the *true* values θ_1, θ_2, $f_1(X)$, and $f_2(X)$. It should be emphasized here that the classifier defined by the equation

$$\hat{B} = \text{sgn}(\hat{\theta}_1 \hat{f}_1(X) - \hat{\theta}_2 \hat{f}_2(X)) \tag{4.19}$$

could be used as an alternative to the NNC. It is a large-store classifier (assuming that $\hat{f}_1(X)$ and $\hat{f}_2(X)$ are large-store estimators). In fact, if $\hat{\theta}_1 = \hat{\theta}_2$ and $\hat{f}_1(X)$ and $\hat{f}_2(X)$ are both 1-NN PDF estimates, (4.19) is an alternative method of stating the NNC decision rule. Cover and Hart (1967) analyzed a related situation, which is applicable to this special case. They found that the error rate of that

NNC, which stores an infinitely large set of locates, selected at random from the parent populations, is equal to some quantity η, such that

$$\beta \leqslant \eta \leqslant \beta(2 - \beta) \qquad\qquad (4.20)$$

Rearranging this we find that

$$\tfrac{1}{2}\eta \leqslant \beta \leqslant \eta, \qquad \eta \ll 1 \qquad\qquad (4.21)$$

By measuring η we could estimate a range of possible values for β. Of course, this is impossible in practice, since η is the error rate of an infinitely large NNC. The author has investigated the practical, finite situation experimentally, using synthetic data which allow the Bayes' error probability β to be measured directly. The author's conclusions must be stated here without justification, since there is insufficient space to give experimental details. It is apparent (from about 25 experiments with varying β) that (4.21) is a good working rule, even if the error rate η is estimated by quite a small NNC (250 locates) (see Batchelor and Hand, 1975).

4.3.3.2. Selecting Typical and Atypical Points

"Atypical" points are those lying in regions of the X-space having low values of the PDF and may easily be detected by simply thresholding the estimated PDF. Ford (1974) and Wilkins and Ford (1972) have emphasized the importance of identifying "strays" (their term). They found that iterative learning schemes, such as those described in the appendix (Section 4.6) to this chapter, can be improved by first removing strays. They found that the learning procedures converged faster as a result. It is also important to identify strays so that they can be given a reduced "weight" during learning. Thus, stray points will have a smaller effect on the movement of the decision surface than will those points lying close to the cluster centers.

It is also important to be able to recognize "atypical" points for another reason. In certain situations, very great importance is placed on the decisions of a classifier. It is then essential to make sure that the classifier is competent to make a decision reliably. This can be achieved using a secondary device (D). Let C be the classifier. D decides whether C is able to make a judgment, while C actually calculates the decision. Hence, C discriminates between two groups (say A and B), while D checks that the pattern is likely to have been derived from their union (A ∪ B). The device D could be a secondary classifier or a PDF estimator followed by a simple threshold. This problem is considered again in Section 4.5.4.

"Typical" points lie close to the cluster centers. Clearly there should be no misclassifications among these points, since this would indicate a gross misplacement of the decision surface. It is valuable therefore to be able to assess the performance of a classifier in terms of a graded list of errors, such as that indicated in the following table.

	Class 1	Class 2
Typical points	$\epsilon_{1,1}$	$\epsilon_{2,1}$
Points with average PDF	$\epsilon_{1,2}$	$\epsilon_{2,2}$
Atypical points	$\epsilon_{1,3}$	$\epsilon_{2,3}$

Clearly, errors at the top are more serious than those at the bottom of this table.

4.3.3.3. Pictorial Representation of Hyperspaces

Suppose that two PDF estimates $\hat{f}_1(\mathbf{X}_i)$ and $\hat{f}_2(\mathbf{X}_i)$ have been calculated for all $\mathbf{X}_i \epsilon S$. A scattergram with axes $\hat{f}_1$ and $\hat{f}_2$ has some interesting properties (Fig. 4.13). "Typical" and "atypical" points can readily be identified. The optimal* decisions can be obtained by referring the points to a line at 45° to the coordinate axes. The overlap between the pattern classes is preserved, although cluster structure *may* not be. By displaying colored points, the interclass relationships will be easily assimilated. In fact, the scattergram is ideal for displaying results, particularly to a nonspecialist audience and a visual display is usually greatly preferred in an application, such as medical diagnosis where a doctor often wishes to receive a little more easily assimilated information than a contingency table can provide.

A display with the same coordinate axes could also be used in an interactive system. An operator sits at a VDU and views a map which displays just one point. The operator has the facility to move a point in hyperspace. The PDF estimators map this into two dimensions, where the subject can observe the effects of the hyperspace movements immediately they occur. This should enable the operator to find areas of high PDF, count them, and find where and how large they are. To be effective this mapping must be fast, which places the technique outside the scope of a general-purpose computer. Suppose that the PDF estimators used are based upon a set containing 10^4 points. Then a parallel hardware system could be built which could calculate these PDFs within 2–20 ms. (We shall see in Section 4.4.7 that this is feasible.)

4.3.4. Cluster Analysis

This section is concerned with procedures which can answer questions (5) and (6) in Section 4.3.1. The subject of cluster analysis is a large one and the reader is referred to the following specialist books: Sokal and Sneath (1963), Jardine and Sibson (1971), Everitt (1974), Patrick (1972), and Fukunaga (1972).

---

*These decisions would be optimal if and only if the PDF estimators were perfect.

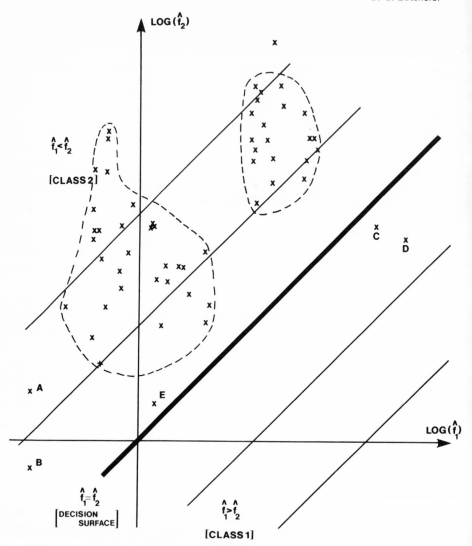

Fig. 4.13. A two-dimensional map, produced using two PDF estimates $\hat{f}_1$ and $\hat{f}_2$. The decision surface on such a map is a straight line at $45°$ to the coordinate axes (heavy line). Contours of constant likelihood ratio $(\hat{f}_1 : \hat{f}_2)$ are parallel to it (faint lines). (Notice that logarithmic axes are used.) The distribution of points on such a map can give the user a "feel" for the problem. For example: (1) Points A and B are "atypical" of class 2, since $\hat{f}_2$ is low. (For clarity, only class 2 points are displayed.) (2) Points C and D are misclassified even though they have moderately large values for $\hat{f}_2$. (3) Point E, although correctly classified, is close to the decision surface and is atypical of class 2. (4) Points falling within the broken curve, which was drawn by hand, can be regarded as being typical of class 2. *Details of the PDF estimators:* 5-NN estimates, 100 points/class, two-dimensional artificial data having multiple clusters in each class.

The border between cluster analysis and PDF estimation is fuzzy. Cluster analysis is concerned with counting and locating *local* peaks in the PDF, dividing the training set into its component clusters. Pattern recognition workers have often been concerned with these problems and have usually sought iterative solutions in preference to the long analysis procedures, required by cluster analysis. To pattern recognition workers the subject is called *nonsupervised learning*, while biologists, psychologists, and other behavioral scientists refer to it as *cluster analysis*. The objectives of these two groups are the same but the sizes of the data sets and subsequently the methods they employ are often different. Apart from this dichotomy, there is a large interest in the problems of visually displaying the "structure" of a set of data.

4.3.4.1. Traditional Cluster Analysis Procedures

The most popular approach, and the only one we can consider here, uses what are called *dendrograms*. These are treelike structures, which display clustering in a hierarchical manner (Fig. 4.14). The lengths of the branches of these trees are taken to represent "similarity," which is often (but not always) measured by distance in the hyperspace. "Adjacency" of two points in the hyperspace is indicated by their being represented as a pair of nodes on the dendrogram with edges to a common node of higher order (nearer to the "root" of the tree). The author has produced trees with 320 terminal nodes ("leaves") which, when drawn on a computer graph-plotter, cover a chart of length 1.6 m. It is doubtful whether the human eye can assimilate this information properly. Perhaps this is the point at which we discard these techniques, since the computation time also becomes prohibitively long. For example, the ICL 1907 computer took 2000 s to compute the distance matrix and perform the cluster analysis on 320 points in 19-space. Furthermore, the computation time rises as the square of the number of points and linearly with dimensionality.

4.3.4.2. Nonsupervised Learning

This area receives its name from the fact that there is no teacher (or, if there is, he has already separated the classes and we are concerned with only one). Adaptive sample set construction is one of the techniques that has been developed. Mattson and Dammon (1965) devised another method. A linear surface is placed so that the data set is partitioned into two relatively compact subsets, which are well separated. (The exact criterion used for fixing the position of this hyperplane offers scope for a multitude of refinements.) Then, the same procedure is applied to these two subsets separately. A hierarchy of partitions is developed by recursive application of the procedure. It then remains to discard any unwanted partitions which do not separate clusters but which divide a single cluster into two parts.

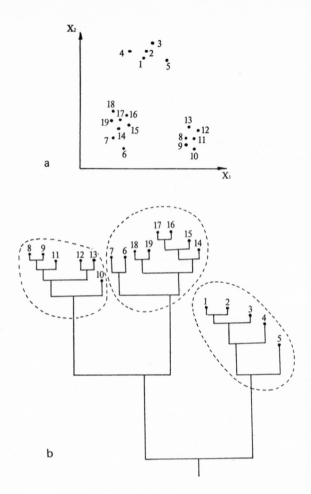

Fig. 4.14. Dendrogram for a simple distribution in 2-space. The power of the dendrogram lies in its ability to represent data structure in a convenient form for human visual interpretation. (a) Descriptor space containing 3 clusters. (b) Dendrogram. Notice how clusters are mapped. (From Batchelor, 1974a.)

Among the fully automated cluster location schemes, there is one called *maximindist* which can be used in conjunction with any convenient PDF estimator (Batchelor, 1974a; Batchelor and Wilkins, 1969; Patrick, 1972). Each point within a set S is scored by estimating the PDF at that point. "Typical" points will, of course, have high values for the estimated PDF. We cannot simply threshold these scores to find the cluster centers, since some clusters will be

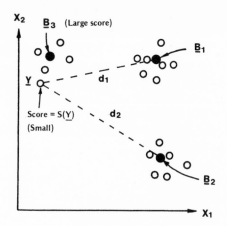

Fig. 4.15. Maximindist. $S(Y)$ is a "score" on point Y (e.g., estimated PDF) B_1 is that point whose score is greatest and is stored first by maximindist. In the situation shown here, a second representative B_2 has been found and stored; the procedure is searching for a third vector to store (which will be B_3). Maximindist calculates $H(Y) = S(Y) \cdot \min_{i=1,2} [D_2(B_i, Y)]$ for each Y in the training set. That vector which maximizes $H(Y)$ is taken to be the next stored representative. Subsequent vectors are added to maximindist's store in a similar manner. (Also see Program 4.5.)

much denser than others. Maximindist identifies clusters as regions of high estimated PDF which are far removed from other such regions. How this is done should be clear from Fig. 4.15. Apart from the preliminary scoring, which can be quite crude, maximindist is a relatively fast procedure.

4.3.4.3. Visual Display of Data Structure

A great variety of techniques exists. One such method is briefly discussed in Chapter 15 and another in Section 4.3.3.3. Sammon (1969) devised a procedure which maps a multivariate space into two dimensions, in such a way that the distances between close neighbors in the two spaces are preserved as far as possible. It is a slow procedure and does not allow new points to be mapped easily. Much less sophisticated procedures have also been applied. The favorite is a map onto the plane containing the two principal eigenvectors of the pattern matrix $[X(\pi_1), \ldots, X(\pi_p)]$. Alternatively, the maximum and second largest diameters of the convex hull of the pattern points in hyperspace may be found by linear programming. These two diameters can be used to define a plane (just as with the principal eigenvector method). All points are then projected onto this plane, which is the required map, showing the data structure.

4.4. IMPLEMENTATION IN HARDWARE

4.4.1. Introduction

However effective in theory, no classifier or data analysis procedure will ever be used if it cannot be realized in hardware or software. It is clear that the distance calculation is central to the CC and NNC, as well as to many of the data analysis procedures discussed in the previous section. We shall therefore pay particular attention to this problem.

Pattern classification does not usually require high-precision computing. In many instances, the descriptors (X_i) are coarsely quantized. It is common to find binary variables in pattern recognition data, which indicates the presence or absence of some features. We frequently encounter coarsely quantized variables, which can be coded in 2 or 3 bits. This is likely to occur in data that have been derived from subjective judgments (Miller, 1967; Martin, 1973). Conventional analog computing methods (0.5% error) and short-byte digital techniques ($\leqslant 8$ bits/descriptor) are normally quite acceptable.

4.4.2. Linear Classifier

A LC may be implemented very conveniently in analog hardware, using at most two operational amplifiers (Fig. 4.16). The weights $W_1, \ldots, W_{q+1}$ are inversely proportional to the computing resistors $R_1, \ldots, R_{q+1}$, which should be close-tolerance, low-drift components.

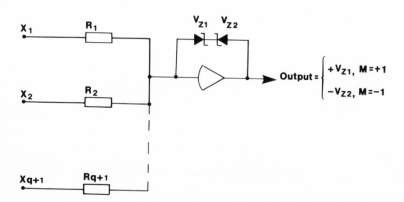

Fig. 4.16. Implementation of a linear classifier in analog hardware. No attempt is made to maximize the speed of operation with phase-correcting capacitors. The circuit implements the equation: $M = \text{sgn}(-\Sigma_{i=1}^{q+1} X_i/R_i)$.

This simple classifier is cheap, reliable, and can calculate a decision in 1 ms. Higher speeds could be achieved by adding phase-correcting capacitors in parallel with the computing resistors.

The LC may also be implemented easily using software. Its digital hardware realization is relatively expensive (or slow), due to the large number of multiplications needed. Multiplication by a constant may conveniently be achieved in digital equipment using a ROM, although this is expensive compared to multiplication in analog hardware using resistors.

4.4.3. Distance Calculations Simplified for Analog Hardware

Before considering the implementation of the CC and NNC let us reformulate the calculations that we are required to perform. The NNC is defined by Equation (4.6).

$$M = \text{sgn}\{ \min_{i=1,\,\ldots,\,N_-} ([D_2(\mathbf{B}_i^-,\mathbf{X})]^2) - \min_{i=1,\,\ldots,\,N_+} ([D_2(\mathbf{B}_i^+,\mathbf{X})]^2)\} \ (4.6)$$

Notice that

$$[D_2(\mathbf{B}_i^+,\mathbf{X})]^2 = \sum_{j=1}^{q} (B_{i,j}^+)^2 - 2 \sum_{j=1}^{q} (B_{i,j}^+ X_j) + \sum_{j=1}^{q} X_j^2 \qquad (4.22)$$

A similar relationship holds for $\mathbf{B}_i^-$. Equation (4.6) may be rewritten in the following form:

$$M = \text{sgn}\{ \min_{i=1,\,\ldots,\,N_-} [A_i^- - \sum_{j=1}^{q} (B_{i,j}^- X_j)] - \min_{i=1,\,\ldots,\,N_+} [A_i^+ - \sum_{j=1}^{q} (B_{i,j}^+ X_j)]\}$$

$$(4.23)$$

where

$$\left. \begin{array}{l} A_i^+ = \dfrac{1}{2} \sum_{j=1}^{q} (B_{i,j}^+)^2 \\[2em] A_i^- = \dfrac{1}{2} \sum_{j=1}^{q} (B_{i,j}^-)^2 \end{array} \right\} \qquad (4.24)$$

The classifier defined by (4.23) produces identical decisions with the NNC defined by (4.6). In this new form it is known as a *piecewise linear classifier* (Nilsson, 1965). Notice that there are no squaring operations at all.* This makes the NNC much easier to implement in analog hardware (Fig. 4.17). Note the following features of this implementation:

---

*A_i^+ and A_i^- are fixed absolute terms in the final classifier.

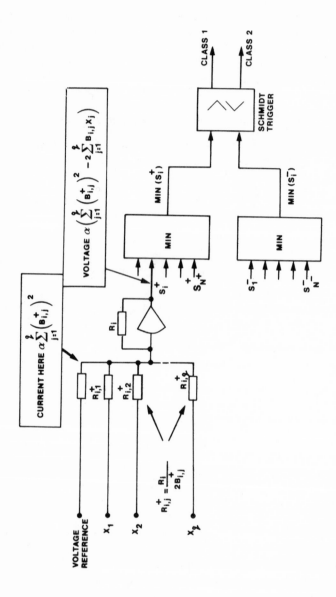

Fig. 4.17. Implementation of a NNC using analog hardware. $S_1^+, \ldots, S_N^+$ and $S_1^-, \ldots, S_{N_-}^-$ are all calculated in the same way as S_i^+. The locates are defined by the resistors $R_{i,j}$. The minimum selector box is a standard analog computing device and only requires one operational amplifier and $(N_\pm + 1)$ diodes.

1. The function sgn($\cdot$) requires a Schmidt trigger circuit.
2. The function min($\cdot$) can be executed conveniently in a circuit similar to a simple diode AND gate.
3. The summation can be performed using a conventional analog summer, and the weights ($B_{i,j}^-$ and $B_{i,j}^+$) can be achieved using fixed resistors.

The CC equation (4.7) may also be rewritten into a form that enables us to use a simple implementation in analog hardware:

$$M = \text{sgn}\{ \max_{i=1,\ldots,N} [E_i + \sum_{j=1}^{q} (A_{i,j} X_j) - \sum_{j=1}^{q} (X_j^2)]\} \qquad (4.25)$$

where

$$\left. \begin{array}{ll} E_i = F_i - \sum_{j=1}^{q} (B_{i,j})^2 & i = 1,\ldots,N \\[2mm] A_{i,j} = 2B_{i,j} & i = 1,\ldots,N; \quad j = 1,\ldots,q \end{array} \right\} \qquad (4.26)$$

The quantity $\sum_{j=1}^{q} X_j^2$ is the only quadratic term in (4.25) involving the descriptors X_j. The squaring operations may be performed conveniently using IC analog multipliers, which are currently available at about \$3 each. The remainder of (4.25) may be implemented using techniques already described in relation to the NNC. The final circuit is shown in Fig. 4.18.

4.4.4. Distance Calculations Using Digital Hardware

In analog circuits, we may very easily multiply a variable signal (voltage) by a constant, whereas squaring is more difficult. The reverse is true in digital hardware, Hence, equations (4.23) and (4.25) are not especially helpful when we are trying to implement the NNC and CC in digital hardware.

Squaring a digital number may conveniently be achieved using a ROM.* If we require a large enough number of squaring operations (all identical) to be performed in parallel, it is economical to employ mask-programmed ROMs. Equation (4.25) would require multiplications by ($N \cdot q$) *different* constants. It would not be economic to use mask-programming here; we must use expensive PROMs.

Locates (and size parameters) could be stored in PROMs, fusible-link ROMs, or hardwiring.

A fast distance-calculator is shown in Fig. 4.19. Its performance figures have already been given in Section 4.1.4.

---

*Remember that we need only provide 8-bits' precision for the X_j.

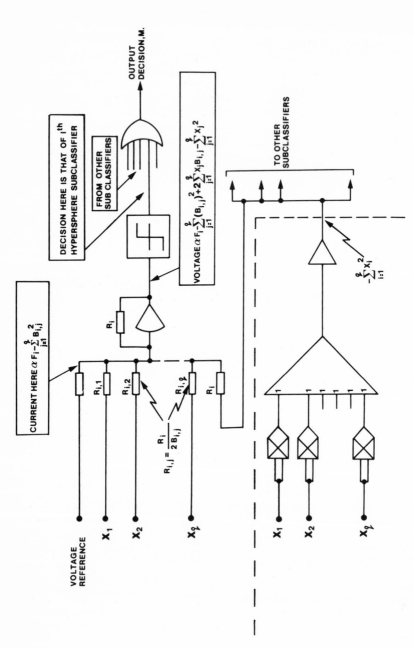

Fig. 4.18. Implementation of a CC using analog hardware. Only the ith hypersphere subclassifier and the common equipment are shown. q multipliers are used, whatever number of subclassifiers is incorporated into the CC. The remainder of the equipment is cheap and robust.

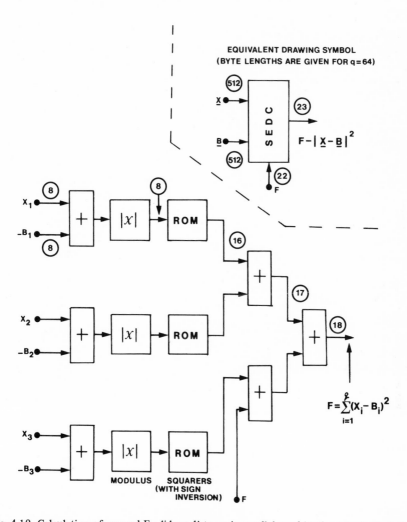

Fig. 4.19. Calculation of squared Euclidean distance in parallel combinational logic. Each X_i and B_i is specified with an accuracy of 8 bits, while F_i requires $16 + \lceil \log_2 [q] \rceil$ bits. Byte lengths are shown as encircled numbers. Each ROM has an 8-bit address and hence possesses 2^8 (=256) storage locations. Its output byte-length is 16 bits. Hence the ROMs must store 256×16 bits each. This can currently be achieved using two IC packages.

4.4.5. Compound Classifier Implemented in Digital Hardware

To achieve very high speed the fully parallel implementation of the CC should be used (Fig. 4.20). This has N distance calculators in parallel. It is important to note that hardwiring could be used for storing the $B_{i,j}$ and F_i, since there is only one locate and one size parameter associated with each distance calculator. In addition, note that Fig. 4.20 implements the CC decision rule given by (4.8).

For medium speed a serial-parallel configuration is best (Fig. 4.21). Locates and size parameters would then be stored in PROMs or fusible-link ROMs.

4.4.6. Nearest-Neighbor Classifier Implemented in Digital Hardware

A (large-store) NNC uses too many locates to allow a fully parallel system to be built. It may conveniently be implemented by a structure similar to that

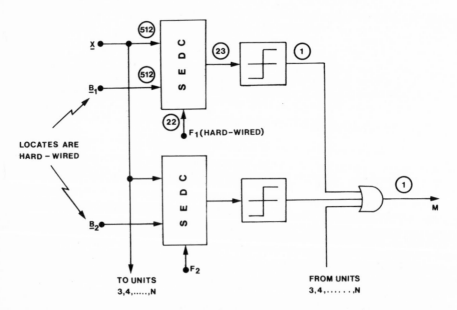

Fig. 4.20. Implementation of the CC in parallel digital hardware. The boxes SEDC are defined in Fig. 4.19. The propagation delay of this circuit is 1.5 μs, using series 74 TTL and MOS ROM squarers in the SEDC; and 150 ns, using series 74 TTL and bipolar PROM squarers. Further speed improvements, at increased cost, could be achieved using ECL or Schottky logic. It is assumed that $q = 64$ and that each descriptor X_i is specified to an accuracy of 8 bits.

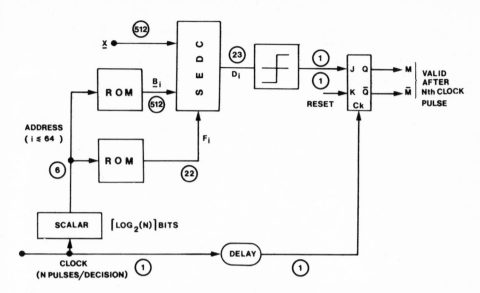

Fig. 4.21. Implementation of a CC in serial-parallel digital hardware. The box SEDC is defined in Fig. 4.19. It is assumed that $q = N = 64$. N locates $(\mathbf{B}_1, \ldots, \mathbf{B}_N)$ are stored in the upper ROM, while the lower ROM stores the size parameters, $F_1, \ldots, F_N$. The $\mathbf{B}_i$ (and F_i) are accessed in sequence by the scalar. The output of the box SEDC is equal to D_i, where $D_i = F_i - \Sigma_{j=1}^{q} (B_{i,j} - X_j)^2$, where i is the output signal from the scalar. The threshold detector computes $\mathrm{sgn}(D_i)$ while the $J-K$ flip-flop forms the logical union of these signals. The delay ensures that the threshold detector has reached a steady state before the $J-K$ bistable device is clocked.

used for the medium-speed CC (Fig. 4.21). The only significant modification is the addition of two minimum-selectors (Fig. 4.22). Such a system can calculate a decision in about $[1.5 \cdot \max(N_-, N_+)]$ ms (the implementation is assumed to use TTL and MOS ROMs).

4.4.7. Auxiliary Data Analysis Procedures

The distance calculator is so fast that supplying it with data is a major problem, although this may be overcome by time-sharing. Let us briefly reconsider k-NN PDF estimation. To estimate the PDF at $\mathbf{X}(\pi)$ we must cycle once through a large set of vectors, S. We shall assume that S is too large to store in electronic memory and is held in a mechanical device such as a disc or drum. The slow access time of these bulk-storage media would be a serious problem, if we were to estimate the PDF at $X(\pi_1)$ on one cycle, at $X(\pi_2)$ on the next cycle, and so on. However, we can *concurrently* compute the PDFs at $X(\pi_1)$, $X(\pi_2)$, ..., $X(\pi_\delta)$, where

$$\delta = \frac{\text{Access time of bulk store}}{\text{Time to calculate distance using the distance calculator}} \qquad (4.27)$$

A system to do this is shown in Fig. 4.23. In recent papers, Batchelor (1974b) and Hand and Batchelor (1974) described a similar system, based upon a MOS feedback shift register.

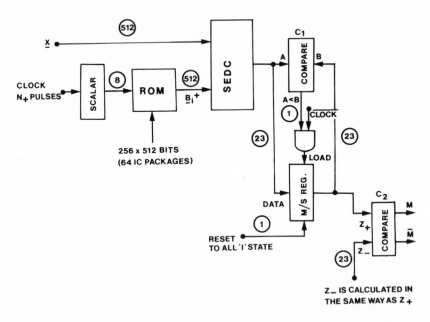

Fig. 4.22. Digital hardware realization of the NNC. The box SEDC is defined in Fig. 4.19. Locates $B_1, \ldots, B_{N_+}$ are held in the ROM and are accessed, in sequence, by the scalar. The signal A is the squared Euclidean distance between X and the currently selected locate. The comparator (C_1) and master–slave register search for the minimum signal appearing on A. After all N_+ locates have been examined, the output of the master–slave register (Z_+) is compared (using C_2) to the equivalent signal (Z_-) for the opposing class. The output (M) is indicated by the smaller of Z_+ and Z_-.

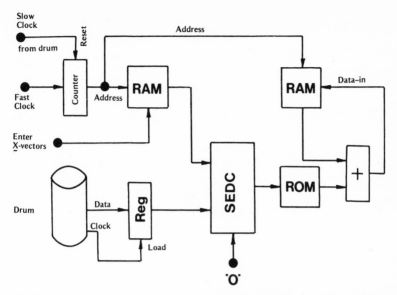

Fig. 4.23. Implementation of the Parzen PDF estimator in digital hardware. Suppose that the box SEDC (see Fig. 4.19) can calculate the squared Euclidean distance between two vectors at H times the speed of the drum clock. Then both RAMs should contain H storage locations. The RAM on the left must be able to store H vectors and that on the right H fixed-point numbers. H input vectors $(\mathbf{X}_1, \ldots, \mathbf{X}_H)$ are loaded into the leftmost RAM. The counter providing addresses for this device is clocked at high speed. At each drum position, the $\mathbf{X}_i$–RAM is cycled once, to calculate partial PDFs for each of the $\mathbf{X}_i$. After one drum revolution the PDFs at each of the $\mathbf{X}_i (i = 1, \ldots, H)$ have been calculated. Hence by using concurrent computing the full speed of the SEDC has been employed. The ROM contains a table of values for the function $\phi(\cdot)$ [see Equation (4.20) in text].

4.5. CURRENT RESEARCH

4.5.1. Introduction

The classification techniques discussed so far in this chapter are all based on the same simple form of pattern description and teacher labeling. It has recently been realized that a large number of applications involve data which do not fit this conceptual model. We shall briefly describe three new types of problems, which illustrate the diversity of current research in pattern classification.

4.5.2. Missing Descriptors

It is not uncommon to find that some elements of the measurement vector have unknown values. For example, we might obtain any one of the following vectors as representing the same pattern:

$$(-6.3, -3.4, ?, ?, -26, 17.6)$$

$$(?, -3.4, 16.7, -3.5, ?, ?)$$

$$(-6.3, ?, ?, ?, -26, 17.6)$$

In most cases the omissions are random, although some descriptors will often be more robust than others.

Missing elements occur in a variety of situations and are especially frequent in data collected using questionnaires. The most significant application yielding these types of data is found when a complex multichannel instrument calculates the pattern vector. When one (or a few) channel fails, the system should continue to operate, even though its error rate will be increased. In other words, the partial failure of the measuring system should not cause complete collapse of the service provided by the machine; there should be a gradual deterioration in the performance as more channels develop faults.

One of the author's research students, D. J. Hand, has recently begun an investigation of this problem. He has considered three methods of solution:

1. Estimating replacement values for the missing descriptors prior to using established classification techniques
2. Storing a decision surface for every contingency
3. Estimating the PDFs in q-space and integrating to approximate the marginal PDFs in lower order spaces

Of these, only method 3 appears to offer any hope of solving the general problem, while method 2 is only feasible in certain special cases. Method 1 does not appear to be very promising (Hand and Batchelor, 1974).

4.5.3. Labeled Sets

The conventional form of pattern labeling is such that patterns are individually assigned by the teacher to one of two classes. A novel type of classification problem has been found in which *sets* of patterns are labeled.

Let G be the set of pattern vectors:

$$(X_1, \ldots, X_N)$$

(N may be different for every set G.) The set G is labeled by the teacher as follows:

$T = +1$: One or more of the $X_i \in G$ also lies within an undefined set H.
$T = -1$: All X_i are known to be outside H.

In X-space the set H has an unknown boundary. The difficulty arises in trying to find an approximation to this boundary.

Batchelor (1974a) shows the importance of this form of pattern labeling in time-series analysis with references to speech recognition and electroencephalography (also see Chapter 15).

4.5.4. Recognizing Familiar Patterns

A group of recognition problems has been discovered in which it is possible to collect a representative sample of patterns from only one class. The other class is so rare or ill-defined that it is not feasible to obtain a set of patterns which properly typify that class. Such problems occur in a variety of situations, for example:

1. Detecting pollution in water
2. Recognizing faults in engines
3. Defining a "reject" class for a conventional classifier
4. Detecting changes in the statistical parameters of the environment (for example, a radar intruder-detector should learn to ignore the reflections from vegetation while being prepared for more sinister activity)

The principal features of this type of problem are:

1. We are given a finite sample taken from only one set of patterns.
2. We are required to define the *boundary* of that class.
3. This boundary should enclose *most* members of the given class.
4. This boundary should not enclose large *voids*, which contain few points from the given class.

The optimal solution is a contour of the PDF for the given class. Most PDF estimators place greater emphasis on the regions of X-space which are associated with high values of the PDF. In this case, low-value PDFs are more important, since the required boundary surface is a low-probability contour. For this reason the problem is, at the moment, unsolved.

4.6. APPENDIX: CLASSIFIER DESIGN METHODS

4.6.1. Introduction

No classifier will ever be used successfully if it does not possess an effective procedure for estimating suitable values for its stored parameters (usually called *learning*). No article on pattern classification would be complete without some discussion of this important topic. It is impossible in a short chapter such as this to select more than one or two items for detailed discussion. Hence we have chosen one learning scheme for the LC, one for the CC, and two for the

NNC. All of these procedures are described in detail elsewhere, and the reader is strongly urged to refer to the literature if he intends to use these learning techniques.

4.6.2. Procedure Definitions Using Iverson Language (APL)

It would not be possible to define the classifier design procedures in such a short chapter without the use of a powerful language. The Iverson language (Iverson, 1962), often called **APL**, was found to offer distinct advantages over other "languages" (ALGOL, FORTRAN, conventional mathematics, flow charts). APL is used both as a practical computing language and as a notation for defining procedures for other people to read. It is unfortunate that APL is not more widely available as a computing language, in view of its ability to manipulate vectors, matrices, sets, and trees as complete units. As a programming language, APL is usually interpreted at run-time, rather than compiled. Hence it is slow in execution, although APL programs can usually be debugged quickly in an interactive mode.

APL allows us to define procedures in a compact, precise form. When first encountered the language is virtually incomprehensible. Even a few hours of study reveal its power and will convince the reader that it represents a significant advance in communicating ideas. It is valuable to note that even a "high-level" flow chart defining the learning scheme for the CC covers three printed pages (Batchelor, 1974a). The equivalent FORTRAN IV program would occupy from 10 to 15 pages! Compare this with programs 3 and 4. Readers who are unfamiliar with APL can find the learning procedures defined (less elegantly!) elsewhere (Batchelor and Wilkins, 1968; Batchelor, 1968, 1973a, 1974a).

The books by Hellerman (1967), Lewin (1972), Katzan (1970), and Martin (1973) offer easier introductions to the language than the definitive work by Iverson (1962). The programs listed below employ only a small subset of a very large language.

4.6.3. Data Selection

We must preface any discussion of learning by some general remarks regarding shuffling of data into random order. The behavior of *any* iterative learning scheme is sensitive to the order in which the data are presented. Ford (1974) demonstrates this for a linear classifier, applied to Fisher's Iris data. Ford found that reversing the order of data presentation between two learning experiments gave quite different results. It is good practice to shuffle the vectors into random order before we begin learning. It is even better to use a random data selector, since this has the effect of continuously changing the order of data presentation during learning. A satisfactory pseudorandom data selector may be produced as shown in Program 4.1.

Program 4.1. Random Data Selector

Step No.	Statement	Comment
1	$r \leftarrow (\perp?(10) + 1) \div 1024$	r is random, uniformly distributed in $[0,1]$
2	$r : 0.5; (>) \rightarrow (7)$	$r \leqslant 0.5$, select class 2 $r > 0.5$, select class 1
3	$t \leftarrow ^-1$	Teacher decision $= -1$
4	$r \leftarrow \lceil \mu(\mathbf{Y}) \times (\perp?(20) + 1) \div (2 * 20)$	r is uniformly distributed in $[1, \mu(\mathbf{Y})]$; r is integer
5	$\mathbf{x} \leftarrow \mathbf{Y}^r$	Select rth vector from $\mathbf{Y}$ (class 2)
6	Return	Exit
7	$t \leftarrow 1$	Teacher decision $= +1$
8	$r \leftarrow \lceil \mu(\mathbf{Z}) \times (\perp?(20) + 1) \div (2 * 20)$	r is a random integer in $[1, \mu(\mathbf{Z})]$
9	$\mathbf{x} \leftarrow \mathbf{Z}^r$	Select rth vector from $\mathbf{Z}$ (class 1)
10	Return	Exit

Random data selector specified in the Iverson language, APL. The $\mathbf{Y}$ and $\mathbf{Z}$ matrices store the training set: $\mathbf{Y}$ contains $\mu(\mathbf{Y})$ vectors from class 2; $\mathbf{Z}$ contains $\mu(\mathbf{Z})$ vectors from class 1. Note that $\nu(\mathbf{Y}) = \nu(\mathbf{Z}) \equiv q$, in the text. The calling sequence for this procedure is RSELECT $(\mathbf{x}, t)$.

4.6.4. Training and Testing a Classifier

Let S denote the set of supplied vectors containing P_1 from class 1 and P_2 from class 2. When $(P_1 + P_2)$ is small ($<10,000$), it is especially important to use distinct data sets for learning (the *training* set) and finding the error rates after learning (the *test* set). To do this, we split S into equal-sized subsets. (This makes subsequent testing easier.) The training and *test* sets should *each* contain $(P_1/2)$ vectors from class 1 and $(P_2/2)$ from class 2. This division of the data should be random to avoid bias. Finally, this partitioning of S must be tested to ensure that it has not introduced bias. To do this we can use the following technique:

Let S_L denote the training set* (both classes) and S_E the test set* (again

*Subscript L is used for learning and E for evaluating.

including both classes). Consider a vector $\mathbf{X} \in S_L$. We find the nearest neighbor to $\mathbf{X}$ from the set $(S_L + S_E - \mathbf{X})$. If the nearest neighbor to $\mathbf{X}$ is also in S_L, we increment a counter C by unity. After we have considered all $\mathbf{X} \in S_L$, this counter indicates the number of vectors (in S_L) which have nearest neighbors also in S_L. If the partitioning of S into S_L and S_E is effective, C will have a binomial distribution with:

$$\text{Mean:} \qquad \frac{P}{2} - 1 \simeq \frac{P}{2} \qquad (\text{for } P_1, P_2 > 50) \qquad (4.28)$$

$$\text{Standard deviation:} \qquad \sqrt{\frac{P(P-2)}{4(P-1)}} \simeq \frac{P}{2} \qquad (4.29)$$

where $P = P_1 + P_2$. If our random partitioning is effective, C should lie within the range:

$$\frac{P}{2} - 1 + \sqrt{\frac{P(P-2)}{4(P-1)}} \simeq \frac{P}{2} \pm \alpha \frac{P}{2} \qquad (4.30)$$

where α determines the confidence interval.

This result provides a basis for testing whether the partitioning is effectively random. If it is not, then we shall usually obtain much larger error rates if we evaluate a classifier using its test set compared to those obtained by reapplying the training set.

4.6.5. Learning in a Linear Classifier

The so-called *fractional-increment error-correction rule* for designing a linear classifier is defined in Program 4.2. If the pattern classes are linearly separable, this procedure will *always* find a solution (Nilsson, 1965). It has been found experimentally that this procedure behaves better than some other rules that have been proposed, when the pattern classes are *nearly linearly separable* (Batchelor, 1968).

The linear classifier is modified in an iterative manner. A modification is made if and only if the linear classifier produces an error ($m \neq t$). The effects of a single modification may be seen from Fig. 4.24.

4.6.6. Learning in a Compound Classifier

The rules for designing a CC are most conveniently separated into two distinct procedures:

1. Program 4.3 includes the classification stage (steps 1–6) and the rules for moving the locates and modifying the size parameters (steps 7–16). The behavior of these rules may be understood with the help of Fig. 4.25.

Program 4.2.   LC Learning

Step No.	Statement	Comment
1	$k \leftarrow 0.001$	Set iteration parameter, k (typical value chosen)
2	$w \leftarrow \bar{e}(\mu(x))$	Set $\mu(x)$ – dimensional vector to all-zero state
3	$n \leftarrow \bar{1}0000$	Initialize loop counter to $-10,000$
4	$n \leftarrow n + 1; (=) \rightarrow (8)$	Loop count; exit loop if $n = 0$
5	RSELECT (x, t)	Program 1
6	$t: 0; (<) \rightarrow (5)$	Look for x vectors from class 1 ($t = +1$)
7	$w \leftarrow w + x; \rightarrow (4)$	Found one! Now add it to w
—	—	w is the weight vector; lines (2–7) provide a "good" initial vector
8	$i \leftarrow \bar{1}0000$	Initialize loop counter
9	$i \leftarrow i + 1; (=) \rightarrow (15)$	Loop count; exit loop if $i = 0$
10	RSELECT (x, t)	Program 1
11	$d \leftarrow (w \underset{\times}{+} x) \div (x \underset{\times}{+} x)$	$d = \sum_{i=1}^{\nu(x)} w_i x_i \Big/ \sum_{i=1}^{\nu(x)} x_i^{\,2}$
12	$m \leftarrow (d > 0) - (d < 0)$	m is the decision of the linear classifier
13	$m: t; (=) \rightarrow (9)$	Compare m and teacher decision, t
14	$w \leftarrow w - k \times d \times x; \rightarrow (9)$	Modify weight vector
15	Halt	Learning complete

APL program for error-correction learning in a linear classifier (fractional increment rule). 1-origin indexing is assumed. $\nu(x) \equiv q + 1$ and $x_{\nu(x)} \equiv 1$ for all patterns. Typical values for k are 0.0001 to 0.05 and the loop counter (i in line 8) should be set initially to an integer in the range -5000 to $-20,000$. This will usually ensure adequate but not excessive learning time.

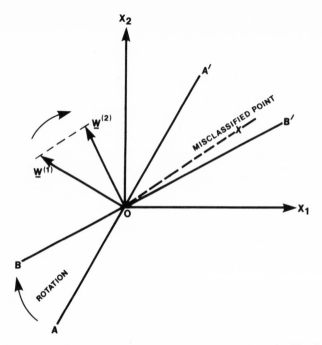

Fig. 4.24. Showing the effect of a single modification to a linear classifier during learning. The vector $W^{(1)}$ denotes the weight vector before the modification and $W^{(2)}$ after. The decision surface before modification is the straight line *AOA'* and after modification it is *BOB'*. The magnitude of the modification is greatly exaggerated for clarity of display.

2. Program 4.4 requires two subroutines, shown in Programs 4.1 and 4.3. A CC of fixed size is subjected to the (iterative) application of the parameter adjustment rules (steps 9–12). Steps 13–22 calculate the index of the "worst" circle (blamed for most errors). A new circle is added close to this "worst" circle and is given the same radius. The addition of the new circle increases the "power" of the CC in that region of X-space which has been found to contribute most errors. Following the addition of the new circle the parameter adjustment rules are reapplied.

These learning rules have been extensively studied by experiment using artificial data* (Batchelor, 1973a, 1974a). It should be emphasized that in this subject *practical* power is much more important than *theoretical* power. Practical power is a function of both theoretical power and *pragmatic* factors such as the computing

---

*Artificial data provide a unique advantage in allowing us to compare the practical classifier with the Bayes optimal classifier. This is possible because the PDFs, which are required by the Bayes classifier, are defined by the user, prior to generating the data.

Program 4.3. Single CC Classify/Modify

Step No.	Statement	Comment
1	$i \leftarrow \mu(\mathbf{B})$	Initialize loop counter
2	$i \leftarrow i - 1; (=) \rightarrow (4)$	Loop count; exit if $i = 0$
3	$\mathbf{d}_i \leftarrow \mathbf{f}_i - (+/((\mathbf{B}^i - \mathbf{x}) * 2)); \rightarrow (2)$	$d_i = f_i - \displaystyle\sum_{j=1}^{\nu(\mathbf{x})} (B_j^i - x_j)^2$
4	$m \leftarrow 2 \times (\nu/(\mathbf{d} > 0)) - 1$	Calculate CC decision, m
5	$m : t; (=) \rightarrow (11)$	Compare m and teacher decision
6	$\mathbf{y} \leftarrow (\epsilon \lceil \mathbf{d})/\iota^1$	Find indices of the largest d_i (may be more than 1)
7	$\mathbf{d} \leftarrow (\mathbf{d} > 0)$	Select all positive d_i
8	$\mathbf{d_y} \leftarrow m$	d will act as a mask later
9	$\mathbf{B^y} \leftarrow \mathbf{B^y} + k \times \mathbf{d_y} \times (\mathbf{B^y} - \mathbf{x})$	Modify locates
10	$\mathbf{f_y} \leftarrow \mathbf{f_y} \times (1 - \mathbf{d_y} \times l)$	Modify size parameters
11	Return	Exit

APL procedure for use as a subroutine in Program 4.4. A vector $\mathbf{x}$ is classified by the CC which has $\mu(\mathbf{B}) = \nu(\mathbf{f})$ locates. The CC is modified iff $m \neq t$. k and 1 are small constants, typically:

$$k = 0.0001 - 0.05$$

$$1 = 0.001 \ \ - 0.05$$

Note that:

$$\nu(\mathbf{B}) = \nu(\mathbf{x}) \equiv q \text{ (text)}$$

Calling sequence: CANDMIX

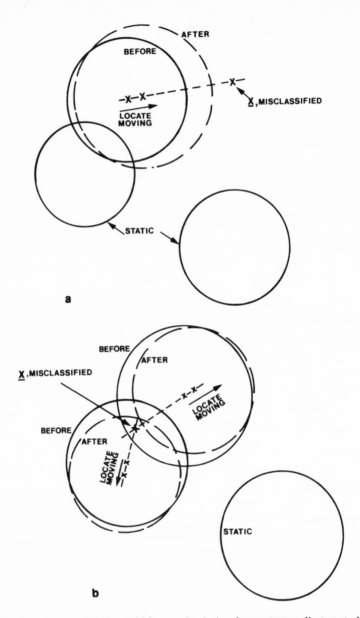

Fig. 4.25. Showing two situations which can arise during the parameter adjustment phase of the CC design procedure (Program 4.3). All changes are exaggerated for clarity of illustration. (a) X should lie inside at least one of the circles but in fact lies outside all of them. The closest circle is made larger and is moved toward X. (b) X should be outside all of the circles, but here it is enclosed within two of them. These two circles are made smaller and are moved away from X. The CC is not modified if the vector X is classified correctly.

Program 4.4.   Growing CC

Step No.	Statement	Comment
1	$\mathbf{B}^1 \leftarrow \bar{\epsilon}(\nu(\mathbf{x}))$	Start learning with one locate set to origin of space*
2	$\mathbf{f}_1 \leftarrow 100.0$	Set size parameter to small value*
3	$u \leftarrow 0$	Initialize loop 1 counter
4	$v \leftarrow 10$	Set upper limit for loop 1*
5	$p \leftarrow -10000$	Store 10,000 vectors* Beginning of loop 1
6	$h \leftarrow -10000$	Initialize loop 2 counter*
7	$k \leftarrow 0.005$	Set iteration parameter*
8	$l \leftarrow 0.001$	Set iteration parameter*
9	$h \leftarrow h + 1$	Beginning of loop 2
10	RSELECTX $(\mathbf{x}; t)$	*Random* data selector
11	CANDMX	See Program 4.3
12	$h: 0; (<) \rightarrow (9)$	Exit loop 2 if $h = 0$
	—	Parameter adjustment learning is complete. Now we add a new locate
13	$c \leftarrow \bar{\epsilon}(\mu(\mathbf{B}))$	Initialize counter c
14	$k \leftarrow 0.0$	Setting k and l to zero "freezes" the classifier in loop 3
15	$l \leftarrow 0.0$	Setting k and l to zero "freezes" the classifier in loop 3
16	$h \leftarrow \bar{p}$	Set loop 3 counter
17	CSELECTX $(\mathbf{x}; t)$	*Cyclic* data selector
18	CANDMX	See Program 4.3
19	$c \leftarrow c + (\mathbf{d} \neq 0)$	c records the frequency that the locates would be modified if we were still learning
20	$h \leftarrow h + 1$	Increment loop 3 counter
21	$h: 0; (<) \rightarrow (17)$	Exit loop 3 if $h = 0$
22	$g \leftarrow ((\epsilon \ulcorner c)/\iota^1)_1$	c_g is largest of the c_i
23	$z \leftarrow 0.01$	z is any small quantity*
24	$\mathbf{f} \leftarrow \mathbf{f}, \mathbf{f}_g$	Add a new circle close to gth circle in old CC. The size of the new hypersphere is $\mathbf{f}_g$
25	$\mathbf{B} \leftarrow \mathbf{B}, (\mathbf{B}_g + z \times ?(\mu(\mathbf{B})))$	Add a new circle close to gth circle in old CC. The size of the new hypersphere is $\mathbf{f}_g$

Program 4.4. Growing CC (continued)

Step No.	Statement	Comment	
26	$u \leftarrow u + 1$	Increment loop 1 counter	⑤
27	$u: v; (<) \rightarrow (5)$	Exit loop 1	
28	HALT	CC design is complete	

Design procedure for a CC. Steps 9–12 form the parameter-adjustment learning rules which optimize a fixed-sized CC. Steps 13–21 calculate the scores c which will be used to select the site for a new circle. This new circle is added very close to (but not coincident with) that circle in the old CC which would be modified most frequently if we continued with parameter adjustment learning. This procedure implements the *growing* rules defined by Batchelor (1973a). Another procedure, called *pruning*, may be used to reduce the size of the CC by eliminating unnecessary locates which do not contribute usefully. Lines indicated by asterisks indicate "typical" values.

Program 4.5. Maximindist

Step No.	Statement	Comment		
1	$i \leftarrow 1001; a \leftarrow 0$	Initialize loop counter (i); set a to zero		
2	$i \leftarrow i - 1; (=) \rightarrow (7)$	Decrement loop counter and test for zero	=	
3	RSELECT $(\mathbf{x}, s, t)$	Random data selector (modified; see legend for explanation)		
4	$t: 1; (\neq) \rightarrow (3)$	Look for a class 1 pattern vector	$\neq$	
5	$s: a; (<) \rightarrow (2)$	Found one! Now look for largest s	<	
6	$a \leftarrow s; \mathbf{B}^1 \leftarrow \mathbf{x}; \rightarrow (2)$	Found it! Now store it in a and store associated x vector in $\mathbf{B}^1$. Repeat, to look for bigger s		
7	$k \leftarrow {}^-51$	Initialize loop counter to -51. Maximindist will find 50 representatives		

Program 4.5. Maximindist (continued)

Step No.	Statement	Comment
8	$k \leftarrow k + 1; (=) \rightarrow (21)$	Increment loop counter and test for zero
9	$j \leftarrow 1001; a \leftarrow 0$	Initialize another loop counter and set a to zero
10	$j \leftarrow j - 1; (=) \rightarrow (20)$	Decrement loop counter and test for zero
11	RSELECT (x, s, t)	Random data selector
12	$t: 1; (\neq) \rightarrow (11)$	Look for a class 1 pattern vector
13	$i \leftarrow \mu(\mathbf{B}^1) + 1; b \leftarrow 10 * 20$	Initialize yet another loop counter and set b to 10^{20}
14	$i \leftarrow i - 1; (=) \rightarrow (18)$	Decrement loop counter and test for zero
15	$d \leftarrow (+/(\mathbf{B}^i - x)*2) \div s$	$d = \sum_j (\mathbf{B}_j^i - x_j)^2 \div s$
16	$d: b; (>) \rightarrow (14)$	Compare d and b
17	$b \leftarrow d; \rightarrow (14)$	Smallest d (so far) found. Now store it
18	$b: a; (<) \rightarrow (10)$	Compare b and a
19	$a \leftarrow b; w \leftarrow x; \rightarrow (10)$	Largest b (so far) found. Store it in a and associated x in w
20	$\mathbf{B} \leftarrow \mathbf{B} \circledast w; \rightarrow (8)$	Add another vector w into the **B** matrix which is the store of representatives
21	HALT	Stop. 50 "typical" representatives are now stored as vectors in **B**

Maximindist, a procedure for choosing typical representatives. 50 representative vectors are found from class 1 by this procedure. Each representative is chosen by considering 1000 class 1 points. The random data selector (Program 4.1) has been modified so that it returns a score, s, for each new x. This is a scalar quantity which measures how "typical" x is of its class. The PDF estimators described in Section 4.3.2 would be suitable for calculating s. (Also see Fig. 4.15.)

requirements. While the theoretical power of a CC is unknown, there is a mass of *experimental* evidence to commend this particular classifier.

4.6.7. Learning in a Nearest-Neighbor Classifier

If the training set contains few patterns ($\leqslant 256$) there is *probably* little point in trying to discard any information. The NNC can then store the complete training set as locates. This simple design procedure will be called the *store-all rule* (Cover and Hart, 1967; Lill and Redstone, 1968).

For larger data sets, it is important to discard redundant information. This is achieved in the *store-when-wrong* rule by only retaining a vector in the locate store of the NNC when it has misclassified that vector. This is an economic solution if the pattern classes are well separated, since most of the stored vectors will lie in the overlap region. The procedure can be improved slightly by starting the store-when-wrong algorithm with a NNC containing several locates (from each class), which have been chosen carefully. Locates which lie close to the cluster centers are best, since they will inhibit the store-when-wrong rule from selecting a lot of points outside the overlap region. Such locates could, of course, be selected using a cluster analysis procedure such as maximindist (see Program 4.5). The store-when-wrong rule is a variant of the *condensed nearest neighbor rule* (Hart, 1968).

REFERENCES

Anderson, G. D., 1969, Comparison of Methods of Estimating a Probability Density Function, Ph.D Thesis, University of Washington, Seattle, Wash.

Batchelor, B. G., 1968, Learning Machines for Pattern Recognition, Ph.D Thesis, University of Southampton, Southampton, England.

Batchelor, B. G., 1971, Improved distance measure for pattern recognition, *Electron. Lett.*, 7:521.

Batchelor, B. G., 1973a, Growing and Pruning a Pattern Classifier, *Inform. Sci.* 6:97.

Batchelor, B. G., 1973b, Instability of the decision surfaces of the nearest-neighbour and potential function classifiers, *Inform. Sci.* 5:179.

Batchelor, B. G., 1974a, *Practical Approach to Pattern Classification*, Plenum Press, London and New York.

Batchelor, B. G., 1974b, Design for a high-speed euclidean distance calculator and its use in pattern recognition, *Proc. Conf. Computer Systems Technol., IEE Conf. Publ.* 121:213.

Batchelor, B. G., and Hand, D. J., 1975, On the graphical analysis of PDF estimators for pattern recognition, *Kybernetes* 4:239.

Batchelor, B. G., and Wilkins, B. R., 1968, Adaptive discriminant functions. In: *Pattern Recognition, IEE Conf. Publ.* 42:168.

Batchelor, B. G., and Wilkins, B. R., 1969, Method for location of clusters of patterns to initialise a learning machine, *Electron. Lett.* 5:481.

Cover, T. M., and Hart, P. E., 1967, Nearest neighbor pattern classification, *Trans. IEEE* **IT-13**:21.

Everitt, B., 1974, *Cluster Analysis*, Heinemann, London.

Ford, N. L., 1974, Pattern Classification and the Analysis of Training Sets, Ph.D. Thesis, University of Southampton, Southampton, England.

Ford, N. L., Batchelor, B. G., and Wilkins, B. R., 1970, Learning scheme for the nearest neighbour classifier, *Inform. Sci.* 2:139.

Fukunaga, K., 1972, *An Introduction to Statistical Pattern Recognition*, Academic Press, New York.

Hand, D. J., and Batchelor, B. G., 1974, A preliminary note on pattern classification using incomplete vectors, *Proc. 2nd Int. Joint Conf. Pattern Recognition*, p. 15.

Hand, D. J., and Batchelor, B. G., 1975, Classification of incomplete pattern vectors using orthogonal function methods, *Proc. 3rd Int. Conf. on Cybernetics and Systems*, Bucharest, Rumania.

Hart, P. E., 1968, Condensed nearest neighbour rule, *Trans. IEEE* **IT-14**:515.

Hellerman, H., 1967, *Digital Computer System Principles*, McGraw-Hill, New York.

Iverson, K., 1962, *A Programming Language*, Wiley, New York.

Jardine, N., and Sibson, R., 1971, *Mathematical Taxonomy*, Wiley, New York.

Katzan, H., 1970, *APL Programming and Computer Techniques*, Van Nostrand-Reinhold, New York.

Lewin, D. W., 1972, *Theory and Design of Digital Computers*, Nelson, London.

Lill, B., and Redstone, L., 1968, A study of learning and recognition algorithms, *IEE Conf. Publ.* **42**.

Martin, J., 1973, *Design of Man–Computer Dialogues*, Prentice-Hall, Englewood-Cliffs, N.J.

Mattson, A. L., and Damman, J. E., 1965, A technique for determining and coding sub-classes in pattern recognition problems, *IBM J.* 9:4.

Miller, G. A., 1967, *Psychology of Communication*, Penguin, Harmondsworth, U.K.

Mucciardi, A. N., and Gose, E. E., 1971, Comparison of seven techniques for choosing sub-sets of pattern recognition properties, *Trans. IEEE* **C-20**:1023.

Mucciardi, A. N., and Gose, E. E., 1972, An automatic clustering algorithm and its properties in high dimensional spaces, *Trans. IEEE* **SMC-2**:247.

Nilsson, N. J., 1965, *Learning Machines*, McGraw-Hill, New York.

Patrick, E. A., 1972, *Fundamentals of Pattern Recognition*, Prentice-Hall, Englewood-Cliffs, N.J.

Rosenfeld, A., and Pfaltz, J. L., 1968, Distance functions on digital pictures, *Pattern Recognition* 1:33.

Sammon, J. W., 1969, Non-linear mapping for data structure analysis, *Trans. IEEE* **C-18**:401.

Sebestyen, G. S., 1962, *Decision-Making Processes in Pattern Recognition*, Macmillan, New York.

Sebestyen, G. S., and Edie, J., 1966, An algorithm for non-parametric pattern recognition, *Trans. IEEE* **EC-15**:908.

Sokal, R. R., and Sneath, P. H. A., 1963, *Principles of Numerical Taxonomy*, Freeman, San Francisco.

Wilkins, B. R., and Batchelor, B. G., 1970, Evolution of a descriptor set for pattern recognition, *Proc. Conf. Technical Biol. Probl. Control*, Instrument Society of America, p. 794.

Wilkins, B. R., and Ford, N. L., 1972, Analysis of training sets for adaptive pattern recognition. In *Machine Perception of Patterns and Pictures, Inst. Physics, Conf. Ser.* 13:267.

Wolfe, J. H., 1967, Normix: Computational Methods for Estimating the Parameters of a Multivariate Normal Mixture of Distributions. (Research Memo SRM 68-2), U.S. Naval Personnel Research Activity, San Diego, Calif. (Defense Documentation Center Ad 656 588).

Wolfe, J. H., 1970, Pattern clustering by multivariate mixture analysis, *Multivariate Behav. Res.*, 5:329.

5

Editorial Introduction

D. A. Bell discusses methods for making decisions, which are based on complex "measurements." In this chapter, measurements are far more complicated than those envisaged by Aleksander or Batchelor, and this gives a different emphasis to the subject. Bell and Aleksander only consider binary-valued measurements, while Batchelor allows continuous, or discrete, variables. Bell and Aleksander differ elsewhere in their approach: Bell contemplates using a conventional computer (von Neumann architecture), while Aleksander assumes that only special-purpose hardware (with distributed storage and processing) will be fast/cheap enough. The difference between their approaches is due to the fact that they are attempting to make different kinds of decisions. Bell assumes that decision-making *per se* forms only a small proportion of the complete calculation. Bell tries to avoid making measurements wherever he can, while Aleksander and Batchelor both assume that all of their (simple) measurements can be obtained in parallel, for every pattern. Bell's decisions are often at a different level of abstraction from those computed by Aleksander and Batchelor. Consider an application, namely, *air traffic control*: our problem is to build an alarm to notify us of potential hazards. This type of decision-making task is extremely difficult for Batchelor's classifier or Aleksander's networks of memory elements, but is ideally suited for Bell's techniques. (For reasons of space, we cannot justify that Batchelor's and Aleksander's techniques are unsuitable.) First of all, we can formulate known hazards, simply by asking a (human) air traffic controller. These data can be used to specify a decision *table*. Bell then transforms this table into a decision-*tree*, and reduces its storage requirements by connecting crosslinks between nodes, to avoid storing common subtrees more than once. The end result is a decision-*lattice*, which requires only a small proportion of the measurements to be evaluated, for any given pattern. The selection of which measurements are to be evaluated is performed as a *sequence of decisions*. Bell's technique achieves its high speed by evaluating only those measurements which are essential to reach the end decision; Aleksander and Batchelor achieve (very) high speed by using parallel computing methods. Returning to the problem of air traffic control, we can see how Bell's program does this. He begins by making a quick, crude evaluation of a wide scene and, on the basis of this, he decides where to concentrate his more refined techniques. Note that hazards, in this context, are both spatially and temporally local. It is this that makes a sequential (Bell) procedure more suitable than a parallel (Aleksander and Batchelor) one, in this type of task. Aleksander and Batchelor's approaches are both better suited to assessing a body of data and making a decision about all of it, and not just a small subset.

5

Decision Trees, Tables, and Lattices

Donald A. Bell

5.1. INTRODUCTION

The use of small built-in processors in pattern recognition machines is becoming much more widespread, with the advent of very cheap processing power in the form of LSI microprocessors and other devices. It is therefore quite reasonable to design systems which evaluate rather complicated features according to the dictates of an "intelligent" computer program and arrive at a decision by branching and sequential feature extraction, rather than by deriving several features in parallel and then classifying by hypersurface separation.

The *decision tree* offers a way of directing the overall strategy of computation in a tidy way, and it should, in principle, permit rapid progress to a final decision, using the minimum of computing resources (Bell, 1973). It will be convenient in the discussion which follows to consider only binary decision trees, but all of the techniques to be described could be extended to cope with three- or four-way branching without much difficulty, apart from the demand for much greater resources in the design phase. Briefly, a decision tree consists of a series of tests to be applied to the pattern which is to be recognized. The tests are situated at the branching points of the tree, and each branch leads either to another branching point or to an endpoint of the tree. (The branching points are

Donald A. Bell · Computer Science Division, National Physical Laboratory, Teddington, Middlesex, England

often called *nodes* and the endpoints *end nodes* or *leaves*.) A typical tree is shown in Fig. 5.1, with the *root* at the top and the leaves at the bottom.

Each letter represents a test to be applied and each digit is an endpoint. Note the following points about such a tree:

1. There are no cycles in the tree, i.e., any endpoint can be reached after traversing four or fewer branching points, or nodes in this case.

2. Each node gives rise to exactly two branches (i.e., it is a *binary tree*).

3. Endpoints with the same label can be reached in a variety of ways.

4. Some endpoints can be reached without evaluating all of the tests. The "3" point only requires A and D to be tested. As a convention, we will draw the *true* branch going to the right and the *false* branch going left, so that "3" is equivalent to *not* A *and not* D, i.e., test A gives the result *false*, so we apply test D and again get the result *false*, leading to the end node "3."

5. There are two identical subtrees in the tree, hanging from the "B" tests.

The pattern recognition situations in which decision trees might be expected to be most useful are those in which it is not reasonable to use very simple features, and a processor with a stored program could evaluate much more reliable measurements, since each test can be precisely defined and qualified. If, in addition, there is a requirement for rapid processing of the patterns, it is important to direct the evaluation of later features according to the results of evaluating earlier ones. This type of situation is particularly relevant to the field of handwritten numeral recognition, and some of the examples in this chapter

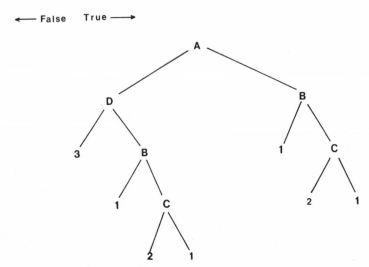

Fig. 5.1. A simple binary tree with 6 nodes and 7 endpoints.

will be drawn from such situations. The handwritten numeral (in contrast to a printed one) is sufficiently "sloppy" to require an intelligent processor to decide whether, for example, a closed loop is present in the lower half of the character. In fact it may not be quite closed or it may be filled in with a blob of ink. For fast processing of handwritten documents in a commercial environment, it is essential to recognize hundreds of characters per second, so we cannot tolerate nonessential calculations.

5.2. FEATURE DESIGN FOR DECISION TREES

If binary trees are to be used, it is clear that each feature evaluated in software must give a yes or no answer (i.e., true or false). The tree will not be able to cope with an equivocal response, and must be designed to be flexible enough to recover even if a feature is evaluated in a misleading way. However, it is much better if each feature can be designed to be as useful as possible in its decisions on the majority of the pattern classes. An example from numeral recognition will make this clear: a curve in the lower right-hand corner of a character is reliably present in the cases of 0, 3, 5, and 8 and reliably absent for 1, 2, 4, and 7. The character 9 may have a curve there or it may be straight, and the loop on the 6 may be so tight that the curve is too small to register. However, this is a very good feature, since 8 of the 10 classes have predictable responses. The same remarks do not apply to a curve in the upper right-hand corner, however, since the characters 0, 3, 8, and 9, which might be expected to show it, are often so carelessly written that it does not appear. Only in the 2 is the curve always present and the feature is reliably absent only in 1, 4, and 6. There must obviously be a trade-off between processing speed and reliability of detection, but, generally speaking, rapidly evaluated features are to be preferred to those which take a lot of time, since they can produce, after all, only a binary decision. As we shall see, it is possible for one feature to cover the mistakes of another if the tree is designed for *robustness*. It is sometimes argued that the weakness of a treelike system is that it takes irrevocable decisions, and that robustness can only be provided by great complexity. Later sections will describe how this complexity can be handled by relatively simple specifications and a highly automated procedure.

5.3. THE DESIGN OF DECISION TREES

The psychological problems in designing decision trees are immense. It is a good rule-of-thumb in computer programming that a subroutine should occupy no more than two written pages, because otherwise it becomes too difficult to comprehend. Unfortunately this is not always possible to achieve, and if it were necessary to restrict the size of a decision tree in the same way, the performance of the pattern recognition machine would be derisory. Several techniques have been devised in an attempt to overcome this problem. The most straightforward is to design the top (or "root") part of the decision tree in an *ad hoc* fashion with inspired guesswork, then to use this *partial tree* to partition the set of training patterns into several sets, one for each "exit" from the partially developed tree and to use each set to develop further subtrees until the task is complete.

Unfortunately, this method is very time-consuming, since every modification at a high point in the tree has unpredictable consequences for tree design further down. Another approach is to concentrate on the pathways which "ought" to be followed by each pattern class, by blanking off most of the tree and noting how often a pattern attempts to follow a forbidden pathway. Again a rather time-consuming operation results, and the repeated iterations may prevent a really small and effective tree from ever emerging.

One of the major problems of tree design is that the features of a pattern may be evaluated wrongly in one or two cases. This results in a "wrong turning" in the tree, so that there must be sufficient redundancy in the tests to make sure that this is not disastrous. Unfortunately, this may result in several copies of the same subtree being present. The designer is now faced with a dilemma: should he combine all the common subtrees into one and turn the tree into a *lattice* where two or more pathways may recombine (Fig. 5.2), or should he leave the common subtrees as they are, on the grounds that it may become necessary to modify one of them at a later date? (The consequences of modifying a subtree which is part of a lattice are extremely difficult to predict.) A few figures may serve to clarify these points. A tree with 200 nodes is quite difficult for one person to manipulate or visualize. The recognition of a hand-

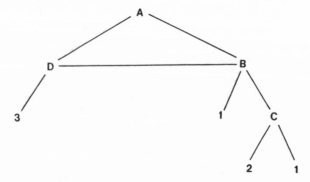

Fig. 5.2. The lattice corresponding to the tree of Fig. 5.1. The common subtree hanging from test B now appears only once, but it can be reached by two different routes.

written numeral takes 10 to 12 decisions in most cases; a binary tree with this number of levels would have 1000 to 4000 nodes. Some rather peculiar character styles must be accommodated if the best performance is to be realized from the pattern recognition machine. This results in some pathways being 15 or 16 steps in length, with the total number of nodes rising to around 30,000. Some of these are visited only rarely, but they must be included.

Faced with this problem, it is necessary to look at the design of trees from the bottom up, and immediately a more hopeful picture arises. Each endpoint of the tree can be expressed as a Boolean function of the tests in the pathway which leads from the root to the endpoint. There will be several endpoints for each pattern class, and if these Boolean expressions could be found, it would be possible to specify the operation of the tree (on very few sheets of paper) and hence to design an efficient order of evaluation of the features. We shall postpone the problem of how to derive these Boolean expressions for the time being, and consider the question of how to turn the set of *decision rules* (for that is what they amount to) into a tree or lattice.

5.4. DECISION TABLES

The COBOL programmer will be quite familiar with the idea of decision tables and several other commercial programming languages make use of them too (Schmidt and Kavanagh, 1964). However, some of the basic concepts must be modified somewhat when they are taken over to deal with pattern recognition problems. A decision table has one row for each test or "condition" which may be evaluated, and one column for each "rule," which specifies the outcome of each condition for the rule to be satisfied. There is usually a "menu" of actions to be performed and the satisfied rule will specify which are to be done, but since, in pattern recognition applications, the actions will be very simple (such as "recognize the character as a 6"), it will be more convenient to number the rules according to the pattern class to which they refer (where several rules specify the same class they are "1a," "1b"). An example will illustrate the general arrangement. Each row has a letter, representing a feature or condition to be evaluated, and each column specifies the pattern class which will be selected. The symbols "Y," "N," and "–" represent YES, NO, and DON'T CARE, respectively. The table following, top of p. 124 is equivalent to the decision tree which was used earlier in the chapter.

There are four rules specifying class 1, two for class 2, and one for class 3. In addition to the above information, we will later make use of the information available on the frequence of application of each rule, and on the relative cost of making each decision.

In passing, it should be mentioned that there are several ways of processing

		Rules						
		1a	1b	1c	1d	2a	2b	3
	A	Y	Y	–	–	Y	–	N
Conditions	B	–	N	–	N	Y	Y	–
	C	Y	–	Y	–	N	N	–
	D	–	–	Y	Y	–	Y	N

the information in a decision table, not all of them involving the conversion to a decision tree. The *rule mask* technique (Kirk, 1965) simply evaluates all of the conditions before looking at the rules, then it picks the first rule to satisfy the pattern of Y and N. The *interrupted rule mask* technique (King, 1966) evaluates only enough conditions to test the first rule, then enough additional ones to test the second rule, and so on. The rule mask technique is clearly not applicable for pattern recognition, since we should like to reach a conclusion without evaluating more than about a third of the available features. The interrupted rule mask technique is useful if one rule is dominant in frequency of application or if a small subset of the conditions will satisfy the bulk of the rules. However, it is also likely to call for the evaluation of too many features, though careful ordering of the rules could keep this within bounds.

Most of the available techniques for converting a decision table into a decision tree use the so-called *table splitting* method (Pollack *et al.*, 1971). The basic operation involves selecting a condition on which to split the table. This condition then becomes a test in the developing tree and the table is split into two subtables. Each of the subtables has all the rows of the original except the one corresponding to the condition which has been transferred to the tree. The so-called *YES subtable* has those columns which did not have "N" against the splitting condition and conversely for the *NO subtable*. Splitting the example table on condition "A" produces the following:

"NO" subtable				
	1c	1d	2b	3
B	–	N	Y	–
C	Y	–	N	–
D	Y	Y	Y	N

"YES" subtable						
	1a	1b	1c	1d	2a	2b
B	–	N	–	N	Y	Y
C	Y	–	Y	–	N	N
D	–	–	Y	Y	–	Y

The left-hand table may be further split on condition "D" to give the subtables:

(A = NO, D = NO)				(A = NO, D = YES)		
	3			1c	1d	2b
B	–		B	–	N	Y
C	–		C	Y	–	N

The left subtable consists entirely of dashes, and so represents an endpoint. The right subtable still requires two tests to distinguish rules 1c and 1d from rule 2b. The conditions for terminating development of a subtable are as follows:

1. *There are no conditions with* Y *or* N *entries* (i.e., there are no more tests left to apply). If more than one class is left in the table, a contradiction has occurred, and the pattern is undecidable. If exactly one class remains, the identification is complete.

2. *There are no rules left.* The pattern does not fit into the set of rules; again it is undecidable.

Those programming languages which incorporate decision tables would cast out tables which could have undecidable patterns of Y and N as either "contradictory" or "incomplete," and this would be flagged as a programming error. In the pattern recognition case, however, the relevant endpoints are marked as "reject" or undecidable patterns.

5.5. TABLE CONVERSION METHODS

Most of the controversy in the literature on converting decision tables to trees centers on the criterion by which the splitting condition should be chosen in a subtable. There are usually two basic aims for efficient conversion, namely, economy of storage, and economy of running. The two requirements do not necessarily go together, though in the small example tables usually quoted in the confines of a paper, it may be difficult to realize the difference. Choosing the "correct" splitting algorithm does not guarantee the absolute minimum in either storage or processing time, and, almost without exception, all of the methods which have been put forward can be shown to be inadequate to deal with some cunningly chosen counterexamples. Those methods which claim to give economy in storage are generally simpler than the ones which seek the best running speed of the *tree traversal program*, since they can afford to ignore statistical information on the relative frequency of application of the rules, or the costs of making the decisions.

Ideally one would like a method which determines which of the conditions in a subtree is likely to yield the maximum information when it is evaluated. If, at the same time, it is possible to look ahead to the subtables likely to result from the split and ensure that they are as small as possible, this would be an added bonus.

Several considerations must be included in any splitting algorithm. For each condition in the table, we ask the following questions:

    1. Are there a lot of DON'T CARE entries against this condition? That is, will the resulting subtables be nearly as big as the current one?

    2. Are there both YES and NO entries in this condition, against rules which specify the same pattern class? In the long run it is not necessary to separate the different rules corresponding to the same class, so that effort which does this is potentially wasted.

    3. Does this condition separate two classes cleanly? That is, are there YES entries against one and NO against the other? Conditions which do this are, of course, very valuable, especially if other conditions merely separate YES or NO from DON'T CARE.

    4. Is one of the subtables resulting from a split on the condition under consideration likely to be very small? If so, it is likely to provide a short route to a "reject" category corresponding to an unrecognizable pattern.

With these considerations in mind, we may now look at the methods which have been proposed, though only a brief survey can be given here.

5.6. TABLE-SPLITTING METHODS

Pollack (1965) described two methods, one for minimum storage and one for minimum average processing time. Only the first will be described:

1. For each rule in the table, count the number of dashes, k. k DON'T CARE's can be satisfied by any one of 2^k patterns of Y and N; 2^k is called the *column count*.
2. For each condition, add up the column counts in the rules for which the condition is DON'T CARE. This is the *dash count*.
3. Choose the condition with the smallest dash count.

The following example will show the principle:

	Rule			Dash count
	1	2	3	
A	N	–	Y	4
B	N	Y	–	2
C	–	Y	N	4
D	–	–	N	8
Column count:	4	4	2	

Hence, condition B should be chosen, giving the tree shown in Fig. 5.3.

Pollack's use of the term 2^k in the dash counts is somewhat questionable, and probably k would do just as well. Press (1965) advocated a "look ahead" principle, by which a condition which had all Y or all N entries was chosen, if possible, so as to yield one subtable which has no columns (i.e., an "undecidable" endpoint). His strategy is therefore:

1. Choose the condition row with the least number of dashes.
2. If this gives a tie, choose the one which is either all Y or all N.

If this does not give a firm choice, look ahead one step and try to choose a condition which would yield a subtable having at least one row of All Y or all N.

Yasui (1971) visualized the decision table as a hypercube, where each decision rule specifies a k-cube (k-dimensional hypercube) of the original hypercube, and the splitting of the table thus results in the partitioning of the hypercube. It is desirable that the k-cubes should not be split, so once again the rule is to choose the condition with the smallest number of dashes.

None of the above methods takes into account the fact that there may be several rules for the same action or pattern class. However, Rabin (1971) does consider the case of several rules for the same action, and proposes a Karnaugh-map approach. The cells of the map may be labeled according to the corresponding action, and the aim of choosing the splitting condition must then be to try to keep together the cells which specify the same action. The condition which splits the minimum number of these "submaps" is then chosen. This is equivalent to selecting the condition with the smallest number of dashes. The algorithm is not quite the same as Yasui's, however, since Rabin insists that the table should not be redundant, i.e., some of the dashes must be filled in with Y

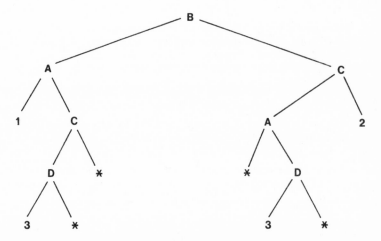

Fig. 5.3. Example tree derived by Pollack's algorithm. Asterisks mean undecidable.

or N where this is possible. He also describes a look-ahead principle which uses blocks of Y or N on one row which exactly parallel similar groups in another row. Schwayder (1971) attempted to use entropy considerations, but did not attach much importance to the undecidable cases. However, he gave a second method, similar to the ones already referred to, which made use of the frequency information:

1. Add the Y and N entries in each column rule.
2. Divide this number into the rule frequency to give the *column weight*.
3. For each condition, add the column weights where a Y or N occurs to give the *row weight*.
4. Select any one of the conditions with the maximum row weight.

Unfortunately none of these methods tackles both the problem of what to do with the rule frequencies and of how to interpret a group of several rules specifying the same action. (Such a group is often called an *action group*.) It is therefore necessary to devise an empirical formula for scoring each condition in order to be able to select the most appropriate one in each subtable. This should avoid splitting the rules belonging to the same action, but, at the same time, give due weight to the expected frequency of each rule. We begin by defining a YES count, f_y, for each action group and condition, given by the sum of the rule frequencies which specify Y for the action and condition; similarly a NO count, f_n, is defined. For example, in the table

Frequency	60%	40%	70%	30%
Action	1	1	2	2
A	Y	N	N	–
B	Y	–	Y	N
C	–	N	–	Y
D	–	N	Y	–

the frequency f_y for condition A in action 1 is 60% and f_n is 40%, whereas for condition B in action 1, f_y is 60% and f_n is zero. Conditions A and B have fewest dashes and therefore might seem to be good candidates, but each of them splits the rules for either 1 or 2. Within each action and each condition, therefore, it is necessary to have a function of f_y and f_n, which gives the desirability of splitting the subtable on that condition. A reasonable estimate is the absolute difference between f_y and f_n abs($f_y - f_n$), but this does not distinguish between $f_y = f_n = 0$ and $f_y = f_n \neq 0$, which is a drawback. The function actually used therefore is

$$F = \text{abs}(f_y - f_n) + E(f_y + f_n)$$

where E is a constant. Empirically it has been found that E should be between $\frac{1}{2}$ and 2, and a value of 1 is usually taken. To choose the condition on which to split the table, the value of F is obtained for each of the actions on each condition. The F values for one condition are summed to give its score, and the condition with the largest score is then chosen. There are, however, two complications. Consider the segment of a subtable

Frequency	30%	40%	30%
Action	1a	1b	1c
A	N	–	Y
B	Y	Y	–
C	–	Y	Y

where there is only one action present, and the rules have been distinguished by small letters. At first sight, the condition to be chosen should be B or C since in each case the f_y score is very high. However, if the table is rewritten as a Boolean expression, it looks like this:

$$\text{action } 1 \equiv A'B + BC + AC$$

where $+$ means OR, and A' means NOT A. It is clear that this expression reduces to $(A'B + AC)$ by the usual rules of Boolean simplification, so that the subtable can be rewritten thus

Action	1a	1b
A	N	Y
B	Y	–
C	–	Y

showing that condition A is now free of dashes (though it does split the rules). To get a true picture of the subtable therefore, it is necessary to do this Boolean simplification for each of the sets of rules which specify common actions, before evaluating the condition scores.

The other complication concerns the treatment of the frequency information when a table is split. Sometimes a rule goes into both of the resulting subtables, because it had a dash against the splitting condition. Should its frequency in the subtables remain as it was in the original or should it be divided by 2? Again an empirical answer must be given: if the frequency is repeatedly divided by 2 on successive splittings of the tables, the rule eventually ceases to influence the choice of splitting condition and therefore a rule which may have started out as highly significant is relegated to the end of a very long pathway in the tree.

The frequency count of a rule in a subtable is therefore kept the same as its parent rule in the original table. There is another reason for this convention; it simplifies enormously the translation of tables directly to crosslinked lattices, which will be discussed shortly, since this process requires the storage of a great many subtables, and the storage of separate frequency counts for each one would be very bulky. The technique for Boolean simplification of the rules will not be described in any detail, but a few observations are in order. The classical techniques of Boolean simplification which involve the canonical expansion of the expression are entirely unsuitable for the tables which are used in practice, since they may have upward of 30 variables in the early stages. However, the heuristic simplification technique of Breuer (1968) has proved entirely adequate for this purpose. It does not expand the expression, but merely tries to eliminate factors and terms from it. It will eventually reduce an expression to prime implicants, though it does not guarantee the smallest number of either terms or factors. McCluskey (1965) gives a good illustration of this point. Consider the expressions

$$f = w'y + xy'z + wx'z$$

$$g = w'y + w'xz + wy'z + x'yz$$

These expressions represent the same function, but they cannot easily be transformed into each other (except by the full canonical expansion or by drawing the Karnaugh map). The full algorithm for transforming the table to a tree may now be given.

1. Set up a stack to hold the subtables and place the original table on it.
2. Take a table from the stack if there is one left, otherwise stop.
3. Examine the table to see if it represents an endpoint, i.e., if it has no rows (other than those full of dashes) or no columns. If this is so, record either an endpoint with a valid action or an undecidable endpoint and return to step 2.
4. If the table is not apparently an endpoint, apply a Boolean simplification to the rules where possible, and examine it again to see if it is an endpoint. If it is, deal with it as in step 3.
5. Split the table into two subtables, by choosing the splitting condition, and load the resulting subtables onto the stack, then return to step 2.

The stack need only be a small one, since there will never be more subtables on it than there are conditions in the main table. If frequency information is not required for each individual subtable, then an M by N main table can have any subtable specified by $(M + N)$ bits, which are not greedy of storage. (Each row and column has a single bit to say "included" or "not included.") This algorithm will develop a tree in a "depth first" manner, following one path to its endpoint before beginning another. The alternative technique, the "breadth first" approach will also be found to be useful; this develops all the available subtables at one level before progressing to the next one. It has the disadvantage of requiring a lot of subtable storage, but it has important implications for the elimination of common subtrees, as we shall see later.

The "depth first" approach is very fast, and can develop a tree of 30,000 nodes in a few minutes on a minicomputer, being limited only by the speed of the output device. When the method was first applied to a problem in the recognition of handprinted numerals, it produced very large trees, which were unlikely to be useful, since they would have demanded the use of a great deal of secondary storage. This phenomenon is not too surprising, however, since it is necessary for a tree to be somewhat redundant in terms of common subtrees, to cater to the possibility of "wrong turnings" at various places. However, it is a great embarrassment, and ways of eliminating subtrees which are identical, or nearly so, must be investigated.

5.7. THE COMMON SUBTREE PROBLEM

Since the table-splitting translation delivers the tree as a string of tests and results, it may be stored in this form as a Polish string. For example, the tree in Fig. 5.4 may be stored as A B C 2 D 3 * D 3 * C 2 D 3 * and does not require any brackets, since each subtree is in the form: (test, right subtree, left subtree).

The first reduction method to be discussed requires the entire tree to be stored in random access memory in the form:

Address	Test	Left subtable address
1	A	11
2	B	8
3	C	5
4	result 2	
5	D	7
6	result 3	
7	result *	
8	D	10
9	result 3	
10	result *	
11	C	13
12	result 2	
13	D	15
14	result 3	
15	result *	

Here, the result of a test specifies whether the next following location should be used (the condition is TRUE) or whether the tests should continue from the address given in the right-hand column.

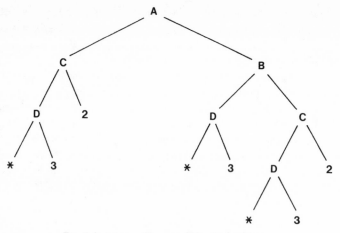

Fig. 5.4. A tree with several identical subtrees.

The reduction algorithm starts at the bottom of the stored tree, i.e., at the highest numbered address and moves upward until a complete subtree has been detected. In a complete subtree the number of endpoints is exactly one more than the number of conditions. In the above case, the subtree in locations 13, 14, and 15 would be found. The algorithm then scans upward through the tree to find a matching subtree as near to the top as possible, at address 5 in this case. Cell 13 is now altered to read "jump to 5" and becomes a kind of endpoint, while cells 14 and 15 may be erased. The subtree beginning in cell 11 now reads (test C, result 2, jump to 5). This may be seen to be equivalent to the subtree beginning in cell 3 which reads (test C, result 2, subtree beginning in cell 5). Cell 11 may thus be altered to read "jump to 3," and 12 and 13 may be eliminated. Finally, cell 8 may be altered to read "jump to 5" with elimination of 9 and 10. Since this leaves a gap in the useful storage, the lower part of the tree may be moved up, and any pointers which pointed beyond the gap (as in cell 1, which pointed to 11) suitably altered. The final form of the tree is thus:

Address	Test	Left subtree address
1	A	9
2	B	8
3	C	5
4	result 2	
5	D	7
6	result 3	
7	result *	
8	jump to	5
9	jump to	3

A second method, rather like the first, may be used where there is insufficient storage for the entire tree, and it must be condensed as it is read in from an appropriate backing store. This is possible provided that a stack is used to accommodate forward links. The nodes are stored as in the first method until a complete subtree has been entered. Note that a single endpoint will, in general, complete several subtrees of different sizes, each a subtree of the bigger ones. At this point, the input is halted until the smallest of the subtrees has been compared with all the other subtrees already present, to look for a match. This is not so difficult as it at first appears, since it is sufficient to compare the triplets

(test, right-branch, left-branch)

with each other. If all possible redundant subtrees have been eliminated, then the entries for "right-branch" and "left-branch" will be unique. If an endpoint completed more than one subtree, and the smaller ones succeeded in finding matches, the larger subtrees must also be tested. This will result in the freeing of some storage, but since it will always be that which has most recently been filled, no movement of subtrees or adjustment of addresses will be necessary.

The third method of tree compaction to be described, and probably the most important, does not develop the whole tree at all, but attempts to insert the necessary crosslinks during the translation from table to tree. It depends for its operation on the observation that identical subtrees must have been derived from logically equivalent subtables, and therefore seeks to detect this equivalence before the subtable is translated. It will be recalled that the first example tree in this chapter had a pair of identical subtrees, which were derived from subtables which were not identical. The first one came from the route (A = yes) and the second from the route (A = no, D = yes). The tables were:

	1a	1b	1c	1d	2a	2b			1c	1d	2b
B	–	N	–	N	Y	Y		B	–	N	Y
C	Y	–	Y	–	N	N		C	Y	–	N
D	–	–	Y	Y	–	Y					

The equivalence between these tables is readily seen if the Boolean expressions corresponding to the various actions are written out: for the table on the left, they are:

$$\text{action } 1 \equiv C + B' + CD + B'D$$

$$\text{action } 2 \equiv BC' + BC'D$$

and for the table on the right:

$$\text{action } 1 \equiv C + B'$$

$$\text{action } 2 \equiv BC'$$

It is now obvious that the expressions for the two actions in the first table reduce to those of the second. Not surprisingly, the tables give rise to identical subtrees. This method does not take into account the frequency information of the subtables which could have influenced the development of the tables into slightly different trees. It is likely to achieve greater compression of large trees than either of the other methods. On the other hand, it might not achieve the best performance in terms of running speed. As it turns out, the storage problem is likely to be the dominant one in practice.

5.8. EQUIVALENCE OF SUBTABLES

The problem of demonstrating the equivalence of subtables is, however, somewhat more complicated than this example would suggest. It has already been mentioned that a Boolean function may be described by more than one expression which is itself irreducible by simple techniques. It is therefore necessary to demonstrate the equivalence of Boolean expressions without reducing them to the same form. The technique is as follows:

1. Take one term from the first expression f and set it equal to 1 (or TRUE). This will impose values on some of the variables; e.g., $w'y = 1$ means $w = 0$ and $y = 1$, but other variables are undefined.
2. Substitute these values in the second expression, g. If g now becomes *identically* TRUE, then the chosen term of f has been shown to be contained in g. If, however, g is found to be other than identically TRUE, the term selected from f is not contained in g, and so the expressions describe different functions.
3. Repeat the operation for each term in f, testing it in g.
4. Then test each term in g against f, in the same way.

The expressions from McCluskey's book (1965) already quoted may be used to illustrate this method. They are:

$$f = w'y + xy'z + wx'z$$

$$g = w'y + w'xz + wy'z + x'yz$$

Substituting the first term of f in g results in immediate equivalence to 1. If the second term is set to 1 then $x = 1$, $y = 0$, $z = 1$, and so $g = 0 + w' + w + 0 = 1$. Similarly, the other terms of f are in g and the terms of g are in f. By contrast, consider the expression:

$$h = x'y + w'y + xy'z + wy'z$$

It is readily seen that each term of g is contained in h, but the first term of h is not contained in g, since, if $x'y = 1$, then $x = 0$, $y = 1$, and so $g = w' + 0 + 0 + z$, which is not identically true.

The demonstration of equivalence of subtables thus involves the following steps:

1. Is the set of actions described by each table identical? If not, then the tables cannot be equivalent.
2. Are the Boolean functions of corresponding actions equivalent?

The problem is thus reduced to one of demonstrating equivalence of Boolean expressions to 1 or TRUE after substitution of some values. Fortunately, in many cases one term of an expression becomes TRUE after substitution of values, but where it does not, the following method, taken from Breuer's paper will suffice. If the expression F, in sum-of-products form is identically true, then F' will be identically FALSE. (Usually F' will first be derived from F by de Morgan's rules, and so will be in product-of-sums form.) Therefore, if F' is expressed in sum-of-products form, every term will be identically FALSE. If this is shown not to be the case, then F was not identically TRUE. F' is formed from F by de Morgan's rules, initially in product-of-sums form, then terms are collected into sum-of-products form and each one tested for equivalence to FALSE. If a term is found where a factor is not immediately balanced by its opposite (to give identity to 0) then the expansion need proceed no further. A simple example will illustrate the method:

Suppose $F = y + y'x' + y'x$, then in product-of-sums form:

$$F' = y'\,(y + x)\,(y + x')$$
$$= y'yy + y'yx' + y'xy + y'xx'$$
$$\equiv 0 \quad \text{and} \quad F \equiv 1$$

If, however, F is changed to

$$F = y + yx' + y'x$$

Then

$$F' = y'\,(y' + x)\,(y + x')$$
$$= y'y'y + y'y'x' + y'xy + y'xx'$$
$$= y'x' \neq 0$$

So that

$$F \not\equiv 1$$

5.9. TABLE TO LATTICE CONVERSION

The conversion of tables directly to lattices may now be described. The method is somewhat constrained by considerations of storage and speed of translation, since, unless some precautions are taken, the problem can become unmanageable. At each stage in the translation process, a subtable will be split to yield a node of a tree and two subtables or endpoints. It is necessary to test immediately whether the subtable is equivalent to another one in the tree, and, if so, to replace it with a suitable crosslink to the appropriate point in the developing tree. Although the storage of a subtable can be greatly condensed, if it is represented by a bit map of the rows and columns of the original master table which it contains, specifying which rows and columns remain, there are likely to be so many subtables created in the course of development of the lattice, that they cannot all be held in memory. A policy of selective throwing away must therefore be adopted. This leads to a "breadth first" development of the tree or lattice, with storage of the intermediate subtables as they arise. At a point midway through the development, there will be some subtables which have already been developed, and some which are still awaiting development. Most of those already developed will be larger than those still waiting, and most of the subtables which are currently being discovered will be more alike in size to the undeveloped subtables than to those already developed. It is therefore possible, when storage for further subtables runs out, to discard the earliest subtables on the grounds that they are unlikely to match any subtable not yet found. At any point in the development, therefore, there will be a "queue" of subtables, some of which have been processed but not yet deleted from the store, since they can still be used in the search for equivalent subtables when a new one is found. When a subtable is split, it will give rise to two more, or to one or two endpoints. If either of these new subtables cannot be equivalenced to an existing subtable, it will join the queue. If the queue is full, it will be necessary to delete an already processed subtable to make room. When the situation arises that the queue consists entirely of unprocessed subtables, then the translation must stop, and a rerun with larger storage should be attempted. Clearly, the more store which can be allocated to the queue the better, since it is possible that a new subtable could have matched one just thrown away. As a rule of thumb, the queue needs to be at least one-third the length of the expected lattice. For example, a tree of 30,000 nodes is likely to be equivalent to a lattice of one-fifth the number of nodes, 6000, demanding a queue capable of holding 2000 subtables.

When the subtables are held in the highly condensed form demanded by the requirement to store a great many of them, much computer time can be wasted by expanding subtables which turn out to be quite different from the one which is attempting a match. It is therefore valuable to store a bit pattern with each subtable giving the pattern classes to which it refers. Only if these bit patterns match is it worth taking the subtable out of its condensed storage.

5.10. THE DEVELOPMENT OF DECISION RULES

Two more problems remain to be discussed: the development of the decision rules and the choice of sets of rules to give the best performance in the overall recognition system. These problems are essentially disjoint from the problems associated with the translation of decision tables; nevertheless, unless some interaction takes place, the final outcome will be a system which is impossibly large or indifferent in its performance (or even both).

Decision rules may be "learned" from scratch by a computer or developed from inspired guesses. That is, the starting point may be good or bad and one would like a method which produced good decision rules either way. The form of a rule is the allocation of a YES, NO, or DON'T CARE status to each of the possible conditions or features in the system. Given a training set of patterns, described according to their reaction to each of the possible tests, it is possible to allocate a score to the performance of a given rule.

It should be mentioned that one rule need not cover all the possible varieties of its class, since the other rules can make up the deficiencies. It must not, however, admit patterns of a different class. The scoring technique must therefore take account of the relative difficulty of each sample of the training set. If two rules for one pattern class already have the capability of recognizing a particular pattern, then a third rule should not be scored highly because it can also recognize it; the pattern is clearly "easy." If, however, the two rules already found do not recognize a pattern belonging to their own class, and a third rule does recognize it, then the reward of a high scoring is deserved.

Decision rules can be "grown" from suitable *seed rules*. These are systematically modified in one of four ways for each condition:

1. a DON'T CARE entry changes to a YES
2. DON'T CARE to NO
3. YES to DON'T CARE
4. NO to DON'T CARE

The other possibilities, NO to YES and YES to NO are rather too drastic to be useful. For example, the "seed" rule on the left in the three conditions A, B, and C can give rise to four more thus:

	seed	1	2	3	4
A	Y	–	Y	Y	Y
B	N	N	–	N	N
C	–	–	–	Y	N

Each of the newly created rules can be tested on the recognition system to see if it does better than the seed rule, and by a process of refinement, some good rules will emerge. Not surprisingly, there are several complicating factors which must be catered for.

Initially all of the patterns in the training set must be processed by all of the feature tests that will be used. If there are N of these, then an N-dimensional binary vector will be associated with each pattern. Several patterns may share the same binary pattern vector. Each distinct vector will have a count associated with it, and the total number of vectors will be less than the total number of training patterns. In the course of development of the decision rules, it will emerge that some patterns are "easy" and some are "hard." A score for the degree of difficulty must therefore be associated with each pattern. To begin with, all patterns are scored "hard" when developing the first rule for a class. Subsequently, those patterns which are correctly dealt with by the rule are scored "easy" when the next rule is being developed. This is necessary to prevent the system repeatedly converging to the same decision rule.

A rule which is intended to recognize class C will probably misrecognize some patterns of another class as belonging to class C. It will also fail to recognize some genuine class C patterns. Both errors are undesirable, but the second is much less serious than the first. If the rule recognizes T patterns from the target class and U patterns from other classes, it may be allocated a score

$$Tp - Uq = S$$

where p is the benefit derived from recognizing a pattern and q is the penalty of making a wrong decision. When the rule is almost acceptable, q will be much larger than p, and U will be much smaller than T. However, when the unrefined rule is in its infancy, U will be as big as T or even larger, since several patterns will be accepted by a rather unselective rule which ought not to be accepted. If at this stage q is larger than p, then the best rule will be the curious one which has $T = U = 0$, not a very promising start! It is necessary, therefore to have a sliding scale of scoring according to the expected level of sophistication of the rule. Such a sliding scale may be readily achieved by setting $q = 0$ initially, and increasing it by one at each generation of rules by the computer. To increase the chances of developing good rules, several promising rules can be kept in a "pool" and the weakest may be discarded when a stronger one is found. Each of these may be developed in turn until no further improvement is observed. When this happens, all of the patterns which are recognized by the best rule are marked "easy," and a further round of rule development may begin.

The selection of the best rule from the pool is not quite straightforward since, by the time no further developments are possible, most of the rules in the pool have very similar characteristics. The choice is therefore made in four stages, eliminating several rules at each stage.

1. Choose the rule(s) with the highest score.
2. From these, choose the rule(s) with the minimum nontarget acceptance.
3. From these, choose the rule(s) with the minimum target acceptance.
4. From these choose the least complex rule.

Step 3 may appear a little strange; it is designed to select a rule which derives its score from recognizing fewer "hard" patterns rather than many "easy" ones. When a set of rules is ready, its performance, in terms of recognition errors and rejects, may be calculated. Then it may be converted from a table of rules to a decision tree.

5.11. CHOOSING THE SET OF RULES

Once a large set of rules has been derived, consisting of several rules for each pattern class, it is necessary to select from the entire set a subset which will give the best performance-to-cost ratio. The addition of a rule to a decision table results in an increase in the size of decision tree or lattice, the growth being approximately exponential. Furthermore, it is quite likely that the removal of some rules from a large set will actually result in an improvement in the recognition performance. The reason for this is that a pattern has a greater chance of being recognized by a larger set of rules, but it has also a larger chance of being misrecognized. A misrecognition will result in a misclassification if the pattern is only recognized by the wrong rule, or in an "undecidable" verdict if it is also recognized by one of the rules for its own class.

To select a set of rules, it is necessary to decide on the relative cost of rejects and substitution errors. Normally a substitution is much more serious than a reject, typically 10 to 100 times more serious. This ratio may be designated E, since it strongly influences the choice of rules for the recognition set.

The entire set of patterns in the training set is submitted to the set of recognition rules. Each rule accumulates a score, S, describing its performance as a member of the set. At the end of a trial, one rule may be eliminated, and the trial repeated on the reduced set until the performance ceases to improve, or until the set of rules is reduced to a manageable size.

If a pattern is presented to the set of rules and none of them recognizes it, then this gives no indication of possible improvements, since removal of a rule will still result in nonrecognition of the pattern. If, however, one or more rules do claim to recognize a pattern, then several cases may be distinguished. The score S of rule R may then be adjusted in one of the following ways:

1. *R is the only rule to recognize the pattern and it gives a correct recognition.* Removal of R would change a recognition into a reject, which is worth one unit, so S is increased by 1 or, in symbolic form,

$$S: = S + 1$$

2. *R is the only rule to respond to the pattern, but it gives an incorrect recognition.* Removal of *R* would thus turn a serious fault into a less serious one, so

$$S: = S + 1 - E$$

3. *R is the only rule of its class to respond to the pattern, and does so correctly, but some other rules, all corresponding to one other class also respond.* *R* has therefore prevented a serious error, though a lesser one remains, so

$$S: = S + E - 1$$

4. *R responds correctly to the pattern, but so do the rules belonging to more than one other class.* Removal of *R* would make no difference so *S* is unchanged.

5. *The same argument applies even if R responded wrongly.*

6. *R is not the only rule to respond correctly to the pattern,* therefore its score does not change.

7. *R responds incorrectly to the pattern:* other rules, belonging to the correct pattern class, also respond. *R* is thus causing an unnecessary reject so

$$S: = S - 1$$

8. *R responds incorrectly to the pattern, but so do some other rules, all of them from another wrong class.* *R* is therefore preventing a substitution, so rather surprisingly its score must be increased substantially

$$S: = S + E - 1$$

One drawback about this set of scoring rules is that if two recognition rules have exactly the same performance on the training set, they will tend to consolidate each other's position in the set of recognition rules, unless they both end up with the lowest scores. Some special action is necessary to detect this case.

REFERENCES

Bell, D. A., 1973, Decision Trees in Pattern Recognition, Ph.D. Thesis, University of Southampton.

Breuer, M. A., 1968, Heuristic switching expression simplification, *Proc. Assoc. Comp. Mach. Natl. Conf.*, p. 241.

King, P. J. H., 1966, Conversion of decision tables to computer programs by rule mask techniques, *Comm. Assoc. Comp. Mach.* 9:796.

Kirk, H. W., 1965, Use of decision tables in computer programming, *Comm. Assoc. Comp. Mach.* 8:385.

McCluskey, E. J., 1965, *Introduction to the Theory of Switching Circuits*, McGraw-Hill, New York.

Pollack, S. L., 1965, Conversion of limited entry decision tables to computer programs, *Comm. Assoc. Comp. Mach.* 11:677.

Pollack, S. L., Hicks, H. T., and Harrison, W. J., 1971, *Decision Tables: Theory and Practice*, Wiley-Interscience, New York.

Press, L. I., 1965, Conversion of decision tables to computer programs, *Comm. Assoc. Comp. Mach.* 8:385.

Rabin, J., 1971, Conversion of limited entry decision tables to optimal decision trees: Fundamental concepts, *SIGPLAN Notices*, Sept., p. 68.

Schmidt, D. T., and Kavanagh, T. F., 1964, Using decision structure tables, *Datamation*, Feb., p. 42; March, p. 48.

Schwayder, K., 1971, Conversion of limited entry decision tables to computer programs – A proposed modification to Pollack's algorithm, *Comm. Assoc. Comp. Mach.* 14:69.

Yasui, T., 1971, Some aspects of decision table conversion techniques, *SIGPLAN Notices*, Sept., p. 104.

6

Editorial Introduction

M. J. B. Duff describes the system architecture of a special-purpose machine for processing two-dimensional data arrays (pictures). He shows that picture-processing algorithms are most naturally implemented on a parallel machine, whereas the conventional serial computer is much slower. A purpose-built parallel processor has two advantages:

1. *Speed*: Although it is expensive compared to a serial processor, a parallel machine is much faster. If a "retina" of $N \times N$ picture points is to be used, the speed improvement is of order N (possibly N^2). By careful design of the system architecture, even this speed improvement may be greatly exceeded. Duff describes a system which allows the user to interact with the processing; the response time is typically a few milliseconds ($N = 96$).

2. *Software*: The "natural" language for a parallel processor is one in which vector and matrix operations can be specified simply, without recourse to the equivalent of FORTRAN DO-loops. The ergonomic advantages of this facility are very great, since the user thinks in parallel-processing terms and can more easily recognize sections of his programs, if they are compact. "Parallel languages" exist, of course, for serial machines (e.g., APL) or can be written by the user, expanding an extensible language (e.g., ALGOL 68).

The advantages of man-processor interaction cannot be overemphasized; the analysis of pictures requires intuitive judgment, to assess the effectiveness of proposed algorithms. It is clear that interaction can only be achieved using a parallel processor.

Many interesting points of comparison exist between this chapter and those by Aleksander (Chapter 3) and Ullmann (Chapter 2). Aleksander, like Duff, uses electronic implementation methods, while Ullmann looks to the future of optical processors. All three authors discuss systems which are well suited to processing visual images or other two-dimensional data arrays. The essential difference between Duff and Aleksander is one of philosophy:

1. Duff begins by trying to implement a certain (very broad) class of algorithms, which he knows to be useful and wishes to apply to data with interaction between the user and the machine.

2. Aleksander begins with the premise that he has a flexible and very convenient hardware module (a RAM) and seeks to apply this to pattern recognition. In colloquial terms, Aleksander poses the question "What can I do with this module which (at first sight) looks very promising?"

Batchelor, like Duff (and Aleksander), emphasizes the importance of parallel machines, and also advocates the greater use of "parallel languages," which can specify vector/matrix operations without DO-loops.

6

Parallel Processing Techniques

M. J. B. Duff

6.1. INTRODUCTION

It is a limiting and perhaps unnatural feature of conventional digital computers with classical von Neumann architectures that they are designed to perform only one calculation at a time. Systems have been developed which allow calculations to proceed at the same time as input–output operations and make possible some measure of interleaving of slow and fast operations. However, at any instant, a computer will be carrying out only one basic arithmetic operation, such as adding together two binary numbers, or subtracting them. Hardwired arithmetic units speed up these operations, but the principle remains the same: instructions involving more than three numbers (multiplier, multiplicand, and product, for example) must proceed sequentially.

In contrast, conventional computers are equipped to store vast quantities of information, some bits representing instructions and others data. In part this is necessary, since the serial nature of the processor results in a requirement for the storage of very long strings of instructions. These strings become excessively long when the data being processed are in the form of large two-dimensional arrays of numbers, such as those representing optical densities at regularly spaced points in gray-tone pictures.

In considering the possibility of using parallel processing techniques for pattern recognition and image processing, two questions must be answered:

M. J. B. Duff   Department of Physics and Astronomy, University College London, England

1. Is the nature of the problem such as would allow simultaneous operations at many points, or even at every point, in the pattern to be processed?
2. Could a special-purpose computer be produced at a reasonable cost which would offer an efficient means for implementing parallel processing algorithms?

6.1.1. Image Digitization

Although parallel processing techniques can also be applied with advantage in speech recognition and other areas where the temporal structure of the data is particularly significant, the following discussions will refer only to operations on spatial (two-dimensional) patterns, or pictures, so as to avoid unhelpful complications in the arguments. As a further simplification, it will be assumed that a picture is presented to the processor as a two-dimensional array of numbers referred to as gray levels, each representing average blackness over a small square region in the picture. The size of the squares, or cells, must be sufficiently small to represent adequately significant small detail in the picture. The square cells form a regular nonoverlapping array covering the entire picture area. Low numbers are usually assigned to dark regions and high numbers to light regions. If the picture is to be represented in binary form, then a threshold is chosen such that grey levels equal to or greater than the threshold are replaced by 1's, lower levels being replaced by 0's. Later in this chapter, for discussions of binary images, black cells are labeled 0 and white cells 1.

6.1.2. Parallel Operations

The gray level assigned to a small region in a picture is of no significance unless there is some information about gray levels in other parts of the picture. Thus a light gray cell may represent a dark spot in a light field if it is surrounded by white cells, but would be a light spot in a dark field if surrounded by black cells. Self-evident though this example may seem, it is nevertheless characteristic of most of the arguments concerned with picture processing. Nearly all processes involve establishing a relationship between the gray level of each cell and that of all or some of its neighbors. The neighbors may be immediate neighbors (of which there are eight in a square array) or may be a larger subset of the whole picture. In extreme cases it may be necessary to consider the whole array in relationship to each cell.

There is no universally accepted formal definition of a parallel process. However, let P be a parallel process which acts on an array A of n numbers producing a new array $A*$, where

$$A = (a_0, a_1, a_2, \ldots, a_{n-1}) \tag{6.1}$$

and

$$A* = P\{A\} \tag{6.2}$$

The process P can be regarded as the simultaneous operation of a subprocess p acting on each component of the array, such that the new value of the component is

$$a_j* = p\,[a_j, s(a_j)] \tag{6.3}$$

and

$$A* = (a_0*, a_1*, a_2*, \ldots, a_{n-1}*) \tag{6.4}$$

Here $s(a_j)$ is the subset of components of the original array A which are taken to be neighbors of a_j. Less formally, we can say that in a truly parallel process, the result of applying the process to a particular cell will not be affected by the processing of one of its neighbors, since it is the *original* state of the neighbors which is used in the process, not the final processed state.

As an example, consider the pattern shown in Fig. 6.1, which shows a straight line of 1's in a background of 0's. This can be taken to represent a white line on a black background. If we apply an algorithm which states that the new pattern will consist of those 1-cells which originally had more than one 1-cell in their immediate neighbor set, then a true parallel-processor would produce a new array in which the shaded cells changed value from 1 to 0 (i.e., the ends of the line would be chopped off). However, if the processor had been designed to operate continuously on the current state of all of the array cells, then the line would rapidly erode until it had completely disappeared. Each removal of an end would in its turn produce a new end which would then be removed by the continuous application of the algorithm.

Thus, the two essential elements of a parallel process would seem to be as follows:

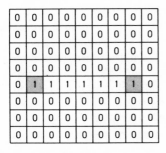

Fig. 6.1. A parallel algorithm for removing line ends.

1. An identical operation must be carried out simultaneously at each point in the array.
2. The operation must involve only the original states of each cell and its defined neighbors.

It will be shown later that these conditions can be slightly relaxed with advantage in some situations.

We can now list operations which can be seen to fit the requirements for parallel processing. In this category, we can include all simple arithmetic relations between two patterns held in two similar arrays, in which the new array is formed by taking the sum, difference, product, or ratio of corresponding pairs of cells in the two original arrays. Similarly, Boolean relationships (AND, OR, etc.) taken cell by cell can be used to form a new array when the original arrays are in binary form. Very simple processes such as these do not involve interactions between cells and their neighbors and can be implemented by means of arrays of electronic microprocessors or by optical means. For example, if two binary transparencies (in which regions in the images are either opaque or clear) are laid one on the other, then the resultant transparency is the AND of the two originals, assigning 1 to clear regions and 0 to opaque regions. If, on the other hand, two gray-tone images are projected onto the same screen, the composite picture so produced has intensities which are the sum of the intensities in the two originals (in the case of binary transparencies the result will be the OR of the originals).

Other simple processes which could be described in terms of parallel operations include shifts, expansions and contractions (in which either every 1-cell surrounds itself by 1's or every 0-cell by 0's), and neighborhood processes in which cells are eliminated or retained depending on the precise state of their neighboring cells. We have already considered one such example in Fig. 6.1.

It should be noted here that the majority of parallel processes require more than one parallel instruction to carry them out. It is not implied that a complete parallel process, whatever it may be, requires only one parallel step. The nature of the parallelism lies in the simultaneous application of the process in each step to all parts of the data field.

Considering more complex processes, it should be possible to carry out two-dimensional Fourier transforms and the Laplace transform in a parallel manner. Indeed, since a simple, image-forming convex lens produces a Fourier transform (see Chapter 2) then we can be sure that it is at least possible to implement the two-dimensional Fourier transform in a parallel, optical processor. This can also be regarded as a single-step parallel process. No doubt an electronic implementation would be considerably more complicated.

At this stage, it becomes instructive to ask what processes cannot be regarded as simple, parallel operations. The most obvious is probably counting.

Suppose an array contains 1's and 0's, and none of the 1's have 1's as neighbors. Thus each 1 can be regarded as an isolated object. The problem is to write a short sequence of parallel instructions which will count the number of 1's in the array. This process will involve a sequence of operations whose length will be at least of the order of n. A similar problem involves the determination of the *area* of an object composed of 1's in a background of 0's. In another example, the aim is to determine whether or not an object (composed of 1's) is *simply connected*, that is to say, does it have holes in it (made of 0's)?

We now come to an important point in the argument. Let us suppose that a process to be carried out on an array is essentially serial in its nature, in that a different subprocess is to be performed at each array point. Could all such processes be performed on a parallel-processing computer? The answer is *yes* and can easily be demonstrated to be so.

In the array of binary numbers shown in Fig. 6.2 the shaded cell is the only one in the array which does not have a neighborhood containing a 1 or a 0 in the West or North positions. It is therefore possible to create in an empty array a 1-cell which will act as a pointer to the shaded cell. Consider the algorithm which states that the new state of a cell is 1 if it has a 1-cell as its West neighbor, and is 0 otherwise. This is a *right-shift* algorithm. It can be seen that the "pointer" can be made to visit, in turn, each cell in the top row of the array. A similar algorithm can be made to produce *downward* shifts, so that the pointer can be made to scan the entire array. By combining an expanded pointer (expanded into the eight immediate neighbor cells, plus the center cell) using a Boolean AND with the original array, each cell in the array together with its neighbors can be lifted out into an empty array. It can then be processed appropriately, using the local algorithm defined for the region in the array indicated by the pointer. The results of the process at each step are then transferred into an output array. Figure 6.3a illustrates the principle employed, and Fig. 6.3b shows the steps in the process. Although this would appear to be very cumbersome, it does illustrate an important feature of parallel-processing programming. Even though an operation may seem to be completely serial, it can always be forced into a series of parallel algorithms and implemented in a parallel computer, provided that the processor is sufficiently general in its design. In the same way, a completely parallel process can always be implemented on a

1	1	0	1	1	0	0	0
1	1	0	1	1	1	1	1
0	1	1	0	0	0	0	0
0	0	0	0	1	1	1	0
1	1	1	1	1	1	1	1
0	0	0	0	1	1	0	0

Fig. 6.2. A parallel algorithm for specifying a single cell.

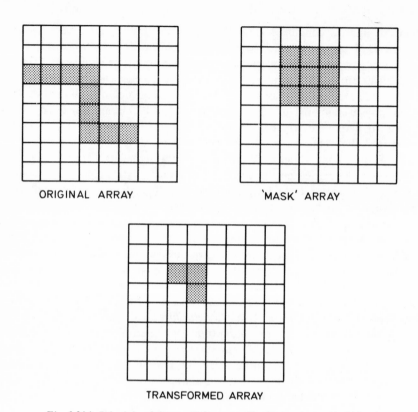

Fig. 6.3(a). Principle of the parallel scanner (1-cells are shown shaded).

general-purpose serial computer. However, in both cases where the parallel/serial implementation does not correspond with the nature of the problem, an uneconomic program structure results and the hardware of the computer is used inefficiently.

It is the purpose of this chapter to suggest that the ratio of the numbers of parallel to serial processes in picture processing is usually high enough to merit the use of specially designed parallel processors, and to suggest that such processors can be sufficiently general in their structure to permit their application to a wide range of problems.

6.2. BRIEF SURVEY OF PROPOSED PARALLEL PROCESSORS

Most techniques for parallel processing fall into two general categories: optical and electronic. A third category, optoelectronic, falls between these two,

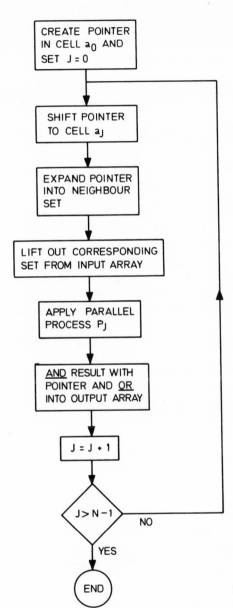

Fig. 6.3(b). Flow diagrams for a parallel scanner.

but represents a largely undeveloped field. Optical and optoelectronic methods will not be treated in this chapter but are reviewed by Ullman in Chapter 2.

It has been implied, although not explicitly stated, that one processor will be required for each cell in the picture to be processed. It can further be deduced that each processor must receive inputs from its corresponding cell and from cells in the neighborhood set. Clearly, any electronic implementation of a processor array will involve large numbers of both components and interconnections. Consequently, and with regard to cost restrictions, the majority of proposals for parallel processor arrays, or cellular logic arrays as they are often called, have not been taken further than a simulation on a conventional serial computer.

In 1958 and 1959 Unger proposed, and later simulated, a square array with 36 X 36 elements and with nine memory registers in each cell. Possible instructions to each cell include left, right, up, and down shift instructions, logic addition and multiplication between the cell accumulator, neighboring cell accumulators, and the cell memories, transfers to memories, input/output, and a special "link" instruction for finding connected sets. Typical character recognition programs required of the order of 300 to 500 instructions per recognition. Some of the features of Unger's proposals were incorporated in the Solomon computer built at the Westinghouse Electric Corporation, Baltimore, Md. by Gregory and McReynolds (1963). Many subsequent researchers have acknowledged inspiration from Unger's proposals, although others have sought rather to simulate parts of biological visual systems and have drawn their inspiration from neurophysiological work such as that carried out by Hubel and Wiesel (1962). Simulated neural nets, and nets of neuron-like elements, have been investigated by many workers, including Aleksander and Mamdani (1968) and Brown *et al.* (1970). Theoretical studies of the properties of cellular arrays have indicated their power and versatility in image processing for operations such as thinning and skeletonizing, image enhancement, and feature extraction. Minnick (1967) has reviewed the use of cellular arrays for these and other tasks. But construction of hardware operating in a parallel manner has not often been attempted. McCormick (1963) constructed the ILLIAC III computer and Levialdi (1968) has built parallel systems for shrinking and counting images of objects and for detection of certain image properties such as closed loops.

The greater part of the study of parallel algorithms which could be run on ILLIAC III has been carried out using simulations, first PAX by Johnson (1972) and, later, XAP by Hayes (1972). Serial simulations are necessarily very slow compared with hardware implementations, and it is likely that the current disenchantment with such simulations stems from the low process speeds obtained.

6.3. THE CLIP PROCESSORS

Logic arrays used for parallel processing of pictures exhibit certain common characteristics. In general, processors will comprise an array of cells, one processing cell being assigned to each picture cell. Each cell will receive inputs from an external data source (usually either a TV camera or a flying-spot scanner) and from neighboring cells. The cell will contain logic gates to process the inputs and a certain number of bits of storage which can store intermediate processed patterns prior to producing a final processed output. Outputs from each cell will connect with neighboring cells and further outputs provide the processed pattern output.

Optimization of the structure of an array must lead to a determination of the interconnection pattern, the internal logic, and the internal storage in each cell. The selection of these three factors will depend on striking a compromise between an expensive cell with a rich interconnection pattern, together with powerful internal logic and a large amount of storage, capable of fast and sophisticated processing, and, at the other extreme, a cheap and simple, sparsely interconnected cell with minimal storage and very little processing capability. The advent of integrated-circuit technology has moved the economically feasible cell much nearer to the first extreme. It should now be possible to construct arrays with useful processing capability at a cost comparable with that for constructing a small to medium-size conventional serial computer.

We will now consider the three main design characteristics in turn and indicate the reasoning behind the optimization selection reached in the CLIP (cellular logic image processor) program at University College London (Duff *et al.*, 1973, 1974).

6.3.1. Array Interconnections

The first decision concerns tessellation: should the array be triangular, square, or hexagonal? If all picture cells are to be the same shape and size, then only these three shapes will provide complete coverage of the picture area. While the triangular array would appear to have no redeeming features, the choice between square and hexagonal depends on several conflicting considerations.

Square arrays are subject to a connectivity paradox. In Fig. 6.4a the line of black cells can be regarded as continuous if diagonal (corner) contact is defined as true connection. Thus the white background must be separated into

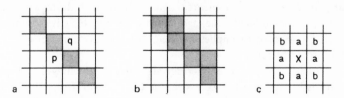

Fig. 6.4(a). Connectivity paradox (see text). (b) A 4-connected line. (c) The square array immediate neighbor set.

two unconnected areas. However, the same criterion applied to white cells, such as p and q, would imply that these cells are similarly connected, so that the white areas are not unconnected.

On the other hand, if we insist that connection between cells must be across flat faces rather than merely through corners (so that the black cells in Fig. 6.4b constitute a continuous boundary), then in Fig. 6.4a, the black line of cells must be regarded as discontinuous, and it follows that the white areas will be connected together. But white cells such as p and q (Fig. 6.4a) will themselves be counted as unconnected on this new criterion; no pair of cells can be found which will connect the two white areas. This paradox can be resolved only by applying different connectivity rules for white and black cells.

A second undesirable feature of square arrays is that, of the eight immediate neighbors to any cell, four (labeled a in Fig. 6.4c) are at unit Euclidean distance from the central cell, whereas four (labeled b) are at distance $\sqrt{2}$. This property obviously complicates the magnitude of the effect on the central cell due to its neighbors when effects between cells are a function of the distance between them.

In the hexagonal array, neither of these features is observed, since all the immediate neighbors are of equal status, being similarly connected to the central cell and at a uniform distance from it.

On the credit side, the square array allows the accurate representation of both horizontal and vertical directions. It may at first sight seem surprising that the lack of a pair of orthogonal array directions in a hexagonal array should prove to be a disadvantage, but it should be remembered that picture processing is closely related to visual perception in which the horizontal and vertical both have special significance.

Once the array tesselation has been determined, it is then necessary to decide the interconnection structure between cells. For complete generality, it should be possible to write a sequence of instructions to the array, such that the value of any selected cell in the array can be transferred into any selected array location. This condition is satisfied if each cell communicates only with its immediate neighbors (4 in a square array, 6 in a hexagonal array). A denser interconnection structure will increase the speed of some algorithms but will considerably aggravate the technical difficulties in constructing the array. In particular, comparing an 8-connected square array with both a 4-connected square array and a 6-connected hexagonal array (see Fig. 6.5), it can immediately be seen that 8-connectivity implies the need for crossing conductors for all the diagonal interconnections.

Since each structure offers certain algorithmic advantages and since the cost penalty is in any case not excessive, the CLIP arrays (from CLIP 3 onward) can be switched under program control into either square or hexagonal connectivity, with each interconnection direction independently gated. The optical

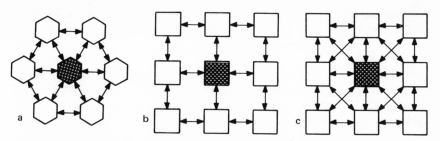

Fig. 6.5(a). 6-connected hexagonal array. (b) 4-connected square array. (c) 8-connected
square array.

scanner can similarly be switched into square or hexagonal operation. For
example, if a vertical shift is required, then the square-array option is effected
and only the vertical connection gates enabled. Alternatively, the array can be
programmed so that every cell receives inputs from any required selection of
directions.

6.3.2. Cell Logic

Restricting the discussion to the processing of binary images, each cell in
an 8-connected square array will receive nine inputs and, in the most complicated
case, could be required to produce nine independent outputs (see Fig. 6.6).

If the internal logic of the cell comprises a "simple" Boolean processor
which can be set to provide all possible states of nine independent outputs, then
the number of functions that the processor must be able to implement is
approximately 10^{1387}. In general m inputs imply $M = 2^m$ input states and n
outputs imply $N = 2^n$ output states. For a given input state, there are N ways of

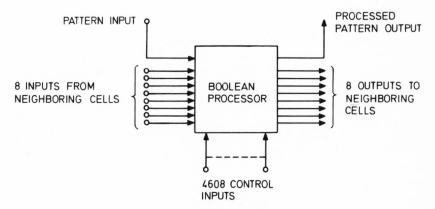

Fig. 6.6. A generalized but overcomplicated logic cell.

selecting the required output state. The total number F of possible functions is therefore given by

$$F = (N)^M$$
$$= (2^n)^{2^m} \tag{6.5}$$

The number of control bits, C, needed to specify the internal logic of a cell to implement all possible functions is

$$C = \log_2 F$$
$$= n2^m \tag{6.6}$$

In the example chosen, this number amounts to 4608. This is clearly an impracticably large number representing an intolerable interconnection problem, since every cell in the array would need to be connected to at least this number of control lines.

An obvious simplification would be to restrict each cell to two independent outputs, one of which fans out to the neighbors. However, even with this simplification, 10^{308} states are possible, requiring 2048 control lines. Since the magnitude of these numbers are dominated by the index m, it is necessary to keep m small in order to obtain a manageable system.

A possible system is shown in Fig. 6.7 in which the Boolean processor has only two inputs and two outputs (therefore requiring eight control inputs). A common output to the neighboring cells is provided and the inputs are individually gated (using eight independent gate inputs) into a preprocessor T. This pre-processor in all cases produces an output which forms one of the two binary inputs to the Boolean processor.

Various forms for the box labeled T can be considered, a very flexible form being the variable threshold gate which sums the enabled inputs and compares the sum with a threshold in the range 0 to 7 (requiring three further control bits to specify the threshold).

It will be noted that an additional feature in Fig. 6.7 is the OR-ing of the T output with a second pattern from B. This facility permits immediate Boolean operations on each pair of cells in two patterns, which is an important function in an array processor. Its operation and application will be discussed later in this chapter.

The logic structure illustrated in Fig. 6.7 has been adopted for the CLIP 3 cell. Experience has shown that the threshold gate in T could well have been replaced by an OR gate with no loss of efficiency for the vast majority of algorithms which have been programmed into CLIP 3. This would also have increased the processing speed, which is a function of the delays in the various gates in the cell, the threshold gate being responsible for more than half the total delay.

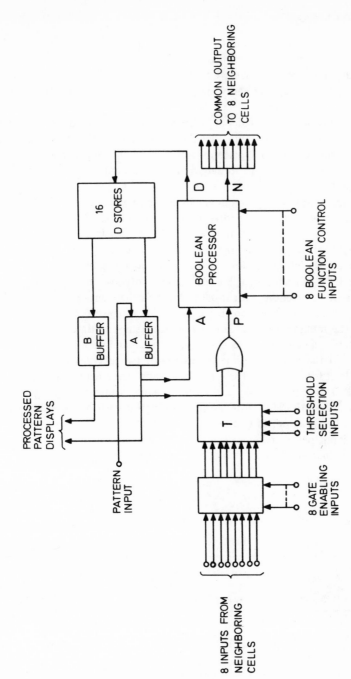

Fig. 6.7. A practical cell structure for logic arrays.

6.3.3. Cell Storage

If the pattern storage is distributed in the array so that logic and data are located together, problems of addressing the stored data do not arise and "fetch" and "store" times are minimized. Calculations of the number of bits of storage required in each cell are not easily made, since the pertinent criteria are not well established.

If we assume that two bits of buffer store are required for the two pattern inputs A and B in Fig. 6.7, and that these will be loaded from an addressable store into which processed patterns and the original input patterns can be directed, then it only remains to determine the optimum number of bits in the addressable store.

The maximum storage requirement arises in the processing of gray-level pictures. Let us assume that the intensities of cells in pictures to be processed are stored in the array cells in b bits representing 2^b gray levels. In a typical program, the intensity of each cell in the picture is replaced by the average of the intensities of its eight neighbors. Before division, the summed intensities will require $(b + \log_2 8)$ additional bits of storage, totaling $(2b + 3)$ bits. At least three more bits are required for storage of intermediate answers in the necessary logical-arithmetic operations, so that a final total of some $(2b + 6)$ bits will be required. The minimum number of bits of storage appropriate for various gray-level inputs is shown in Table 6.1. A more demanding operation, involving calculation of the product of the intensities of two gray-level pictures would require at least $(4b + 3)$ bits. These figures are also tabulated in Table 6.1.

CLIP 3 is interfaced to a television camera resolving eight gray levels. Sixteen bits of addressable store are provided within each cell in addition to the two bits of buffer store. In CLIP 4, which will be expected to handle information from input peripherals resolving up to 64 gray levels, 32 bits of addressable store will be incorporated in each cell. These figures are compatible with those in Table 6.1.

Table 6.1.   Cell Storage

No. of gray levels in the input	b	No. of bits of cell storage required for averaging programs	No. of bits of cell storage required for multiplication programs
8	3	12	15
16	4	14	19
32	5	16	23
64	6	18	27

6.4. PROGRAMMING A CLIP ARRAY

It is not the purpose of this chapter to give a detailed description of the CLIP mnemonic programming language, nor, for that matter, to suggest that all parallel processors must, of necessity, look like our own CLIP array. However, since considerable experience has now been gained in the use of the CLIP arrays, it is convenient to use them to illustrate those features of parallel processing which will probably prove to have some generality.

To facilitate the presentation of programs, a simplified form of the CLIP language will be adopted, ignoring the available sophistications in the language where they provide more confusion than help in explaining the parallel algorithms.

6.4.1. CLIP Instructions

It will be assumed that patterns reside in the array stores addressed as D_0, D_1, D_2, etc. The two bits of buffer store in each cell are referred to as A and B (see Fig. 6.7). A LOAD instruction is of the form:

$$LD\ X, Y, Z$$

and loads bit A from store D_X, bit B from store D_Y, and addresses the processed pattern resulting from the next process instruction into store D_Z (X, Y, and Z are all octal numbers). Alternatively, the A or B bits can be cleared by writing C in the appropriate field.

The only other type of instruction which will be used is the PROCESS instruction which is of the form:

$$D = \theta(1, m, n, \ldots)\ B_N\ (P, A),\ B_D\ (P, A),\ E.$$

In these instructions, θ is a number in the range 0 to 7, representing the threshold in the variable threshold gate (T in Fig. 6.7), 1, m, n, etc. are enabled directions for the interconnections (see Fig. 6.8), and B_N and B_D are the Boolean

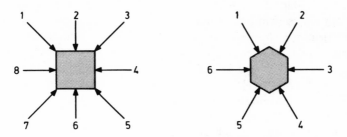

Fig. 6.8. CLIP array interconnection directions.

functions of P and A (the two inputs to the Boolean processor) which define N and D, respectively (the interconnection output and the processed pattern output from the processor). The addition of the letter E at the end of the instruction implies that all interconnection inputs entering cells from outside the array are held at 1 rather than 0.

Since most iterative processes should be terminated when the pattern reaches a required state (for example, a thinning operation should be halted when further application of the algorithm would produce no further change in the pattern), some form of pattern "inspection" is needed. The CLIP array achieves this by taking the D outputs from every cell to a common AND-gate. setting to 1 a bit of storage (&) after each process instruction, only if every D output is 1. The (&) bit is used to control certain branch instructions. A typical use of this branching technique is illustrated in Section 6.4.3. Other branch instructions are conditioned by the state of four sense switches (S1, S2, S3, S4) on the control panel. In the following descriptions, branch and I/O instructions are stated in words rather than by using the mnemonic instruction codes, thus avoiding consideration of peripheral-dependent coding. As a further simplification, only square-array operations will be discussed here.

6.4.2. Simple CLIP Programs

Since the implementation of a parallel algorithm in a parallel processor does not involve a cell-by-cell scanning operation, many simple processes can be achieved by means of a single pair of LOAD and PROCESS instructions. For example, the following program removes isolated white (1) points from a pattern stored in D_1:

Line No.

0   LD 1,C,1

1            $D = 0(1 \rightarrow 8) A, PA$

Note that $(1 \rightarrow 8)$ implies all interconnection directions from 1 to 8 are enabled. White cells produce a 1-output at N, since the Boolean function defining N is $N = A$. Only those white cells which receive an interconnection input from any of their eight neighbors appear as 1 cells in the D output which had been addressed into D_1 (the zero threshold gives a 1-output from any T gate with a 1 on any of its 8 N-inputs; $D = PA$ is the Boolean function which therefore removes isolated white cells). Note also that clearing the B buffers (as indicated by the C in the LOAD instruction) implies that $P = T$ (rather than $P = B + T$).

As another example, consider a program which extracts as a pattern of 1-cells, those 0 (black) cells which are totally enclosed by 1 (white) cells in the input pattern (see Fig. 6.9). In the figure, 1-cells are shown shaded. The program is:

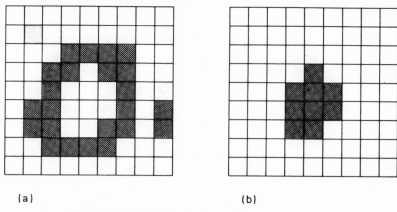

(a)   (b)

Fig. 6.9. A hole-finding algorithm. (a) Input. (b) Output.

Line No.

0   LD 1,C,1

1                    $D = 0(1 \rightarrow 8)P\overline{A},\overline{P}\,\overline{A},E$

At the edges of the array a propagation signal flows into all interconnections (E) and is transmitted only by black cells ($N = P\overline{A}$). Black cells enclosed by white cells therefore do not receive the interconnection signal and appear as 1's in the output ($D = \overline{P}\,\overline{A}$). Although a single instruction-pair implements this algorithm, the process instruction involves propagation through all the black "background" cells. In CLIP 3, propagation takes 0.1 μs/cell, comparing with 1.0 μs for non-propagating instructions.

Propagation can present two hazards which therefore must be guarded against. The first is the possibility of a "race" condition. This could arise only if the Boolean function $B_N(P,A)$ were to invert the propagation signal (so that N would be a function of $\overline{P}$ rather than P). Clearly, this can be prevented by disallowing such functions which are, in any case, of little programming value.

The second hazard is the possibility of the introduction of a spurious propagation signal which is self-maintaining once introduced. Careful physical layout and elimination of electrical noise completely overcome this potential difficulty.

Many of the more complex programs involve SHRINK and EXPAND processes which strip off one or more layers from an object and fill in small holes. The following program (SHRINK – EXPAND – EXPAND – SHRINK) removes white and black noise and removes small detail in the input pattern stored in D_1. The "cleaned-up" output pattern is then stored in D_2.

Line No.

0	LD 1,C,2	} SHRINK	}	removes small
1	$D = 0(1{\to}8)\bar{A},\bar{P}A$			white detail
2	LD 2,C,2	} EXPAND		
3	$D = 0(1{\to}8)A,P + A$			
4	LD 2,C,2	} EXPAND	}	
5	$D = 0(1{\to}8)A,P + A$			removes small
6	LD 2,C,2	} SHRINK		black detail
7	$D = 0(1{\to}8)\bar{A},\bar{P}A$			

All the above operations are isotropic. As a final example, a directional process will be described. The program is:

Line No.

0	LOAD A PATTERN INTO D_0
1	LD 0,C,0
2	$D = 0(2)A,P$
3	BRANCH TO 6 WHEN ARRAY IS EMPTY
4	DISPLAY D_0
5	BRANCH TO 1
6	END

The display of D_0 shows the pattern sinking step-by-step through the lower edge of the array since each execution of the directional propagation process in instruction 2 causes a shift downward by one cell. The branch in instruction 3 is conditioned by testing the (&) bit described above.

6.4.3. Longer Programs

In order to perform useful image processing, much longer sequences of operations than those illustrated in the previous section must be carried out. A few examples will give some idea as to how more complex programs are constructed.

Suppose it is required to find, in a sequence of binary images, a binary image in which only one connected object is present.

Line No.

0	LOAD NEW IMAGE INTO D_0
1	LD 0,C,1
2	$D = 0(2,4)P + A,P + A$
3	LD 1,C,1
4	$D = 0(2,8)A,\bar{P}A$
5	LD 0,1,1

Line No.

6	D = 0(8)P$\bar{A}$,PA
7	LD 0,1,0
8	D = 0(1→8)PA,P + $\bar{A}$
9	BRANCH TO 11 IF ARRAY IS FULL OF 1'S
10	BRANCH TO 0
11	END

The process is illustrated in Fig. 6.10. Instructions 1 and 2 spread the binary image to the left and downward. Instructions 3 and 4 identify the top left-hand point of the spread image, being the only 1-cell without another 1-cell either above or to the left of it. This cell can then be loaded into the B store so as to initiate a propagation signal traveling to the right along 0 cells in the original image (which is still stored in D_0). The propagation ceases at the first 1-cell in the image which can therefore be extracted, since it will be the only 1-cell receiving a propagation input. It is then loaded into the B store and instruction 8 initiates propagation from this cell through connected 1-cells in the original image. The D Boolean function outputs, as a pattern of 1's, cells which were 0 cells or which received the propagation signal. Clearly, any disconnected white object in the image will not be reached by the propagation signal and will appear as 0's in the output. The "full array" test is then used to decide whether to terminate the program or to return for another image input.

The program described above represents a "natural" application of parallel processing to images. It is less obvious that parallel processors can be used effectively for arithmetic calculations. Greater advantage is obtained when the calculation is such as can permit simultaneous operations on large arrays of numbers. Although a cellular logic array can be used to multiply together a single pair of numbers, representing each number as the leftmost column in two arrays, the process is slow in comparison with the speeds obtainable using the hardwired arithmetic unit in a serial computer. The fast processing speed of these units requires a high-speed electronic technology which would be far too expensive to duplicate in a large array.

Histogram construction can be carried out very economically in a cellular logic array. Suppose, in the course of an analysis program, that the size of each object is counted and a single 1-cell in the bottom row of an otherwise empty array D_0 is stepped one place to the right. At the end of the count giving the area measurement the array D_0 containing the single cell is combined with a histogram array D_1, increasing by one the count in the column containing the single cell. The process is then repeated for each object being measured. The histogram array D_1 then displays an area histogram for objects in the input image, the columns representing binary numbers with their least significant digits in the bottom row of the array. The program is illustrated in Fig. 6.11 and the coding

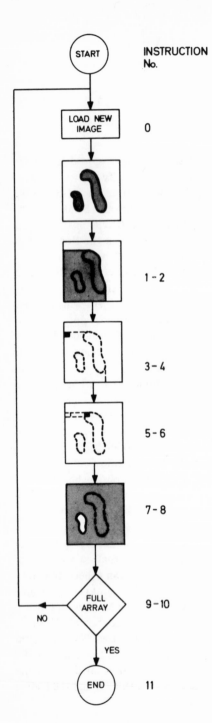

Fig. 6.10. Schematic flow diagram illustrating
the single-connected object program.

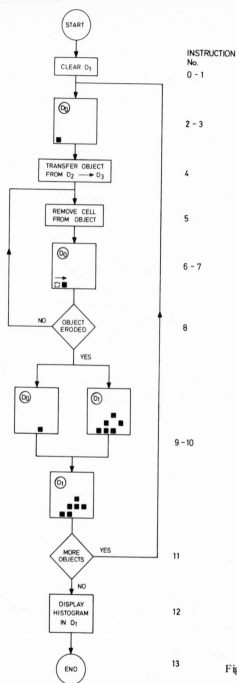

INSTRUCTION
No.

0 - 1

2 - 3

4

5

6 - 7

8

9 - 10

11

12

13

Fig. 6.11. Schematic flow diagram illustrating
the area histogram program.

is as follows:

Line No.

0	LD C,C,1	} Clear the histogram array D_1
1	D = A	
2	LD C,C,0	} Create a pointer in the bottom left-hand corner of D_0
3	D = 0(6,8)$\bar{A}$,$\bar{P}$	
4	TRANSFER AN OBJECT FROM D_2 TO D_3	
5	REMOVE ONE 1 CELL FROM THE OBJECT IN D_3	
6	LD 0,C,0	} Shift pointer one place right in D_0
7	D = 0(8)A,P	
8	BRANCH TO 5 IF D_3 STILL CONTAINS PART OF AN OBJECT	
9	LD 1,0,1	} Add 1 to the count in the appropriate column of D_1
10	D = 0(6)PA,P⊕A	
11	BRANCH TO 2 IF THE ARRAY D_2 CONTAINS ANOTHER OBJECT	
12	DISPLAY HISTOGRAM ARRAY D_1	
13	END	

The technique for removing whole objects to another array (instruction 4) is a slightly modified version of the algorithm illustrated in Fig. 6.10. A further slight modification, in which the propagation instruction (number 8) of the connected object program is changed to a point deletion instruction, enables the same program to effect instruction 5 (object erosion for area measurement) of the present program.

Processing of gray-level pictures is achieved by bit-plane arithmetic. Instead of representing the gray levels as binary numbers in columns of a single-dimensional array of storage (as in the histogram program), several planes are assigned to the picture and, in any particular cell, the appropriate gray level will appear as a binary string in D_0, D_1, D_2, etc. Thus all the most significant bits for the whole array will be in, say, D_0, the next significant in D_1, and so on (see Fig. 6.12). Arithmetic processes between adjacent cells are implemented by sequences of single shifts in any one plane and Boolean operations between the contents of adjacent planes. Figure 6.13 shows the displays from six CLIP 3 programs illustrated a wide range of contrasting processes.

6.5. PROCESSING LARGER IMAGE AREAS

Obviously, the small size of the present CLIP 3 array, representing only 192 image points, restricts its use as an image processor. Two courses of action are possible: either a hybrid system must be developed in which the small array processor is scanned across a larger image area, or a larger array must be built. The first solution is comparatively cheap but results in a processor which is several orders of magnitude slower than a pure array-processor. The second

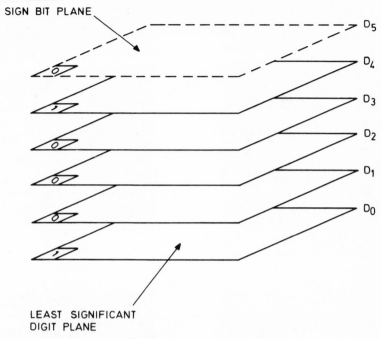

SIGN BIT PLANE

D_5

D_4

D_3

D_2

D_1

D_0

LEAST SIGNIFICANT
DIGIT PLANE

Fig. 6.12. Bit plane storage showing binary 17 stored in the bottom left-hand corner of the stack.

solution is only economically feasible if large-scale integration of the cell electronics is carried out. The CLIP development program at University College London is proceeding in both these directions.

6.5.1. The Hybrid Processor

In the hybrid processor, the CLIP 3 array is surrounded by edge registers which store the propagation signals emanating from the edges of the array, and re-present these signals to the array as adjacent sectors of the image are processed. The total image area which is covered is 96 X 96 cells.

During the first scan, which proceeds from left to right, row by row, starting in the top row (see Fig. 6.14), the process starts with all edge registers R1 and R4 cleared. In principle, after the first sector has been processed, the contents of R4 are transferred to R3 and R4 cleared and the contents of the left end of R2 shifted into R1. The array is now ready to process the data in the second sector which is to the right of the first sector and the sequence of operations is repeated until all sectors in the top row have been processed. At this point, R2 contains all the propagation bits from the first row and processing of the second row commences.

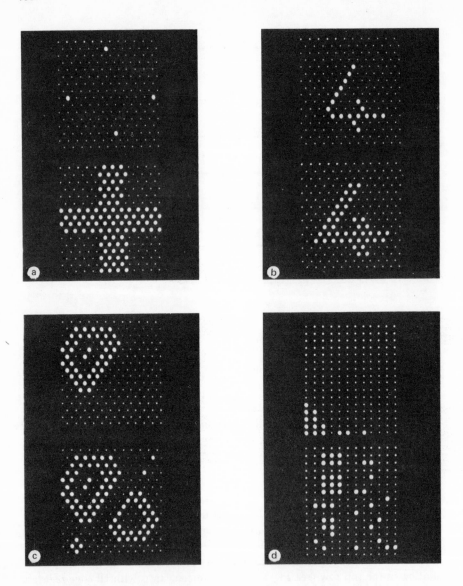

Fig. 6.13. Examples of CLIP 3 programming. In each case the input image is in the lower display and the output appears in the upper display. (a) Thick line ends. (b) Skeleton: the skeleton is of single cell thickness and preserves the connectivity of the original image. (c) Nucleated cell: objects which are not closed loops containing at least one unconnected object are rejected. (d) Perimeters histogram: the perimeter of all objects are extracted and the area histogram program in Fig. 11 then applied. (e) Smoothing: the output gray level of each cell is the average of its 4-connected neighbors. (f) Gradients: edges of gray regions are displayed where the gray level gradient exceeds a chosen threshold.

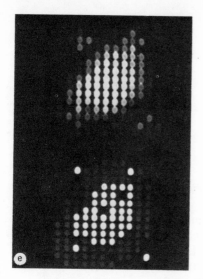

Fig. 6.13 (*continued*)

When the array process involves propagation, it is necessary to divide the process into two subprocesses. The first of these produces an output pattern in which all cells in the propagation path appear as 1-cells; the second uses this propagation path and forms the appropriate Boolean function of the path with the original image. Suppose that we wish to implement the following instruction sequence in the 96 × 96 cell array:

Line No.

0	LD 0,C,1
1	D = 0(1→8)A,P + A

then we must implement the following sequence of instructions in the hybrid array:

Line No.

0	LD 0,C,1
1	D = 0(1→8)A,P
2	LD 0,1,1
3	D = P + A

It should be noted that the 96 × 96 patterns cannot be stored in the 16 × 12 array-D stores. Suitable external storage must be provided. A further complication is that the scan must be followed by a complementary scan in the reverse direction (right to left, bottom to top) to allow for propagation in all directions,

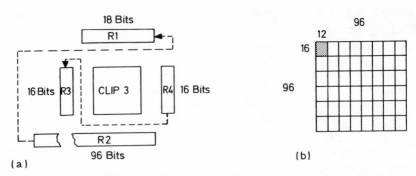

Fig. 6.14. Hybrid system. (a) Edge register configuration. (b) 48 sector scan pattern.

and a check scan is always required to determine the completion of propagation. Figure 6.15 shows the complete CLIP 3 hybrid system which enables a wide range of optical inputs to be processed in CLIP 3 via a television camera input. The PDP 11/10 computer is interfaced to the array so as to allow convenient program preparation and flexible control.

Figure 6.16 shows three photographs of the television monitor during the processing of a biological section, input via the microscope — television camera peripheral. Figure 6.16a shows the unprocessed camera output (occupying the central linear third of the frame); Fig. 6.16b is the display following binary thresholding at one of eight gray levels; Fig. 6.16c shows figure edges resulting from the application of the simple process represented by the CLIP process instruction

$$D = 0(1{\rightarrow}8)A,P + A$$

6.5.2. Large Arrays

CLIP 3 was constructed using commercially available TTL medium-scale integrated circuits. The logic circuits for 4 cells occupy one printed-circuit board and the RAM storage occupies another board, so that the 16×12 cell array fills three 8-in. high, 19-in. wide standard card frames. Using the same method of construction, a 96×96 cell array would therefore fill some 16 racks, each 6 ft. tall, without making any provision for power supplies.

In contrast, by using an MOS large-scale, integrated circuit chip comprising 8 cells (each cell having 32 bits of storage and a logic structure similar to CLIP 3) it will be possible to build a 96×96 array and house both the array and its power supplies and controller in one 7-ft. rack. Component and construction costs are therefore greatly reduced. This array, to be known as CLIP 4, is scheduled for operation by the end of 1977.

Unfortunately, the CLIP 4 array will be rather slow in comparison with CLIP 3, due to the unavoidable use of MOS technology. At the time of writing,

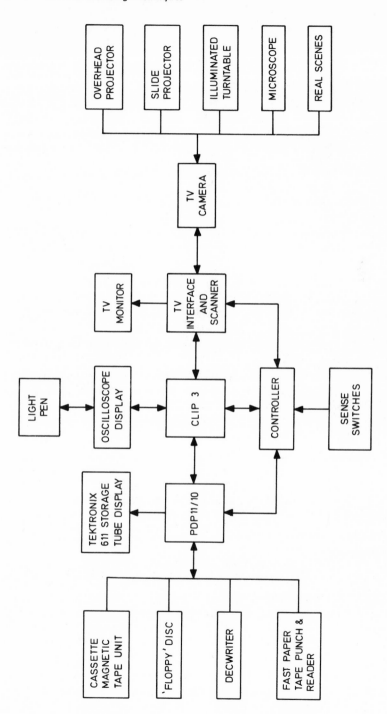

Fig. 6.15. The CLIP 3 hybrid system.

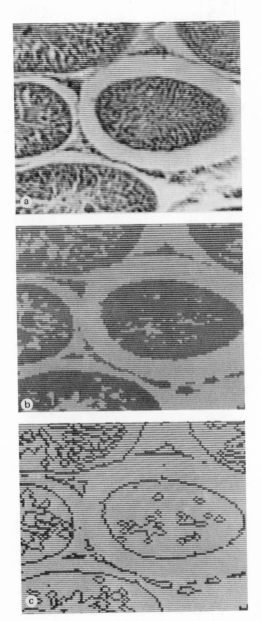

Fig. 6.16. Edge finding program for
the CLIP 3 hybrid system. The photographs show the central region of the
television monitor display. (a) Original
video output. (b) Binary output following gray level thresholding.
(c) Edges.

faster technologies are not available for chips of the required size and complexity. As an example, it has been estimated that the execution time for a LOAD/PROCESS pair of instructions will be about 10 μs compared with 2 μs in CLIP 3. A more serious disadvantage is that the propagation speed is expected to fall by at least a factor 10 to about 1 μs/cell. Nevertheless, overall process times will still compare very favorably with those for the hybrid system.

6.6. FUTURE TRENDS

A goal which is currently quite unobtainable is to produce an LSI chip embodying a complete array, together with optical inputs and outputs. Disregarding the I/O devices, the probable area of CLIP 4 will be in the region 2 mm^2/cell, so that the chip area for a 96 $\times$ 96 array would be of the order 1.84 $\times$ 10^4 mm^2. This figure is very approximate since losses at the edges of the CLIP 4 chip must account for an appreciable fraction of the area, but a chip with sides of about 14 cm is quite impossible. Nevertheless, the possibility of bonding chips onto a single large substrate to form an array does not seem to be excluded. This method of construction would also facilitate the incorporation of semiconductor optical components.

A more likely development would be the addition of more arithmetic and data transfer facilities within the cell logic and memory structure. CLIP 4 will include further gates to permit its use as a full adder; other suggestions which would be of interest in image processing applications would involve sophistication of the D-store structure to improve the bit-plane arithmetic capability. In applications requiring greater arithmetic precision, the amount of storage in each cell would have to be substantially increased.

A further sophistication which could influence the design of future arrays might be the provision of facilities for variable architecture and variable cell logic, both on a local basis, so that one area of the array could function in a quite different manner to other areas at any given instant. The inclusion of "trainable" elements in the cell logic might well be thought to make such variability feasible. However, the difficulties likely to be encountered in programming such an array are expected to be more daunting than the constructional problems, so that a considerable amount of thought and experimentation will be needed before arrays of this type can be put to effective use.

Cellular logic arrays are still comparatively unexplored and their potential as image processors largely unappreciated. It remains to be seen in the next few years whether their promise can be fulfilled.

ACKNOWLEDGMENTS

The author gratefully acknowledges the contributions of Dr. D. M. Watson, Mr. T. Fountain, and other members of the Image Processing Group to the research program discussed in this chapter which is supported by the Science Research Council in collaboration with University College London.

*REFERENCES**

Aleksander, I., and Mamdani, E. H., 1968, Microcircuit learning nets: Improved recognition by means of pattern feedback, *Electron. Lett.* **4**:425–426.

Brown, D., Hall, M., and Lal, S., 1970, Pattern transformation by neural nets, *J. Physiol.* **209**:7p.

Duff, M. J. B., Watson, D. M., and Deutsch, E. S., 1974, A parallel computer for array processing, *Inf. Processing 74 (Proc. IFIP Cong.)*, pp. 94–97.

Duff, M. J. B., Watson, D. M., Fountain, T. J., and Shaw, G. K., 1973, A cellular logic array for image processing, *Pattern Recognition* **5**:229–247.

Gregory, J., and McReynolds, R., 1963, The SOLOMON computer, *Trans. IEEE* **EC-12**:774–780.

Hayes, K. C., 1972, XAP: an 1108 file-oriented picture management system, *Tech. Rpt. TR-213*, Computer Science Center, University of Maryland, College Park, Md.

Hubel, D. H., and Wiesel, T. N., 1962, Receptive fields, binocular interaction and functional architecture in the cat's visual cortex, *J. Physiol.* **160**:106–154.

Johnson, E. G., 1972, The PAX user's manual, *Computer Note CN -10*, Computer Science Center, University of Maryland, College Park, Md.

Levialdi, S., 1968, "CLOPAN": A closed-pattern analyser, *Proc. IEE* **115**:879–880.

McCormick, B. H., 1963, The Illinois pattern recognition computer – ILLIAC III, *Trans. IEEE* **EC-12**:791–813.

Minnick, R. C., 1967, A survey of microcellular research, *J. Assoc. Comp. Mach.* **14**:203-241.

Unger, S. H., 1958, A computer orientated toward spatial problems, *Proc. IRE* **46**:1744–1750.

Unger, S. H., 1959, Pattern detection and recognition, *Proc. IRE* **47**:1737–1752.

---

*Further and more detailed information concerning the CLIP system is published in the internal reports of the Image Processing Group, Department of Physics and Astronomy, University College, London from whom copies may be obtained.

7

Editorial Introduction

J. A. G. Halé and P. Saraga present a broad review of the motivation and techniques for processing digital images. They set visual pattern recognition into perspective, among other closely allied subjects. As they point out, the problem of feature extraction is one of the most difficult areas of pattern recognition. The fact that numerous workers (outside "pattern recognition") are busy finding reduced-"bandwidth" techniques for storing, transmitting, and displaying data should not be ignored, since the same methods may be applicable to feature extraction. Many of the topics they introduce are developed in more detail by other contributors, for example:

1. Detail enhancement and noise reduction (Chapters 6, 11, and 12)
2. Edge detection (Chapters 6 and 8)
3. "Skeletonization" (Chapters 6 and 12)
4. Tomography (Chapter 11)
5. Complex scene analysis (Chapters 8 and 16)
6. Optical implementation (Chapter 2)
7. Electronic implementation (Chapters 3, 4, and 6)
8. Spectral analysis (Chapter 11; see also Chapters 13 and 15)
9. Industrial inspection and assembly (Chapters 2, 8, and 10)

7

Digital Image Processing

J. A. G. Halé and P. Saraga

7.1. INTRODUCTION

There has been a very rapid growth in the field of image processing in the past two decades, a growth which shows every prospect of accelerating. One witness to the current interest is the number of surveys and special issues of journals being devoted to the topic (Rosenfeld, 1969; *IEEE Transactions – Computers*, 1972; *Proceedings of the IEEE*, 1972a, 1972b; Graselli, 1969; Huang *et al.*, 1971; Harmon and Knowlton, 1969; Andrews *et al.*, 1972; *Pattern Recognition*, 1973; Hall *et al.*, 1971; *Proceedings of Seminar on Image Information Recovery*, 1968; *Proceedings of Conference on Machine Processing of Remotely Sensed Data*, 1973; Arps, 1974). This growth is occurring both in the range of applications and in the variety of the techniques being employed.

The aim of this chapter is to give a broad survey of the whole area of work which can be loosely grouped under the heading of image processing. It is not our intention to give a comprehensive state of the art report. We will examine the motivations behind the work in the field. We believe that these motivations can be summarized in terms of three main requirements:

1. The presentation of images to human beings
2. The presentation of images to machines
3. The economic handling of image data

*This chapter was originally published in *Opto-Electronics*, Vol. 6 1974, pp. 333–348. Permission to republish was kindly supplied by Chapman & Hall Ltd.

J. A. G. Halé and P. Saraga • Mullard Research Laboratories, Redhill, Surrey, England

We will also examine the position within the field of the various "disciplines" such as pattern recognition, scene analysis, image enhancement, etc., which are concerned with the handling of images and pictures. We will analyze the tasks involved in each discipline and give examples of the techniques currently being used to accomplish these tasks. We will also give some examples of the range of applications that fall within the scope of each discipline, trying to give a brief indication of the state of the art. We will concentrate on two specific examples: optical character recognition and the application of scene analysis to industrial automation.

In the final section of the chapter we will describe the typical configuration of a digital image processing system. We will consider the various system components: input, processing, storage and output, and attempt to indicate some current bottlenecks.

7.2. MOTIVATIONS FOR IMAGE PROCESSING

7.2.1. Presentation of Images to Human Beings

Visual images are probably the most important means by which men experience their environment. However, the information required by the human being is often only a small part of the data in the image. The human visual system has a huge channel capacity [about 10 MHz (Schade, 1956)] and a very powerful processing capability. Nevertheless it can, under some circumstances, be extremely inefficient in extracting information. Thus there is a demand for techniques that process the image after capture but before presentation to the human being, in a manner that makes it easy for him to extract the information he requires. The satisfaction of this demand must involve processes for the enhancement and simplification of images and a means of displaying a high-quality image to the viewer.

Because human beings find it easy to absorb visual information there is considerable scope for techniques that can convert nonpictorial data and mathematical abstractions into an image form. The boom in computer graphics is just one example of the importance of this type of transformation.

7.2.2. Presentation of Images to Machines

A human who is continuously performing a boring and repetitive task can become frustrated and error-prone. Consequently we increasingly require machines to replace human beings in mundane work, for which they are ill-equipped. For example, we want machines that can "read" printed text and transfer it to a computer. We also want machines that can automatically inspect factory products for manufacturing defects.

Other applications arise when machines are required to work in dangerous, dirty, or remote environments, for example, the automatic inspection of underwater drilling rigs. The accomplishment of such tasks requires much more than the enhancement of an image. Information must also be extracted from the image and on the basis of this information decisions must be taken. Techniques are therefore required to extract specific types of information from images and to present them to a machine in such a way that the machine can take the appropriate decision.

7.2.3. Economic Handling of Image Data

Because visual information is so important to human beings it is not surprising that there is a need to transmit and to store it. Obvious examples occur in facsimile, videophones, space exploration, aerial surveys, voiceprints, biomedical images, fingerprints, engineering drawings, etc.

Images contain large amounts of data and in most of the applications mentioned there are very many images to be transmitted or stored. The channel capacity required to transmit the data is often far greater then the capacity available. There is therefore considerable incentive to reduce the quantity of data to be transmitted.

7.3. IMAGE PROCESSING DISCIPLINES

From the examples discussed in the previous section it is clear that image processing could also be categorized in terms of the input and output material of the image processing operation. One advantage of this categorization is that it relates closely to the main "disciplines" involved in the processing of images. As shown in Fig. 7.1 we identify five different transformations involving images:

1. Image to image
2. Image to data
3. Data to image
4. Image to name
5. Image to description

The conversion of one image into another "better" one is generally characterized by terms such as "image enhancement," "image preprocessing" or "image filtering." As mentioned in Section 7.1, an image may be transmitted or stored before being presented to a human being. The process of conversion of an image to data involves disciplines such as "signal theory" and "picture coding." The reverse operation, the creation of an image from other data or mathematical abstractions, is generally covered by disciplines such as "map-making" and "computer graphics."

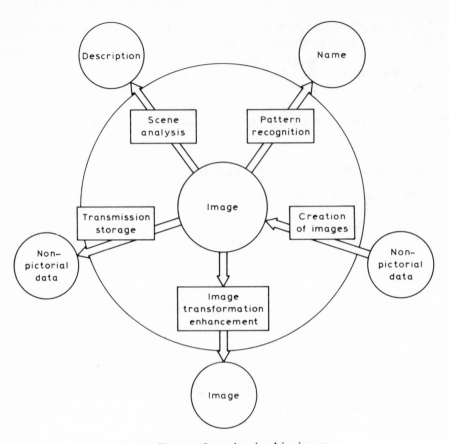

Fig. 7.1. Five transformations involving images.

One of the most fruitful areas of image processing has been pattern recognition. Where the input pattern is in the form of an image, pattern recognition involves the conversion of the image to a name. If rather than a name, the output of the process is some kind of description, then the process is usually referred to as scene analysis.

In the following sections we consider each of the five transformations in more detail.

7.3.1. *Image to Image Transformation*

The process of image capture can introduce considerable distortion. When for example a planet is photographed from a spacecraft, the picture obtained may be degraded by perspective, atmospheric turbulence, aberration in the optical system, and the motion between the camera and the planet. It is obviously important to minimize this type of degradation by clever design of the imaging system. Huang *et al.* (1971) describes this as "*a priori* image improvement." *A priori* techniques alone, however, will not always produce an acceptable image. *A posteriori* image enhancement techniques can be grouped under three main headings as illustrated in Fig. 7.2: spatial transformation, spectral transformation, amplitude transformation.

Spatial transformation such as the correction of geometric distortion is particularly important in the processing of space photographs (Billingsley, 1972). An example of projection correction is shown in Fig. 7.3. The image stretching is done by defining to the computer the new desired location of known points, and intermediate points are determined by interpolation between the defined points.

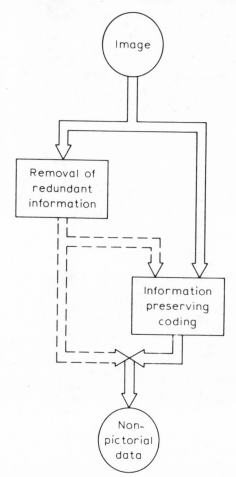

Fig. 7.2. Image to image transformations.

Fig. 7.3. (a) JPL building photographed from a low angle. (b) Rectified version using computer stretching. (From Billingsley, 1972.)

The two most common and useful transformation procedures are those that make the photograph sharper and increase its contrast. Image-sharpening involves spectral transformation, i.e., changing the spatial frequency content of the image, and this may be achieved by applying an orthogonal transformation (e.g., Fourier or Hadamard) (Andrews, 1970; Harris, 1966; Sawchuk, 1973) to the image, high pass filtering, and retransforming. This can be particularly useful in the processing of biomedical images as shown in Fig. 7.4. Other modifications to the spatial frequency can also be used to provide a subjective improvement in the visibility of image features. These modifications need not involve transformations in the frequency domain. Local area numerical filters (Nathan, 1968) can be applied to the original image to achieve similar results. Typical examples of local operator filtering are edge enhancement and the removal of high-frequency spatial noise (Fig. 7.5).

Dramatic changes in apparent contrast can be achieved by amplitude transformations (Stockham, 1972). For example, when one uses a display device which can render a fixed number of grey levels, one might normally arrange these levels so that they are spread linearly over the amplitude range in the image. The contrast can sometimes be improved, however, by arranging the grey levels so that an equal number of picture elements (pixels) are displayed at each level. This is similar to what the photographer does in his darkroom when he brings up details in dark regions.

Image interpretation can also be improved by the use of pseudocolor displays in which different grey levels are displayed as different colors (Lamar *et al.*, 1972).

Amplitude transformation can also be used to detect motion (Billingsley, 1972; Leese *et al.*, 1970). If the difference between successive images of the same scene is displayed, only objects that have moved will be visible. Tasto (1973) has used a sequence of x-ray images to detect motion of the heart wall as part of a program to identify certain types of heart defects. Genuine "real-time," "on-line" processing can also be very important especially in military reconnaissance, surveillance, and navigation.

7.3.2. Image to Nonpictorial Data Transformation

As previously mentioned, data reduction is one of the main problems in image transmission and storage.

One can apply two types of data reduction. First, any visually redundant data can be removed (psychovisual coding), and second, the remaining data can be coded more economically (statistical coding); see Fig. 7.6.

If the transmitted image is to be displayed to a human observer it is possible to take advantage of the characteristics of the human visual system and to remove a considerable quantity of data without introducing noticeable

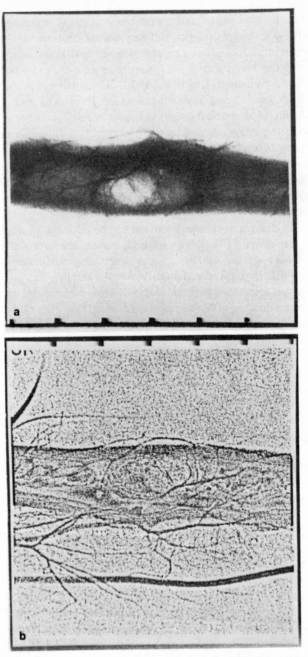

Fig. 7.4. (a) Radiograph of an arm bone containing a tumor. (b) After computer processing which removes gross shading, blood vessel detail may be better seen.

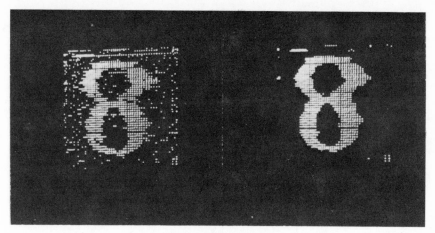

Fig. 7.5. Noise removal.

distortion. For example, one can usually save about 1 bit per sample by using a logarithmic rather than linear quantization scheme (Huang, 1972). There are many different schemes which fall into this group of "psychovisual" coding methods (Stockham, 1972; Huang, 1972; Landau and Slepian, 1971; Harmon, 1971). The particular scheme used depends on the application. A dramatic example of the power of psychovisual coding is shown in Fig. 7.7. The original picture shown in Fig. 7.7a contains 4 bits per pixel, while the coded picture in Fig. 7.7b contains only black and white elements (i.e., 1 bit per pixel) (Brown and Halé, 1972). Despite the data reduction, the face remains recognizable.

Once the data contains only useful information it can be subjected to an efficient information-preserving coding scheme. A further data reduction is achieved by removing the redundancy which exists because not all brightness levels are equally likely, and because the brightness of any point is likely to be similar to that of its immediate neighbors. Many schemes exist for coding images in a way which removes some of this redundancy and hence reduces the number of bits required to transmit the picture (Rosenfeld, 1969; Arps, 1974; Huang, 1972; Landau and Slepian, 1971; Kobayashi and Bahl, 1974; Kobayashi, 1971; Duan and Wintz, 1973; Cherry *et al.*, 1963). One example is run length coding in which instead of the intensity of each point being transmitted, the length and intensity levels of runs of consecutive points having the same intensity are transmitted. The average bit rate is further reduced if those intensity and run length values which occur most frequently are coded with short codes and vice versa. Statistical coding of this type does not remove any information and the original picture can be reconstructed.

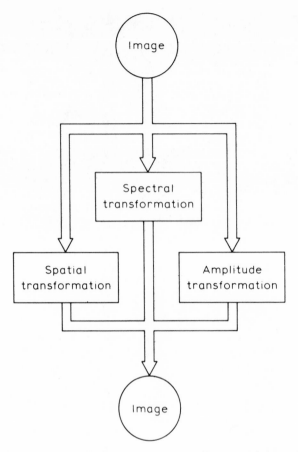

Fig. 7.6. Image to nonpictorial data transformations. (For transmission and storage.)

7.3.3. Nonpictorial Data to Image Transformation

This operation obviously takes place at the receiver of an image transmission system or when a coded image is retrieved from storage and is displayed to a human being. In these cases the transformation required is essentially the inverse of the coding used at the transmitter, although further processing such as the creation of a pseudogray level image from coded data may be relevant here (Roberts, 1962, Schroeder, 1969; Klensch, 1970; Knowlton and Harmon, 1972).

There are many examples, however, where data that have never existed in the form of an image could be more easily understood if it were converted into an image format (Fig. 7.8) (Saunders, 1972). The original data could have come

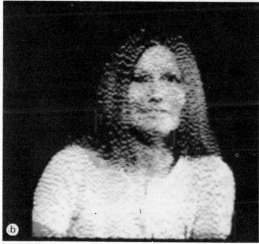

Fig. 7.7. Creation of pseudogray level images with 1 bit per pixel. (a) Line printer picture (4 bits per pixel). (b) Blurred plasma display picture (1 bit per pixel).

from many different sources. A typical example is the ppi radar display in which the amplitude of the received signal is plotted in time and angle coordinates. Other examples occur in all types of map making (Harmon and Knowlton, 1969; Fukunaga and Olsen, 1971; Yoëli, 1967). It is much easier to predict the weather from a picture of isobars and fronts than from lists of measurements from weather stations.

Pictures can also be created from mathematical abstractions. The huge

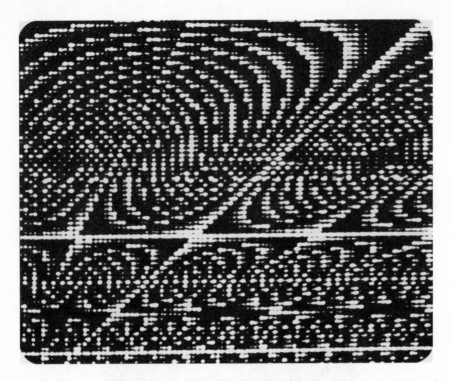

Fig. 7.8. Density of each spot is proportional to the value of the third digit in the division of
X by Y [based on an idea by Saunders (1972)].

growth of computer graphics in the last decade is a demonstration of the value
of visual presentation of information (Newman and Sproull, 1973; Sutherland,
1965). The visual evaluation of mechanical and electrical designs has led to a
great saving of time and effort.

A special case of image reconstruction from projections (Gordon and
Herman, 1971) occurs in x-ray tomography (Grimmert *et al.*, 1973; Bolwell,
1972). This first involves creating a 3D "image" from a number of 2D views
of an object. It is then possible to synthesize any 2D slice of interest from the
3D information.

7.3.4. Image to Name Transformation

Image recognition is part of the wider problem of pattern recognition. The
input to a pattern recognition process could come from many sources
(Kazmierczak, 1973; Ullmann, 1973; Sklansky, 1973; *Proceedings of the 1st
Joint Conference on Pattern Recognition*, 1973; Institute of Physics Conference
Series, 1972; Holdermann and Kamierczak, 1972). It could be a printed

character or equally a set of medical symptoms. In this chapter we will restrict the discussion to images.

An image recognition system is often preceded by a "preprocessing" stage, usually aimed at improving the image by operations such as thresholding and noise removal (Holdermann and Kazmierczak, 1972). Techniques applicable to these tasks have been described in Section 7.3.1. Pattern recognition itself is assumed to start with a preprocessed image. The classical structure of the pattern recognition transformation is shown in Fig. 7.9. The first task, often called segmentation, is to locate and isolate the object to be recognized. Having isolated the object the next task is to choose which measurements or features to extract and then to extract them. Finally, the object is classified according to the measurements made in the previous stage of the process.

The choice of features is often the most important part of the design of a recognition system. Although the problem of measurement and feature selection

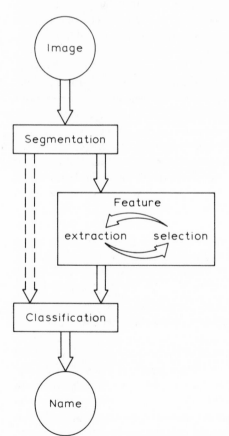

Fig. 7.9. Image to name transformation (pattern recognition).

has received considerable attention (Uhr and Vossler, 1961; Levine, 1969; Kittler and Young, 1973), the general methods that have been developed are not particularly useful in solving practical problems. The measurement and pre-processing techniques employed in the solution of a particular pattern recognition problem tend to be highly dependent on the type of patterns to be recognized. Classification techniques, on the other hand, are mainly dependent on the distribution of the patterns in the measurement space. They are only indirectly dependent on the type of raw data being used, and are beyond the scope of this chapter.

Major applications of pattern recognition techniques can be found in archiving or screening large collections of pictures such as bubble chamber photographs; microscope slides of blood cells, chromosomes and other biological specimens; fingerprints; human faces; cloud and aerial photographs; and speech spectrograms. Character recognition, however, is probably the most commercially successful application. Systems exist which will recognize fairly poor quality printed or typewritten alphanumeric characters. Machines that can recognize fairly well-constrained handprinted numerals have also been available for some years (Ullman, 1973; Romland et al., 1968).

The recognition of handprinted numerals is a useful vehicle to illustrate typical measurement and feature extraction techniques. In contrast to machine-print recognition, in which masking techniques are generally used, handprint systems usually employ some form of stroke analysis. This is reasonable if one considers the way in which the characters are produced. Machine-print characters are "hammered" on to the paper to produce black areas in a fixed spatial relationship. Handprint characters, on the other hand, are composed of strokes in loosely specified directions and positions. Scanned local area masks are used in handprint systems but mainly for noise-reduction and as an aid to stroke analysis, but not for complete recognition.

It is generally considered that sufficient information to recognize hand-printed characters is contained in either the center line of the strokes or in the edges of the characters. Many methods have been proposed for extracting edges and edge directions from both binary and analog images (Rosenfeld, 1969; Saraga and Wavish, 1972; Hueckel, 1973; Saraga et al., 1967; Freeman, 1961). Subsequently, features such as straight lines, cusps, line ends, junctions, angles and loops can be derived from the coded edge of the character (Romland et al., 1968; Saraga et al., 1967).

A very simple scheme for edge extraction from binary images is illustrated in Fig. 7.10. An edge point is defined as a black point, of whose four nearest neighbors (horizontal and vertical) at least one is white. All the edge points in Fig. 7.10a are shown in Fig. 7.10b.

The extraction of topological features such as line ends, junctions, and loops can also be facilitated if the strokes of the character are first reduced to

Fig. 7.10. Edge extraction and thinning: (a) original; (b) edge; (c) thinned character.

lines of a single-point thickness (Rosenfeld, 1969; Hilditch, 1969), Fig. 7.10c. Most stroke-thinning schemes use local operators.

Instead of first thinning the character, some systems follow the original character stroke or fit elemental features to the unprocessed character (Parks, 1969). These local features can then be combined to obtain more global features.

7.3.5. Image to Description Transformation

Image recognition is only one aspect of the larger problem of image analysis and description (Guzman, 1968; Clowes, 1972; Miller and Shaw, 1968; *Proceedings of the 3rd International Joint Conference on Artificial Intelligence, 1973*). The structure of the scene analysis task is illustrated in Fig. 7.11. The problem of describing scenes was first tackled by considering scenes containing polyhedral blocks. The scene is generally viewed with a TV camera. The resulting image is usually cleaned and areas of interest are isolated. The image is then typically reduced to a line diagram, each line corresponding to an edge of a block. This line drawing is analyzed into topological and geometric features, which may include a reconstruction of the original blocks. The scene can then be described in terms of these features, the reconstructed objects, and their inter-relationships. In the naive example given in Fig. 7.12, the table indicates that the eye (⊙) is below (B) the hat (⌂). The machine may well possess additional knowledge of the "world" in which it is working. It can use this knowledge to help with all stages of the analysis process.

More recently scene analysis techniques have been applied to more complicated scenes such as those found in medicine, industry, aerial reconnaissance and space research (Ambler *et al.*, 1973; Winston, 1972; Shirai and Inoue, 1973, Ejiri *et al.*, 1972; Olsztyn *et al.*, 1973; Harlow and Eisenbis, 1973). There has been a particular interest in developing visually controlled machines for tasks such as automatic assembly, and the navigation of self-controlled vehicles. This work has been carried out by university departments primarily concerned with academic problems of artificial intelligence, as well as by industrial research laboratories.

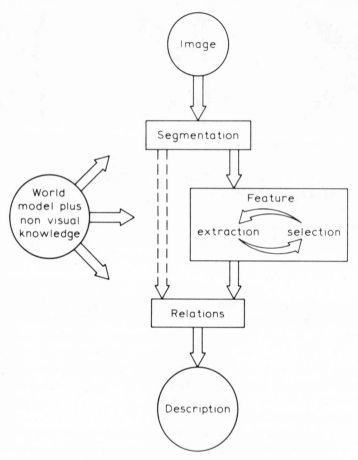

Fig. 7.11. Image to description transformation (scene analysis).

A team at Edinburgh University has developed a versatile assembly system using two TV cameras, and a computer-controlled arm and table. Their system can be taught to make simple assemblies such as a peg and rings or a toy car, Fig. 7.13. This process involves separating parts from a heap and recognizing them by matching a model of the object with the description obtained from the scene.

Similar work has been carried out at MIT, Stanford University, Stanford Research Institute, and Hitachi (Winston, 1972; Shirai and Inoue, 1973; Ejiri, *et al.*, 1972). The work of Hitachi is a good example of the progress in this field being made by Japanese industry. Systems they have developed include a visual image processing system for the automatic detection of defects in complicated patterns like printed circuit boards, integrated circuits, and shadow masks.

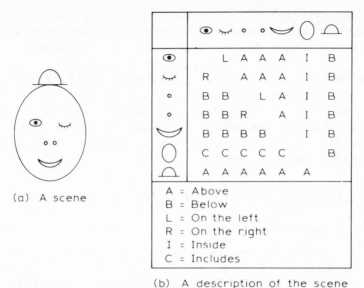

	◉	✕	∘	∘	◡	◯	⌂
◉	L	A	A	A		I	B
✕	R		A	A	A	I	B
∘	B	B	L	A		I	B
∘	B	B	R		A	I	B
◡	B	B	B	B		I	B
◯	C	C	C	C	C		B
⌂	A	A	A	A	A	A	

A = Above
B = Below
L = On the left
R = On the right
I = Inside
C = Includes

(a) A scene

(b) A description of the scene

Fig. 7.12. A naive example of scene analysis.

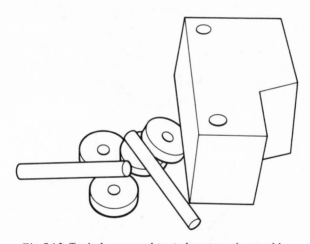

Fig. 7.13. Typical scene used to study automatic assembly.

Another intelligent robot, actually in production, tightens bolts on molds employed in the manufacture of concrete piles. This machine has both visual and tactile feedback.

The Stanford University system has been used to assemble a Model T Ford water pump. Tasks in this assembly include the visual location of the pump body and its components prior to assembly.

Olsztyn (1973) from General Motors has described an experiment in which a numerically controlled machine tool was modified to perform the simulated assembly task of automatically mounting wheels on to automobile hubs. Both the wheel and hub were scanned by a TV camera, and the images were analyzed by computer to determine the positions of the studs on the hub and the stud holes in the wheel. The wheel could then be placed on to the hub. The experiment demonstrated that a visually controlled automatic system was capable of performing this task, but that it was too slow for a production line. This problem will probably be overcome by using special-purpose electronics to perform some of the image processing.

7.4. IMAGE PROCESSING EQUIPMENT

7.4.1. System Structure

The essential components of an image processing system are illustrated in Fig. 7.14. Examples of typical input, output storage, and transmission (Kamierczak, 1973; Vernot, 1973) systems are listed in Table 7.1. The most interesting component, however, is probably the processor itself.

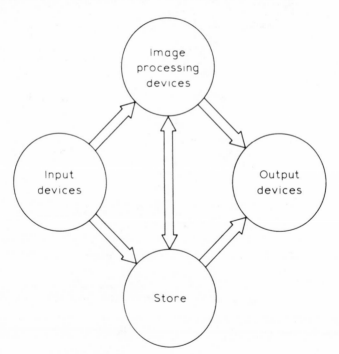

Fig. 7.14. Structure of an image processing system.

Table 7.1.   Image Processing Equipment

Input devices	Output devices
Image dissector	TV display
TV camera	Plasma display
Solid state array	Flying spot scanner + photographic film
Flying spot scanner	Electron beam + film
Radiation detectors	Matrix printer
Radar scanner	Line printer
	Graphics display
Storage devices	**Transmission channels**
Magnetic disc	Laser beam
Magnetic core	TV channel
Magnetic tape	Data link
Paper tape	Hi-fi channel
Cards	Voice telephone
Storage tube (silicon target)	Human reader
COM/CIM	Teletype
Video disc and tape	
Holographic memory	

7.4.2. The Processor

Both optical and digital techniques are widely used to process patterns and there are conflicting claims as to their comparative value. The high speed, parallel processing, and large storage capability of optical systems makes it hard for digital computers to compete in the implementation of processes involving Fourier transformation such as image deblurring (Stroke, 1972). Digital computers, however, are capable of applying a wide range of complicated algorithms, some of which would be difficult to implement optically. Other important advantages include the high precision and repeatability of calculations, and the flexibility and transferability of software. The use of interactive processing, particularly using CRT graphics terminals, is also an increasingly applied and powerful technique.

These advantages have led to a marked preference for digital techniques in pattern recognition and scene analysis. The overall trend in image processing as a whole is also toward digital processing. Quoting from Preston (1972): "A general trend favoring digital electronic systems is evident but it appears that special purpose optical analog systems will have an advantage in certain limited areas of application for the next 10 to 30 years."

Some of the disadvantages of the general purpose digital computer have been overcome by the use of special purpose digital processors. Figure 7.15 attempts to illustrate the range of processor types. The processors are arranged

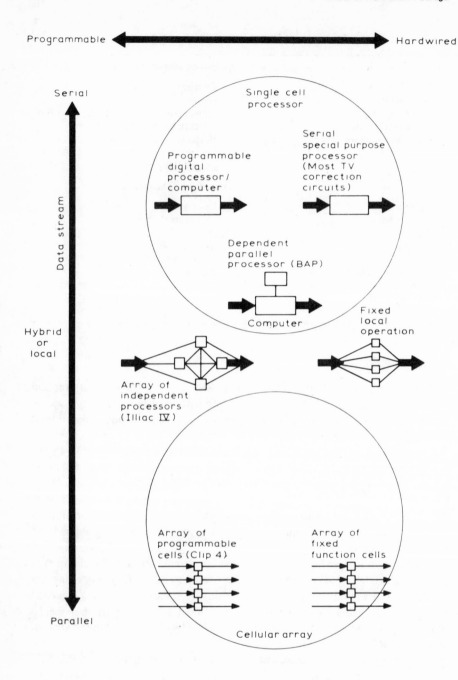

Fig. 7.15. Types of digital image processors.

both according to their degree of programmability and to the degree of parallelism in the data flow. The most widely used processors are the general purpose digital computer, the serial hardwired processor for digital filtering and the special purpose hardwired processors which process small areas of the image in parallel. Multiprocessor computers such as ILLIAC IV (Barnes *et al.*, 1968) are very expensive and cellular array processors in which the same operation is performed simultaneously on each pixel are still in the research phase (Unger, 1958; Preston, 1971; Duff *et al.*, 1974; McCormick, 1963).

7.4.3. Data Rates in Image Processing

The need for special-purpose processors becomes apparent when one examines the data rates which can be achieved in the various parts of an image processing system. This is illustrated in Fig. 7.16 which shows the maximum rates in bits s^{-1} which can be achieved by various image production, transmission, and storage components and compares them with the processing speeds of computers. It can be seen from Fig. 7.16 that the processing is the bottleneck in the overall system since even a high-speed computer can only achieve a data rate of approximately 10^8 bits s^{-1}. This means it can perform a single instruction in the time a TV camera requires to produce a few pixels. To execute a complete image processing operation however, whether a local operator or an orthogonal transformation, requires many instructions per pixel (typically between 10 and 100). The inefficiency of the general-purpose computer is partly because its structure is not well matched to the type of computations involved in image processing. A large part of the time is consumed in locating, fetching, indexing, and storing the picture elements being processed.

In order to obtain a better feel for the data rates shown in Fig. 7.16, we consider the application of various types of digital processor to two generic image enhancement problems. We assume that an image is obtained from a single TV frame and sampled into $N \times N$ elements, each element containing 4 bits (16 grey levels). In the first process each pixel is thresholded at a fixed value and the resulting binary image is subjected to the edge extraction operation illustrated by Fig. 7.10. The second process considered is a two-dimensional Hadamard transform using a standard Fast Hadamard Transform algorithm (Pratt *et al.*, 1969; Jagadeesam, 1970). The approximate times (in seconds) required to execute the two processes for values of N of 64 and 512 are given in Table 7.2. In all cases the time to load the store of the processor is not included Provided that the processing time is less than the data acquisition time of one TV frame (i.e., 4×10^{-2} s), it should be possible to perform real time processing with a single-frame delay. It can be seen from Table 7.2 that for both processes, general purpose computers are too slow and that special purpose processors are required (Bergland, 1969).

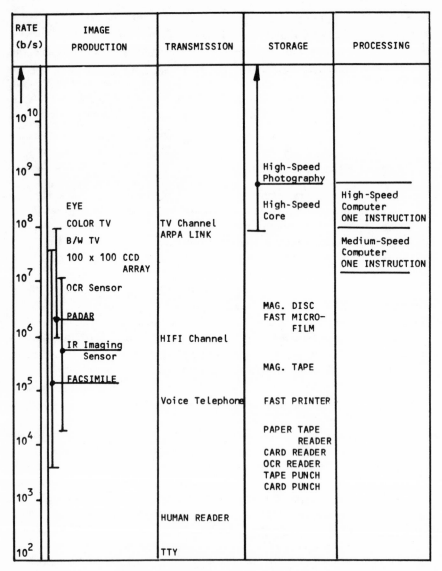

Fig. 7.16. Data rates in image processing.

Table 7.2 also demonstrates that with present technology some kind of parallel processing is required if large images are to be processed at television rates. The figures for parallel arrays are very approximate. Few if any truly parallel systems of any scale have been built. The cost of a hardwired fully parallel 512 × 512 Hadamard transform processor is still prohibitive. It is in the

Table 7.2.   Typical Image Processing Times

Picture size	Edge extraction		Fast Hadamard transform	
	64 × 64	512 × 512	64 × 64	512 × 512
Minicomputer	10^{-1}	10	1	100
Special purpose programmable processor	10^{-2}	1	10^{-1}	10
Sequential hardwired processor	10^{-3}	10^{-1}	10^{-3}	10^{-1}
Programmable parallel array	10^{-5}	10^{-5}	10^{-2}	10^{-2}
Hardwired parallel array	10^{-6}	10^{-6}	10^{-6}	10^{-6}

field of high-speed orthogonal transformations that optical processing seems much more appropriate.

7.5. CONCLUSIONS

In this paper we have attempted to impose a structure on the field of image processing. Although the structure is inevitably oversimplified we hope it is useful in clarifying the relationship between the many disciplines involved with images.

There is a common feeling that any machine used to replace a human function must be as good as or better than the human, before it can be useful. In image processing this feeling is expressed as a demand for machines which can process images at higher rates, or in larger quantities, than can be achieved by human beings.

We have seen how machines have been designed which can capture, transmit and store pictures at data rates approaching that of humans. Technological advances, however, will be required before large images can be effectively processed at the same speed.

ACKNOWLEDGMENTS

We gratefully acknowledge the valuable discussions we have had with P. R. Wavish and Dr. A. Turner-Smith. We would also like to thank the Institute of Physics and Professor F. C. Billingsley for permission to reproduce Figs. 7.3 and 7.4, and the A.I. group at Edinburgh University for permission to reproduce Fig. 7.13.

REFERENCES

Ambler, A. P., Barrow, H. G., Brown, C. M., Burstall, R. N., and Popplestone, R. J., 1973, *A Versatile Computer-Controlled Assembly System*, School of Artificial Intelligence, University of Edinburgh (Experimental Programming Report, No. 29).

Andrews, H. C., 1970, *Computer Techniques in Image Processing*, Academic Press, New York and London.

Andrews, H. C., Tescher, A. G., and Kruger, R. P., 1972, Image processing by digital computer, *IEEE Spectrum* 9:20–32.

Arps, R. B., 1974, Bibliography on digital graphic image compression and quality, *IEEE Trans. Inf. Th.* IT-17:120–122.

Barnes, G. H., Brown, R. M., Kato, M., Kuck, D. J., Slotnick, D., and Stokes, R. A., 1968, The Illiac IV Computer, *IEEE Trans. Comp.* C-17:746–757.

Bergland, D., 1969, Fast Fourier Transform Hardware Implementation: A Survey, *IEEE Trans. Aud. Elect.* AU-17:109–119.

Billingsley, F. C., 1972, Digital image processing for information extraction In: *Machine Perceptions of Pattern and Pictures*, Institute of Physics, Conf. Ser. No. 13, pp. 337–362.

Bolwell, A. J., 1972, Computerized x-rays aids investigation, *EMI Design News* 27:32–33.

Browne, A., and Halé, J. A. G., 1972, British Patent Application (1972), No. 56526/72.

Cherry, C., Kubba, M. H., Pearson, D. E., and Barton, M. P., 1963, An experimental study of the possible bandwidth compression of visual image signals, *Proc. IEEE* 51:1507–1517.

Clowes, M. B., 1972, Scene analysis and picture grammars In: *Machine Perception of Pattern and Pictures*, Institute of Physics, Conf. Ser. No. 13, pp. 243–256.

Duan, J. R., and Wintz, P. A., Information preserving coding for multispectral data, *Proc. IEEE Conf. on Machine Processing of Remotely Sensed Data*, 4a–28–35.

Duff, M. J. B., Watson, D. M., and Deutsch, E. S., 1974, A Parallel Computer for Array Processing, *Proc. IFIP Conference, Stockholm* (Aug., 1974), pp. 74–97.

Ejiri, M., Uno, T., Yoda, H., Goto, T., and Takeyasu, K., 1972, A prototype intelligent robot that assembles objects from plan drawings, *IEEE Trans.* C-21:161–170.

Freeman, H., 1961, On the encoding of arbitrary geometric configurations, *IRE Trans. Electron. Comp.* EC-10:260–268.

Fukunaga, K., and Olsen, D. R., 1971, A two-dimensional display for the classification of multivariate data, *IEEE Trans. Comp.* C-20:917–923.

Gordon, R., and Herman, G. T., Reconstruction of pictures from their projections, *Com. ACM* 14:759–768.

Grimmert, H., Groh, G., and Tiemens, U., 1973, Synthesis of *tomograms*, *Proc. Int. Congr. Radiology*, Madrid.

Automatic Interpretation and Classification of Images, Grasselli, (ed.,), 1969, Academic Press, New York.

Guzman, A., 1968, Decomposition of a visual scene into three-dimensional bodies, *AFIPS Conf. Proc.* 33:291–304.

Hall, E. L., Kruger, R. P., Dwyer, S. J., Hall, D. L., McLaren, R. W., and Lodwick, G. S., 1971, A survey of preprocessing and feature extraction techniques for radiographic images, *IEEE Trans. Comp.* C-20:1032–1044.

Harlow, C. A., and Eisenbis, S. A., 1973, The analysis of radiographic images, *IEEE Trans. Comp.* C-22:678–689.

Harmon, L. D., 1971, Some aspects of recognition of human faces, In: *Pattern Recognition in Biological and Technical Systems* (D. J. Grusser and R. Klinke, eds.), Springer-Verlag, New York, 196–219.

Harmon, L. D., and Knowlton, K. C., 1969, Picture processing by computer, *Science* 164:19–20.

Harris, J. L., Sr., 1966, Image evaluation and restoration, *J. Opt. Soc. Am.* **56**:569–574.

Hilditch, C. J., 1969, Linear skeleton from square cupboards, *Machine Intelligence* **4**:430–420.

Holdermann, F., and Kazmierczak, H., 1972, Preprocessing of grey-scale pictures, *Computer Graphics Image Processing* **1**:66–80.

Huang, T. S., 1972, Bandwidth compression optical images, In: *Progress in Optics*, Vol 10, (E. Wolf, ed.), North-Holland Pub., Amsterdam-London, pp. 3–44.

Huang, T. S., Schreiber, W. F., and Tretiak, O. J., 1971, Image processing, *Proc. IEEE* **59**:1586–1609.

Hueckel, M. H., 1973, A local visual operator which recognizes edges and lines, *J. ACM* **20**:634–647.

Special issue on 2-Dimensional Signal Processing, IEEE Transactions–Computers, 1972, *IEEE Trans. Comp.* **C-21**:633–836.

Institute of Physics, Conf. Ser. No. 13, 1972, *Machine Perception of Patterns and Pictures*, London.

Jagadeesan, M., 1970, N-dimensional fast fourier transform, *Proc. 13th Midwest Symp. Circuit Theory III*, **2**:1–8.

Kazmierczak, H., 1973, Problems in automatic pattern recognition, *Proc. Int. Comp. Sym.*, Davos, *Switzerland*, pp. 357–370.

Kittler, J., and Young, P. C., 1973, A new approach to feature selection based on the Karhunen-Loève expansion, *Pattern Recognition* **5**:335–352.

Klensch, R. J., 1970, Electronically generated halftone pictures, *RCA Review* **31**:517–533.

Knowlton, K., and Harmon, L., 1972, Computer-produced grey scales, *Computer Graphics Image Processing* **1**:(1972) 1–20.

Kobayashi, H., 1971, A survey of coding schemes for transmission or recording of digital data, *IEEE Trans. Com. Tech.* **COM-19**:1087–1100.

Kobayashi, H., and Bahl, L. R., 1974, Image data compression by predictive coding (2 Parts), *IBM J. Res Dev.* pp. 164–179.

Lamar, J. V., Stratton, R. H., and Simac, J. J., 1972, Computer techniques for pseudocolor image enhancement, *Proc. 1st USA-Japan Comp. Conf.*, pp. 316–319.

Landau, H. J., and Slepian, D., 1971, Some computer experiments in picture processing for bandwidth reduction, *Bell Syst. Tech. J.* **50**:1525–1540.

Leese, J. A., Novak, C. S., and Taylor, V. R., 1970, The determination of cloud pattern motions from geosynchronous satellite image data, *Pattern Recognition* **2**:279–292.

Levine, M. D., 1969, Feature extraction: A survey, *Proc. IEEE* **57**:1391–1407.

McCormick, B. H., 1963, The Illinois pattern recognition computer Illiac III, *IEEE Trans. Electron Comp.* **12**:791–813.

Miller, W. F., and Shaw, A. C., 1968, Linguistic methods in picture processing: A survey, *AFIPS Conf. Proc.* **33**:279–290.

Nathan, R., 1968, Picture enhancement for the moon, mars and man, In: *Pictorial Pattern Recognition*, (C. Cheng, R. S. Ledley, D. K. Pollock, and A. Rosenfeld, eds.), Thompson Book, Washington, D.C., pp. 239–266.

Newman, W. H., and Sproull, R. F., 1973, *Principles of Interactive Computer Graphics*, McGraw-Hill, New York.

Olsztyn, J. T., Rossol, L., Dewar, R., and Lewis, N. R., 1972, An application of computer vision to a simulated assembly task, *Proc. 1st Int. Joint Conf. Pattern Recognition*, Washington, D.C., Pub. IEEE, New York, pp. 505–513.

Parks, J. R., 1969, A multilevel system of analysis for mixed font and hand-blocked printed character recognition, In: *Automatic Interpretation and Classification of Images* (A. Graselli, ed.), Academic Press, New York.

Pattern Recognition, 1973, Special issue on Picture Processing, **5**.

Pratt, W. K., Kane J., and Andrews, H. C., 1969, Hadamard transform, image coding, *Proc. IEEE* **57**:58–68.

Preston, K., 1972, A comparison of analog and digital techniques for pattern recognition, *Proc. IEEE* **60**:1216–1231.

Preston, K., Jr., 1971, Feature extraction by Golay hexagonal pattern transforms, *IEEE Trans.* C-**20**:1007–1014.

Proceedings of Conference on Machine Processing of Remotely Sensed Data, 1973, Purdue IEEE Cat. No. 73 CHO/834-2 GE.

Proceedings of the 1st Int. Joint Conf. on Pattern Recognition, 1972, Washington, D.C., Pub. IEEE, New York.

Proceedings of the IEEE, 1972a, Special issue on Digital Picture Processing, **69**:766–922.

Proceedings of the IEEE, 1972b, Special issue on Digital Pattern Recognition, **60**:1117–1242.

Proceedings of Seminar on Image Information Recovery, 1968, Philadelphia, Pa., S.P.I.E. California.

Proceedings of the Third Int. Joint Conf. on Artificial Intelligence, 1973, Stanford Research Inst., California.

Roberts, L. G., 1962, Picture coding using pseudo random noise, *IRE Trans. Inf. Th.* IT-**8**:145–154

Romland, W. S., Traglia, P. J., and Hurley, P. J., 1968, The design of an OCR system for recording handwritten numerals, Fall Joint Computer Conf., *AFIPS Conf. Proc.* **33**:1151–1162.

Rosenfeld, A., 1969, *Picture Processing by Computer*, Academic Press, New York.

Saraga, P., and Wavish, P. R., 1972, Edge tracing in binary arrays In: *Machine Perception of Patterns and Pictures* Institute of Physics, Conf. Ser. No. 13, pp. 294–302.

Saraga, P., Weaver, J. A., and Woollons, D. J., 1967, Optical character recognition, *Philips Tech. Rev.* **28**:197–203.

Saunders, R. P., 1972, A description and analysis of character maps, *Computer J.* **15**:160–169.

Sawchuk, A. A., Space variant system analysis of image motion, *J. Opt. Soc. Am.* **63**:1052–1063.

Schade, O. H., 1956, Optical and photoelectric analog of the eye, *J. Opt. Soc. Am.* **46**:721–739.

Schroeder, M. R., Images from computers, *IEEE Spectrum* **6**:66–78.

Shirai, Y., and Inoue, H., 1973, Guiding a robot by visual feedback in assembling tasks, *Pattern Recognition* **2**:99–108.

Sklansky, J., 1973, *Pattern Recognition*, Dowden, Hutchinson & Ross, Pennsylvania.

Stockham, T, G., Jr., 1972, Image processing in the context of a visual model, *Proc. IEEE* **60**:828–841.

Stroke, G. W., 1972, Image enhancement by holography, developments in electronic imaging techniques, *Proc. Soc. Photo-Optical Instr. Engr.*, **32**:93–106.

Sutherland, I. E., 1965, *Sketchpad: A Man-Machine Graphical Communication System*, MIT Lincoln Lab. TR296.

Tasto, M., 1973, Guided boundary detection for left ventricular volume measurement, *Proc. of 1st Int. Joint Conf. on Pattern Recognition* pp. 119–124.

Uhr, L., and Vossler, C., 1961, A pattern recognition program that generates, evaluates and adjusts its own parameters, *Proc. of the Western Joint Computer Conf.*, pp. 555–569.

Ullman, J., 1973, *Introduction to Pattern Recognition*, Butterworths, London.

Unger, S. 1958, A computer oriented towards spatial problems, *Proc. IRE* **46**:1744–1750.

Vernot, R. D., 1973, Imagery storage techniques, *Soc Inf. Display J.* **10**:17–21.

Winston, P. H., 1972, The MIT robot, *Machine Intelligence* **7**:431–463.

Yoëli, P., 1967, The mechanisation of analytical hill shooting, *Cartographic J.* **4**:82–88.

8

Editorial Introduction

This chapter by C. M. Brown and R. J. Popplestone (and Chapter 16 by E. A. Newman) lays the basic foundations for subsequent chapters describing the analysis by computer of visual scenes. They point out that much of the research has been rather academic. The reason is that real-world scenes (trees, roads, etc.) are very complex and, until simpler images are better understood, there seems to be little hope that complicated ones can be handled effectively. Hence, scenes containing matt-gray polyhedra, in rooms having uniform illumination, have been a favorite subject of study for many years and are now quite well understood. This type of scene, although still very restricted, comes close to some interesting applications in automated packing, and the handling of postal parcels. However, research has progressed considerably beyond this and more complex objects, such as cups, spectacles, tubes, and human faces, are being investigated now. (Newman discusses face recognition in detail.) Halé and Saraga (Chapter 7), Parks (Chapter 10), and Rutovitz *et al.* (Chapter 12) effectively build on the foundations laid here, while Newman returns to the philosophical problems again in a fascinating epilogue. Chapter 14 by Ainsworth and Green on speech recognition, makes use of the same concepts of hierarchy and heterarchy which are developed here.

8

Cases in Scene Analysis

C. M. Brown and R. J. Popplestone

8.1. INTRODUCTION

In this chapter we will try to give insight into some historical and intellectual developments in the analysis of scenes by computer. To do this we will study the approaches taken in some important scene analysis programs. We use the word "scene" variously to mean an arrangement of things in space, their representation as input (e.g., a black and white picture or a line drawing), and their representation as partially interpreted inside a computer.

What is scene analysis, and how is it different from pattern classification? Pattern classification involves making one right choice of alternatives, while scene analysis often involves constructing a description in terms of primitives, their properties, and the relations between them. This is not to say which problem is easier. A seemingly hard problem is to classify correctly the two handprinted latin characters A and A . An easy problem (Selfridge, 1955) might be to recognize words in the string of characters $C A T$; here we have used such powerful cognitive machinery that we may not even notice solving the hard classification-like problem as a subpart of the easy scene analysis-like problem.

A second characteristic of scene analysis, then, is that scenes are always taken from some domain, obeying rules which are formalizable and to which all scenes from the domain conform. Scene analysis depends on a knowledge about

C. M. Brown • University of Rochester, Rochester, New York. *R. J. Popplestone* Department of Artificial Intelligence, University of Edinburgh, Edinburgh, Scotland

the world explaining how such a scene could come about, and providing a context within which the scene may be understood. It is the lack of such wider context that makes the above classification hard; the context makes the difficulty vanish. It would, however, be unwise for a scene analyst to decry the classification problem as being just a "toy problem" or "puzzle," without being prepared for the observation that the analysis of an array of numbers (which is where most current scene analysis programs start) may be needlessly hard and perhaps just a puzzle when compared with what humans do. As they move through the world their cognitive processes, sensory inputs, and data-gathering strategies may well be much richer than a static, coarse, digital array can support.

In that it is concerned with describing a complex sensory input in terms of meaningful primitives and their relationships, scene analysis has much in common with speech analysis; in both fields, even if the difficulties of dealing with real data (or worse, real-time data) are removed, very hard problems still remain. Understanding a typed-in sentence in any meaningful way (to translate it into another language or ask questions about it, say) is difficult, as is understanding a line drawing. Generally the development of scene analysis has paralleled that of other artificial intelligence topics; there has been a growing conviction that one cannot rely on a computer's high speed and perfect memory to generate intelligent behavior by brute force, and that a usable model of the world coupled with some problem-solving capabilities may be needed for correct interpretation of sensory inputs. The Artificial Intelligence community now tend to regard understanding sentences and scenes as constructive acts; the development of this feeling is mirrored in the evolution of scene analysis programs.

The cognition-driven perception idea has reached quite a high pitch in influential quarters (Minsky, 1974). Not surprisingly it has elicited a compensating reaction in the direction of giving the lower processing levels in the visual system more credit (Marr, 1974; Horn, 1973). Perhaps the inclusion of so much high-level knowledge and expectation in scene-analysis programs has only been necessary because the input and low-level processing were of relatively low quality. These questions are open; we know distressingly little about the allocation of computational resources between the various levels of visual processing.

A major problem for workers in scene analysis is to find domains which are rich enough to provide insight and constrained enough to allow progress. In 1975, it seems that many workers are at least as concerned with finding meaningful questions as with finding effective answers.

8.2. SCOPE OF SCENE ANALYSIS

Scene analysis is usually taken to begin in 1963 with the work of Roberts (1965). By then the development of computer hardware and software, and of organizations dedicated to using them to accomplish tasks of artificial intelligence, had advanced enough that it was possible to think about fitting a large picture into the memory of a computer and performing operations of sufficient logical complexity to allow a convincing attack on the understanding of scenes.

Since then, there has been a proliferation of philosophies, domains, computational methods, and input devices for scene analysis, all of which interact and influence each other within any single project. The usual input to a scene analysis program dealing with real data is an array of numbers; where they came from, what they represent, and how they are to be used are questions that may be answered in many ways, each of which has implications for a particular project (placing limits on the types of descriptions which may be generated is an obvious example). An "intensity map" can result from digitizing a black and white picture, or directly from a TV camera or a vidissector, perhaps with the scene viewed through a colored filter. It may be analyzed to find lines or regions of slowly varying intensity, related colors, or identical texture. It may result from structured light, such as a square grid pattern or parallel lines, projected onto the scene. The numbers in the map may represent not intensities but ranges, giving distances from some range-finder to surfaces in the scene. The sensor parameters (location, focal length, color filter, etc.) may be fixed or under program control. Range and intensity maps may be used together. There are many possibilities.

The work reported in this chapter was all motivated more by a desire to shed light on issues of intelligence and perception than to achieve economic or military results. For this reason, domains for scene analysis have been chosen for their intellectual appeal, in one or another sense. Polyhedra were favored initially because of the well-defined way they project onto the retina and the ease with which they may be described and displayed. They also present obvious features which we humans represent when we make a line drawing of them (edges, vertices, faces) and which look detectable in an intensity map. Finally, quite complex shapes and scenes may be built of simple polyhedral blocks. Gradually a theory of polyhedral vision has emerged which is concerned not with intensity maps of polyhedra, but with the line drawings which could be the output of a successful lower-level program. This theory has been extended to deal with simple curved objects as well.

Besides intensity and depth maps of polyhedral scenes and the line drawings of them, line drawings representing scenes including such things as houses, people, automobiles, etc. have been analyzed. Other domains which have provided intensity or depth maps for analysis have been outdoor scenes of beaches, trees, and sky, of roads, cars, grass, and trees, and indoor scenes of offices. Also, biologically curved objects such as human faces, horses, or dolls, or simpler curved objects such as coffee cups and spectacles have been used.

The descriptions generated by these programs have also been diverse.

There has been agreement and cooperation among polyhedral vision theorists as regards descriptions, but generally descriptive mechanisms are of interest in their own right, and different workers tend to use different ones. Often the descriptions are designed to be useful in a later process, such as matching descriptions, or manipulation of the described objects.

In the following sections we will mention three important choices for domains in scene analysis. Within each we will discuss programs that do comparable things, and note changes in methodology which have occurred. The three areas of concern are: the lines in a polyhedral scene, the meaning of lines in a line drawing of a polyhedral scene, and the regions in scenes of curved and irregular objects. It is hoped that this collection will indicate general evolution of style in scene analysis as practiced by artificial intelligence workers. Though we only present a sample of work from a large collection of important old and new projects, we hope it is diverse enough to justify our reluctance to make a statement about the nature of all scene analysis. For more complete descriptions of the individual projects, the reader is referred to the original papers; a good book on many aspects of scene analysis is that of Duda and Hart (1973).

8.3. LINE FINDING

In this section we give accounts of two programs which derive line drawings from intensity maps (Roberts, 1965; Shirai, 1972). Roberts' program was a pioneering one, and in this section we discuss only its first phase (see Section 8.4). Shirai's program stopped after finding and interpreting the lines. Our object is to point out the difference between the "hierarchical" approach of Roberts, which goes from data to lines using no knowledge of the domain and without letting "higher level" steps influence "lower level" ones, and the "heterarchical" approach of Shirai, which alternates between finding lines and interpreting them. This interpretation can in turn guide the line-finder, and the general lesson seems to be that one can often gain much by allowing such feedback between program levels.

In polyhedral vision, a line-finder locates the image of the edges of the polyhedra, where two faces meet or a surface is obscured. The typical case (as in Roberts and Shirai) is that the polyhedra are painted with a matt paint, neither very light nor very dark in color. While different polyhedra may be different colors, the color of a polyhedron, or at least of a face, will be uniform. Lighting is diffuse, so that sharp-edged shadows do not occur. Many processes in computer vision give rise to noise, or discontinuities in intensity which are randomly spatially distributed; the edges of polyhedra, however, tend to give rise to discontinuities which are organized into straight lines. One possible paradigm for a line-finder is to differentiate spatially the intensity map to emphasize discontinuities, select feature points from the differentiated map to be those with values over some threshold, and to find a small set of lines that includes most of the feature points. This approach in its simplest form is purely hierarchical, with the data processed in several independent passes.

A sample scene of the usual type is shown in Fig. 8.1. When it is digitized in the computer, the intensity map may be displayed and could appear as in Fig. 8.2. It has been shown (Herskovits and Binford, 1970) that the features our eyes detect as edges, in scenes such as we have described, are variations in the intensity map. There are several types, including one which is only a change in *intensity gradient*; there may be no intensity discontinuity at the edge at all. These conclusions mean that no single linear detector will react the same way to all edge-causing intensity variations, and in particular that a simple differentiation operator will fail to detect the feature points of some edges.

8.3.1. Finding Lines Using Hierarchical Control

Roberts' program takes as input a 256 × 256 array of digitized values from a black and white photograph of a scene such as could give rise to the line drawing of Fig. 8.3. Using the simple paradigm outlined above, it applies a linear discontinuity detector based on first differences. The differentiated map is next divided into 4 × 4 squares; within each, the maximum-valued element is selected as a feature point if it is over some threshold. The data around these feature

Fig. 8.1. Photograph of polyhedral scene.

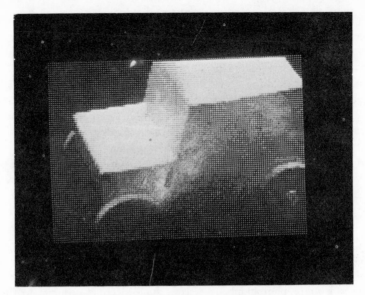

Fig. 8.2. Digitized polyhedral scene.

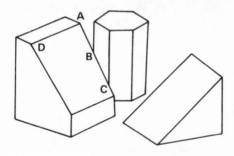

Fig. 8.3. Line drawing of polyhedral scene.

points are then correlated with five-long line elements of slopes 0, 1, −1, and ∞ centered on the feature point. If the ratio of the best to the worst fit is greater than a threshold (about 3) then the feature point and the slope of its line element are remembered. These processes have all been local and hence fairly cheap; they are designed to eliminate from consideration all feature points which do not have a fairly convincing excuse for existence in terms of belonging to some line.

The next pass over the data links up the local evidence into global constructs. Feature points are linked together if they occupy neighboring 4 × 4 squares and if the slope of either of their line elements is within 23° of the slope of the line connecting them. This linking operation gives chains of linked points, which may contain gaps and small networks of connections (Fig. 8.4). The redundant points in the network of connections are eliminated, and lines are extended across gaps; finally a least-squared error fit of remaining points is made and taken to be the line which was desired.

It is the feeling of some present workers that as it is described Roberts' program is probably incapable of finding correctly all lines in a scene which is not very carefully set up (Mackworth, 1974). Nevertheless, the paradigm of generating and grouping local evidence which Roberts seems to have been the first to use has remained to this day the standard approach, with the flow of control and specific techniques being adjusted to improve performance.

8.3.2. Finding Lines Using Heterarchical Control

An example of a less rigidly pass-structured approach is that of Shirai, whose guiding principle is to extract the most obvious information using knowledge so far acquired, to interpret the new information, thus obtaining more knowledge, then to go back for more information from the data, etc. The process starts by using a priori knowledge.

Shirai deals with polyhedral scenes, and starts with approximately 100 X 100 points in his intensity map. His goal is to find lines, and to interpret them as "contour" (between bodies and the background), "boundary" (delineating bodies), or "interior" (within bodies). All contour lines delineate bodies, so are boundary lines also. In Fig. 8.3, line *AB* is contour and boundary, *BC* is boundary, and *AD* is interior. As it happens, his program fails in some of its interpretations for a large class of nonconvex polyhedra, but this only marginally affects the line-finding process.

Shirai's a priori knowledge is that he can reliably detect contour lines; his bodies are light ones on a dark background, so the contour is assumed findable without error. One theoretical observation which helps the interpretation process is that if a boundary line is extended into a scene it remains a boundary line. (In Fig. 8.3, *AB* is contour so must be boundary, therefore *BC* must be boundary.) This observation is true for convex polyhedra. Heuristic rules to aid in line-finding are used, such as the rule that a concavity in a boundary of a body or scene usually indicates lines inside the boundary at that point, and that blocks are usually parallelepipeds.

The program uses 10 different techniques, or jobs, in finding lines; they are ordered into 10 levels of priority according to the amount of information

Fig. 8.4. Linked feature points.

they provide. The highest-priority job which is applicable is always performed next. Each job usually leads to a new line, which is interpreted; the resulting scene is examined to see which job to do next. The highest-priority jobs look for lines which are the most restricted in their possible directions and locations. One high-priority job is to extend a line which is part of a boundary concavity (such as *AB* in Fig. 8.3) into the scene (say to *C*); a lower-priority job is to search around a vertex to see if more lines emanate from it.

Since the program knows or determines the direction of lines to be extended, it can use a sensitive detector which responds to lines in a definite direction. The program knows the characteristics of the line it is extending, and uses the theory of Herskovits and Binford to interpret the results of the detector. Much slimmer evidence is sufficient to extend a line than to find it in the first place with no prior knowledge of its characteristics, and the program can track hanging lines through extremely "barren wastes" of equal intensity in an attempt to join them up to distant vertices.

The priority-ordering of jobs means that each scene is considered afresh, the most informative and reliable job being done at any given time as opposed to doing the jobs in some fixed order. Comparisons indicate that this heterarchical approach yields much better results than a hierarchical program using the same basic operators. The program has built-in knowledge about its domain which affects the priority it gives to jobs, and which saves it much work. However, it is so much driven by its expectations, and so afraid of doing undirected and useless work, that it will never find edges it has no reason to expect. There is nothing in the outline of a scene such as is drawn in Fig. 8.5 to suggest internal lines, and so the small wedge, hole, and cracks will all be missed.

8.4. POLYHEDRAL VISION

In this section we report on a long series of work on perfect line drawings of polyhedral scenes. A theory emerges which is based on the constraints

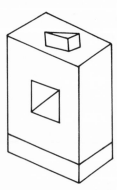

Fig. 8.5. Some lines not found by Shirai's program.

imposed on the drawing by the physics of the spatial world and the geometry of the projection process. We will see how constraints on the meaning and relations of scene parts can allow us to understand the scene as a representation of the world.

8.4.1. Matching Topological Features

In Section 8.3.1. we discussed Roberts' line-finder; in this section is outlined his method for recognizing polyhedra. It should be noted that we are making no reference at all to several other basic contributions that were made in this extraordinary paper.

Having obtained a line drawing such as Fig. 8.3, Roberts desires to interpret it in terms of a small set of polyhedral models, shown in Fig. 8.6. Simple polyhedra in the scene (such as the two rightmost in Fig. 8.3) are regarded as incarnations of transforms of these three models, where a transform may involve scaling along the three coordinate axes, translation, and rotation. *Compound polyhedra*, such as the leftmost in Fig. 8.3, are regarded as being simple polyhedra "glued together." The goal of the program is to derive from the scene the identity of the models used to construct it (including details of the construction of compound polyhedra), to discover the transformations the models underwent before appearing as incarnations, and finally to demonstrate a correct understanding of the scene by being able to construct a line drawing of it as seen from any viewpoint, using its derived description.

To understand a part of the scene, Roberts must first decide which model it resulted from, and then derive the transformation the model underwent in order to appear as it does in the scene. This is done by matching topological features of the scene with those of the models; matching features induces a match between points, and at least four noncoplanar points are needed to derive a transformation. Tentative topological matches are checked by a metrical process which determines whether a model can allowably be transformed into the required shape, and if so whether it lies completely inside the observed polyhedra. Since the same image can result from a close small scene or a distant

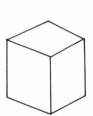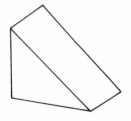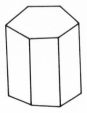

Fig. 8.6. Roberts' polyhedral models.

large one, assumptions about the location of the supporting surface and how objects are placed on it are used to fix the distance.

The three models all have convex polygons as faces, which project onto the image as convex polygons. The faces all have 3, 4, or 6 sides, so faces which have not suffered occlusion or merging with another face while forming a compound polyhedron will appear convex, will have 3, 4, or 6 sides, and will have no sides that are the uprights of "*T*-vertices" (which result from occlusion). Polygons which pass these three tests are "approved," and are remembered on a list of possible model faces (Fig. 8.7).

In searching for points to identify between the scene and the model, Roberts looks for topological structures in decreasing order of their efficacy; at this stage, his program behaves like Shirai's in that it extracts the highest-quality information, reinterprets the scene, and searches again. The most helpful structure is a vertex surrounded by three approved polygons, which are then matched with model faces on the basis of their number of sides. Matching this structure associates seven scene points with model points if the model is a cube (Fig. 8.8a). The second best structure is a line with approved polygons on each side; for a cube, this matches six points (Fig. 8.8b). Third, an approved polygon with a line extending from a vertex, matching five points (c), and last, a vertex with three lines emanating from it, matching four points (d).

When a transformed model is identified in the scene, it is notionally removed, the resulting new visible lines are filled in, and the new scene is analyzed; for compound polyhedra, this process unglues the model and leaves the rest of the compound. The matching is not infallible (Mackworth, 1974,

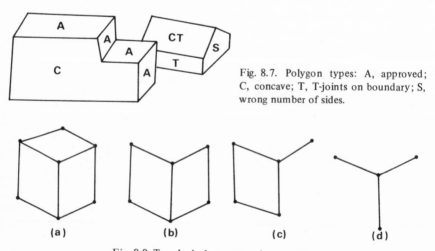

Fig. 8.7. Polygon types: A, approved; C, concave; T, T-joints on boundary; S, wrong number of sides.

(a)   (b)   (c)   (d)

Fig. 8.8. Topological structures for matching.

gives examples of how it fails on some rather simple compound polyhedra), but it forms a starting point for our investigations.

8.4.2. Grouping Regions into Bodies

A program by Guzman (1969) takes as input a drawing of a polyhedral scene which may be quite complicated (Fig. 8.9). The lines divide the drawing into a number of polygonal regions, and the goal of the program is just to group these regions into sets, each set corresponding with one body. He is satisfied with one description which is reasonable out of many possible ones a scene might have. One could say Guzman was addressing a polyhedral version of the general question as to how humans segment the world into objects.

The idea once again is to accumulate local evidence from the scene, and then to group polygons on the basis of this evidence. The evidence takes the form of "links," which link two regions if they may belong to the same body; links are planted around vertices, which are classified into types, each type always planting the same links (Fig. 8.10). No links are made with the background region.

To interpret the links once planted, one idea might be to group together into bodies all regions which are linked once, but it turns out that this scheme is too easily fooled by accidents of visual alignment into clumping regions into nonbodies (Fig. 8.11a). If we insist on two links, then the scene fragment of Fig. 8.11b is called three bodies, identified with regions (*AB*), *C*, and *D*. We would probably not agree with this interpretation, so the next adjustment to the method calls for first grouping regions with two links, and then accepting into the group any regions with two links into the *group*. This produces two

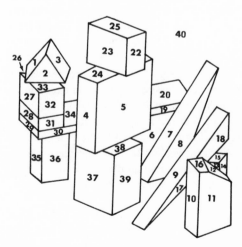

Fig. 8.9. A scene considered by Guzman.

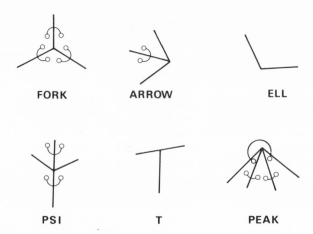

Fig. 8.10. Guzman's vertex types and links.

bodies from our scene fragment, (*ABC*) and *D*. The last tuning of the method is to accept into a group any region with a link into the group and no links with any other group; this version correctly groups the scene fragment into one body, (*ABCD*), and is basically Guzman's method. Unfortunately, as stated the method groups the scene of Fig. 8.11c into one body, a result that seems disagreeable, though possible.

To cope with this latter problem, Guzman recognized situations in which links should be inhibited. For example, suppose a line has an arrow vertex at one end, and the line itself is one of the barbs of the arrow (not the central shaft). This is very strong evidence that the line is the boundary between two bodies; even if a same-body link would otherwise be placed across the line by the vertex at its other end, that link-planting is inhibited. This case inhibits the two links joining the *L*-bar to the base in Fig. 8.11c, which is then correctly interpreted.

The final form of the program performs reasonably well on scenes without accidents of visual alignment, but it is a maze of special cases and exceptions, and seems to shed little light on what is going on in polyhedral line-drawing perception. One might well ask where the links come from; no justification of why they are correct is given. Further (Mackworth, 1974), we note that Guzman can accept as one body the two regions in Fig. 8.12a, and one feels a little dissatisfied with a scheme that just answers "one body" to a scene like Fig. 8.12b, instead of answering "pyramid on cube" or "two wedges," for example.

Despite its problems and its baroque form, it is interesting to consider Guzman's work in relation to Roberts'. In a world of convex isolated *trihedral*

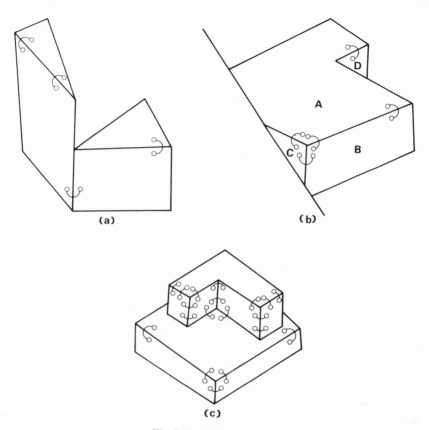

Fig. 8.11. Links in scenes.

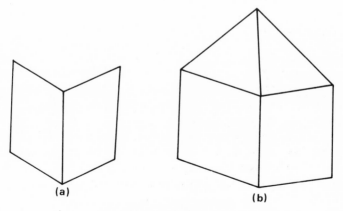

Fig. 8.12. Troublesome scenes.

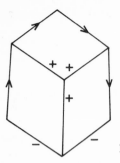

Fig. 8.13. Line labels.

polyhedra,* we will only see fork, arrow, and ell vertices; other types result from occlusion or nontrihedral corners. We will see a fork when and only when we see three faces of one polyhedron meeting in a point. Guzman's most powerful link planter, the fork, is based on this implication. Though it fails to hold strictly with nonconvex bodies and occlusion, it is likely to be correct. All four of Roberts' topological structures can occur when we see three faces around a point (Fig. 8.8); the first structure occurs when there is no occlusion. The second most common link planter, the arrow, occurs in the restricted world of convex isolated trihedral polyhedra when and only when two faces are visible of the three which meet at each corner. Applying this reasoning to the unrestricted world again gives likely conclusions, and Roberts' last three structures can all occur in this situation.

Thus Guzman's method derives from a theory of a very restricted class of scenes, which is extended by *ad hoc* adjustments and evidence-weighing to deal with a much broader range of scenes, albeit not very elegantly. Further progress in the line drawing domain came about when attention was directed at the three-dimensional causes of the different vertex types.

8.4.3. Labeling Lines

Huffman (1971) and Clowes (1971) independently concerned themselves with scenes similar to Guzman's, not excluding non-simply-connected polyhedra, but excluding accidents of alignment. They desired to say more about the scene than just which regions arose from single bodies; they wanted to ascribe interpretations to the lines. We follow Huffman's line-labeling scheme here. Figure 8.13 shows a cube resting on the floor. Lines labeled with a + are caused by a convex edge, those labeled with a − are caused by a concave edge, and those labeled with a > are caused by matter occluding a surface behind it. The occluding matter is to the right of the line looking in the direction of the >, the

---

*That is, polyhedra in which no more than three faces meet in a point.

occluded surface is to the left. If the cube were floating in the air, we would label the lowest lines with < instead of with −.

We can make a systematic investigation of the types of lines we can possibly see around a trihedral corner; such corners can be numbered by how many octants of space are filled by matter around them (1 for the corner of a cube, 7 for the inside corner of a room, etc.). By considering all possible trihedral corners as seen from all possible viewpoints, Huffman and Clowes found that, without occlusion, just four vertex types and only a few of the possible labelings of lines meeting at a vertex can occur. Figure 8.14 shows views of one- and three-octant corners, which give rise to all possible vertices for these corner types. The vertices appear in the first two rows of Table 8.1, which is a catalogue of all possible vertices, including those arising from occlusion, in this restricted world of trihedral polyhedra. It is easy to imagine extending the catalogue to include vertices for other corner types.

It is important to note that there are four possible labels for each line (+ − > <), and thus $4^3 = 64$ possible labels for the fork, arrow, and T and 16

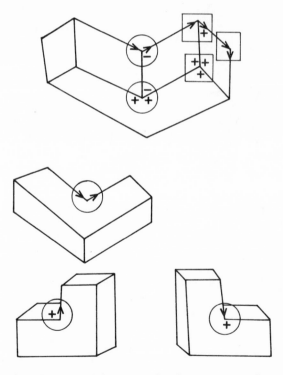

Fig. 8.14. Vertex labels generated by views of a 1-octant corner (in squares) and of a 3-octant corner (in circles).

Table 8.1.   Catalog of Meaningful Vertex Labels

Visible Surfaces / Octants Filled	3	2	1	0
1				—
3				—
5			—	—
7		—	—	—
Occlusion				

possible labels for the ell. By counting the vertices in the catalogue we see that only 5/64, 3/64, 4/64, and 6/16, respectively, of the possible labels actually occur in the catalogue; only a small fraction of possible labels can occur in a scene.

The main observation that lets line-labeling analysis work is that in a real polyhedral scene, *no line may change its interpretation (label) between vertices.* This we may call the coherence rule. For example, what is wrong with scenes like Fig. 8.15 is that they cannot be coherently labeled. The lines in drawings of

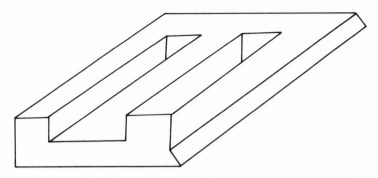

Fig. 8.15. Three-rib, one-slot blivit.

real scenes can be interpreted quickly, because the small percentage of meaning-ful labelings interacts with the coherence rule to reduce drastically the number of explanations for the scene.

Let us see how line labeling relates to Guzman. A labeled-line description clearly indicates the grouping of regions into bodies, and will also reject scenes like Fig. 8.12a, which cannot be coherently labeled with labels from the catalog. We can explain the origin of Guzman's links this way: consider again the world of convex polyhedra; the only labels from the catalog which occur are shown in Fig. 8.16a. Further, it is clear that a convex edge has two faces of the same body on either side of it, and an occluding edge has faces from two different bodies on either side of it. A convex label means we should therefore link the regions on either side of it; we have derived Guzman's link-planting rules (Fig. 8.16b). The inhibition rules are a further corollary of the labels; they are to suppress links across an edge if evidence that it must be occluding is supplied by the vertex at its other end (Fig. 8.16c). When vertices at both ends of a line agree that the line is convex, Guzman would have planted two links; this is in fact the strongest evidence we can get that the regions are part of the same body. If just one vertex believes the edge has a link, a decision based on heuristics is made; the coherence rule is being used implicitly by Guzman, and the same physical and geometric reality is driving both Guzman's scheme and that of Huffman.

The labeling scheme here explained still has problems: syntactically non-sensical scenes are coherently labeled (Fig. 8.17a). Scenes are given geometrically impossible labels (b), and scenes which cannot arise from polyhedra are easily labeled (c). It is very hard to see how a labeling scheme can detect the illegality of scenes like (c); the problem is not that the edges are incorrectly labeled, but that the faces cannot be planar.

Concern with this last-mentioned problem has led to a program (Mackworth, 1974) which can obtain information about a polyhedral scene equivalent to labeling it, and also can reject nonpolyhedra as impossible. There

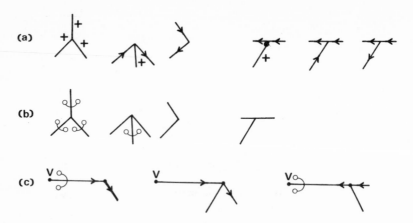

Fig. 8.16. Line labels and region linking: (a) Huffman's labels; (b) Guzman's links; (c) vertex at right inhibits link at V.

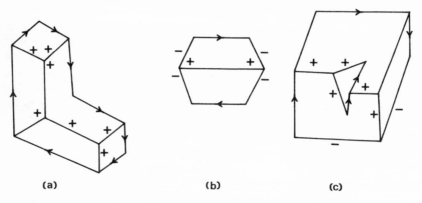

Fig. 8.17. Nonsense labelings and nonpolyhedra.

has also been an exciting denouement to the line-labeling idea in the work of Waltz (1972) and Turner (1974).

Waltz extends the line labels to include shadows, three illumination codes for the faces on each side of an edge, and the separability of bodies in the scene at cracks and concave edges; this brings the number of line labels possible up to just below 100. He also extends the possible vertex types, so that many vertices of four lines occur. He can deal with scenes like the one shown in Fig. 8.18.

The combinatorial consequence of these extensions is clear; the possible vertex labelings multiply enormously. The first interesting thing Waltz discovered was that, despite the combinatorics, as more information is coded into the lines, the smaller becomes the percentage of geometrically meaningful labels for a

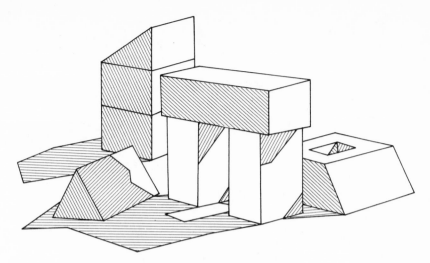

Fig. 8.18. Scene considered by Waltz.

vertex. In his final version, only approximately 0.03% of the possible arrow labels can occur, and for some vertices the percentage is approximately 0.000001.

The second interesting thing Waltz did was to use a labeling algorithm which very quickly eliminates labels for a vertex which are impossible, given the neighboring vertices and the coherence rule. The small number of meaningful labels for a vertex imposes severe constraints on the labeling of neighboring vertices, by the coherence rule; the constraints are passed around the scene from each vertex to its neighbors. To his surprise, Waltz found that for scenes of moderate complexity, eliminating all impossible labelings left only one, the correct one. The labeling process, which might have been expected to involve much search, usually involved none.

Waltz also points out how scenes with missing lines may be labeled; one merely adds to the legal vertex catalog the vertices which result if lines are missing from legal vertices. This idea has some obvious drawbacks, but can be useful.

Turner extends the vertex labels to include scenes with cylinders, cones, spheres, tori, and other simple curves. The number of vertex types and labels grows explosively, and the coherence rule must be modified to cope with the fact that lines can change their interpretation between vertices and can tail off into nothing, and that one region can attain all three of Waltz's illumination types. Turner deals with scenes such as appear in Fig. 8.19. Further, he was able to use the labeling of vertices in a working, integrated system for the analysis of television pictures.

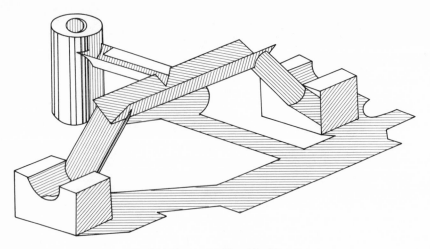

Fig. 8.19. Scene considered by Turner.

8.5. REGION FINDING

While finding and interpreting lines is natural in some domains, it can be an artificial and unhelpful approach to many scenes containing irregular (say biological) objects. The idea soon appeared (Brice and Fennema, 1970) of finding regions of similar intensity in intensity maps; in many scenes the description in terms of scene areas and their relationships is useful, and there is no commitment nor need to impose any strict mathematical constraints on the shape of all areas.

In this section we describe two region-finding programs (Barrow and Popplestone, 1971; Feldman and Yakimovsky, 1975). The earlier one might be compared to Roberts' work, as reported in Section 8.3; the primitives (regions) are found by a hierarchical nonsemantic method, and are then combined into a description which is used to recognize the object being viewed. Here we must ignore the interesting questions of description matching. The later work might be compared to that of Shirai (cf. Section 8.3) in that a strongly semantic component of the program guides the finding of the primitives, and the final product is a set of primitives with some interpretation placed on each. These comparisons should of course not be taken too seriously; we merely desire to point out the similarities of evolution of approach.

8.5.1. Finding Regions without Using Knowledge

Barrow and Popplestone's program works in the domain of isolated irregular bodies, such as a cup, spectacles, hammer, ball, torus, wedge, tube, etc. Its input is a 64 × 64 array of 16 gray-levels; 256 of these in a 16 × 16 array are used to start the analysis. Taking them in sequence the program steps toward the edge of the picture looking for intensity changes greater than ±1 from the starting value. If no such change is found, it is assumed that this starting point is in the background region, and the next point is tried. If a discontinuity is found, it is assumed that a region boundary has been crossed, and the intensity contour is traced; it will either go off the picture or will return to its starting point. In the latter case it is considered as a possible new region. Its properties (e.g., its area, location of centroid, average brightness contrast around boundary) are computed (cf. Chapter 12). Since more than one starting point may be within a region, any new region is checked to see if its properties match one found before. If it was previously unknown, it is added to a list of regions. Regions do not have holes; only their outer boundaries are known. For these objects, up to approximately 50 regions can be found in the first stage, depending on the quality and content of the input; an average number was about 20.

The problem now is to merge these many small regions into a few significant ones. The program has just one rule; merge two adjacent regions if the average contrast across their common boundary is less than a threshold (the notion of common boundary must be slightly adjusted here). This merging is done iteratively, the merged regions being replaced in the region list by their merge after every iteration. Finally, small regions and regions with comparatively weak boundaries are discarded.

With luck, after this process regions correspond to the surfaces of an object, and boundaries correspond to edges in the scene. There are a number of ways the process can fail, but any error in finding regions is irrecoverable.

The final regions are assigned properties, including brightness, compactness, and shape. Relationships are established between each pair of regions, such as relative size, the fraction of boundary in common, distance, etc. The values for both properties and relations are taken from a continuum. The final description of the scene is a graph, with nodes representing regions labeled with values of the region's properties, and the arcs representing relations labeled with their values.

8.5.2. Finding Regions Using Knowledge

The program of Feldman and Yakimovsky has been used on two domains so far, outdoor road scenes and left ventricular angiograms. In the former, the input is a 200 × 300 array containing intensity and color information. Unlike the uninterpreted regions of Barrow and Popplestone, here regions have six interpretations: sky, trees, road, roadside vegetation, car, and shadow of car. The regions have properties with discrete values, instead of the continuous-valued ones of Barrow and Popplestone; some are intensity, color hue, shape, size, and position. The boundaries between regions are given properties as well; with six types of region there are 18 types of boundary, and these have properties such as relative size, relative intensity, relative color, shape, and orientation, and relative position.

In the outdoor domain, a region-finding algorithm based on the absolute properties of regions would fail; both the trees and sky can generate many regions, which among themselves vary in properties more than do any other regions in the scene. Also, the roadside grass has widely varying colors, including the colors of some cars. Last, shadows often fall across the road.

This program guides the merging and interpretation of regions with semantic information about the meaning of the regions a scene can contain and the expected properties those meanings imply. A naive semantic system might give a high probability to regions interpreted as grass being green and as sky being blue, for instance. The boundary properties are used as well, to provide context information; the claim is that, while regions with different interpretations may be similar in properties, their contexts, in terms of the neighboring regions and the boundaries between them, are not likely to be identical as well; a yellow patch surrounded by grass is more likely to be grass than it would be if it were surrounded by road, for instance. The interpretation and merging are done within a framework of mathematical decision theory, with the program trying to maximize a utility function expressing the probability of a global interpretation for the scene, given the measurement made on the scene and the a priori context knowledge about the scene domain.

There are many established methods for maximizing such a utility function, but they are too slow; the program finds an approximate maximum. First the scene is divided into small regions on the basis of nonsemantic criteria (as in Barrow and Popplestone, 1971). For the outdoor scenes, this first operation yielded approximately 100 regions; for the angiograms, somewhat more. The program iteratively merges these initial regions, on each iteration merging two regions with a common boundary. The semantic component of the program is used to decide which regions to merge and when to stop merging.

In controlling the merging, the semantic information is used to measure the boundary strength between two regions. In each iteration, the regions with the weakest boundary are merged. The measure of boundary strength is taken to be the probability that the boundary really exists there, *and* that the regions it separates have different interpretations. In deciding when to stop merging, the semantic system is used after each merge to approximate the utility of the best possible interpretation of the regions; merging is continued until this utility drops off sharply (the best of interpretation becomes less probable). Then the program unmerges back to the point of maximum utility and computes more exactly the interpretation for the set of regions which gives maximum expected utility.

The semantics-controlled merging of regions into a probable interpretation is somewhat like the constraint satisfaction that occurs in the Waltz line-labeling algorithm. In both cases the semantic theory is guiding the interpretation of scene elements, and in both cases the connections given by the neighboring primitives are sufficient to propagate the constraints. (Barrow and Popplestone compute relations between each pair of primitives, but this may be unnecessary for many domains.) In polyhedral scenes, the semantics is definite and tight, in road scenes more probabilistic, but the same idea seems to be at work.

As in Shirai's work, an interpretation stage is interpolated after every primitive-generating operation in this region-finder, so that the next operation may be performed under conditions of maximum knowledge about the scene as so far understood.

A significant idea in this work is the separation of the domain-specific knowledge from the program which uses it. The same region finding and merging routines can be efficaciously applied to an entirely new domain by driving them with different domain-dependent knowledge.

REFERENCES

Barrow, H. G., and Popplestone, R. J., 1971, Relational descriptions in picture processing. In: *Machine Intelligence 6* (B. Meltzer and D. Michie, eds.), Edinburgh University Press, Edinburgh.

Brice, C. R., and Fennema, C. L., 1970, Scene analysis using regions, *J. Artificial Intelligence*, 1:205–226.

Clowes, M. B., 1971, On seeing things, *Artificial Intelligence* 2:79–116.

Duda, R. O., and Hart, P. E., 1973, *Pattern Classification and Scene Analysis*, Wiley, New York.

Feldman, J. A., and Yakimovsky, Y., 1975, Decision theory and artificial intelligence: A semantics-based region analyser, *Artificial Intelligence*, 5, 4, 349–372.

Guzman, A., 1969, Decomposition of a visual scene into three-dimensional bodies. In: *Automatic Interpretation and Classification of Images*, (A. Grasselli, ed.), Academic Press, New York.

Herskovits, A., and Binford, T. O., 1970, On boundary detection, *MIT AI Memo 183*, Massachusetts Institute of Technology, Cambridge, Mass.

Horn, B. K. P., 1973, On lightness, *MIT AI Memo 295*, Massachusetts Institute of Technology, Cambridge, Mass.

Huffman, D. A., 1971, Impossible objects as nonsense sentences. In: *Machine Intelligence 6* (B. Meltzer and D. Michie, eds.), Edinburgh University Press, Edinburgh.

Mackworth, A. K., 1974, On the Interpretation of Drawings of Three-Dimensional Scenes, *Ph.D. Thesis*, University of Sussex.

Marr, D., 1974, An essay on the primate retina, *MIT AI Memo 296*, Massachusetts Institute of Technology, Cambridge, Mass.

Minsky, M., 1974, A framework for representing knowledge, *MIT AI Memo 306*, Massachusetts Institute of Technology, Cambridge, Mass.

Roberts, L. G., 1965, Machine perception of three-dimensional solids, In: *Optical and Electro-Optical Image Processing*, Massachusetts Institute of Technology Press, Cambridge, Mass.

Selfridge, O. G., 1955, Pattern recognition and modern computers, *Proc. Western Joint Computer Conf.*, Los Angeles, Calif.

Shirai, Y., 1972, A heterarchical program for recognition of polyhedra, *MIT AI Memo 263*, Massachusetts Institute of Technology, Cambridge, Mass.

Turner, K. J., 1974, Computer Perception of Curved Objects using a TV Camera, *Ph.D. Thesis*, University of Edinburgh, Edinburgh.

Waltz, D., 1972, Generating Semantic Descriptions from Drawings of Scenes with Shadows, *Ph.D. Thesis*, MIT AI Laboratory, Massachusetts Institute of Technology, Cambridge, Mass.

9

Editorial Introduction

In their paper on the visual control of a printed-circuit-board drilling machine, Halé and Saraga show that complex and intelligent activity can be achieved by combining together simple, almost naive, operations. Sophistication is not always necessary and the success of their project lies in the realization of this fact. In a restricted environment, such as that seen on top of a drilling table, there is no need to make elaborate provisions for a multitude of rare events. To do so would merely be wasteful; instead the machine simply calls the operator if the unexpected occurs. Neither is there any other virtue in avoiding this restriction on the environment; Halé and Saraga have controlled the environment of their machine whenever they found it expedient to do so. This is "just" good engineering practice. Strict constraints on the size of store and processing power were imposed. The final result is a control strategy which could conveniently be implemented using an inexpensive modern microcomputer (costing about 20% of the total system!). The effectiveness of their solution can best be judged by watching the machine running for a few minutes. This is a fascinating experience, since the machine operates at about the same speed as it would with a human controller and correctly drills almost all of the required holes. (I did not witness any false drilling and very few omissions.) This application has its origins in engineering; there is a "bottleneck" in the manufacture of short runs of printed-circuit boards, at the digitizing or drilling state. The effective solution to a practical problem such as this does credit to pattern recognition, which has suffered for too long from abstract theorizing.

9

The Control of a Printed Circuit Board Drilling Machine by Visual Feedback

J. A. G. Halé and P. Saraga

9.1. INTRODUCTION

It is becoming increasingly difficult for industry to find people who are willing to do dangerous or monotonous jobs. Those who are performing boring or repetitive tasks are liable to become frustrated and error-prone. Problems of this type together with the increasing cost of manpower provide strong incentives for industry to automate their production lines.

   With the advent of computers, traditional fixed automation is being selectively replaced by programmable numerically controlled machines. These machines nevertheless follow a fixed set of instructions, and in general they cannot cope with a change in their environment. As computers have become cheaper and more powerful it has become possible to introduce into factories more intelligent machines (Radchenko and Yerevich, 1973) which are aware of their environment and which are capable of reacting to changes in that environment. Examples of this type of machine include Hitachi's visually controlled bolt tightener, and General Motor's system for placing wheels on hubs (Hitachi, 1973; Olsztyn *et al.*, 1973). In this chapter we are concerned with machines which examine their environment with visual sensors and which we call visually

J. A. G. Halé and P. Saraga · Mullard Research Laboratories, Redhill, Surrey, England

controlled machines. Much of the discussion, however, would apply to machines equipped with other types of sensor.

9.1.1. Justification for Visually Controlled Automation

It can be argued that there should be no need for visually controlled machines in the ideal automatic factory. Consider, for example, the problem of automatic assembly. If the orientation and position of all components were preserved from the point of initial fabrication, where they are well known, to the final assembly into the completed product, then "blind automation" would be quite satisfactory. In practice, however, there are many reasons why this desirable goal cannot be achieved. It may often be necessary to "let go" of parts (for example, when small sheet metal pressings are plated or deburred), and to store them between manufacture and assembly. If the components are stored in a "loose" state, then either a person or a machine is required to feed the component to the automatic assembly machine. In many cases, this problem is solved by ingenious mechanical designs such as bowl feeders. Some components, however, provide harder problems for the mechanical designer, and it is in these circumstances that a visually controlled machine is useful.

Thus visually controlled machines may be cost-effective when frequent product changes make complete programming expensive, when it is important to detect drifts in the product or the machine (Hitachi, 1973; Heginbotham, 1973, 1974; Olsztyn 1973), or when it is uneconomical to keep components in jigs either because the jigs are too costly or because this type of storage takes up too much space.

It is also worth noting that the application of visual feedback can remove the need for high absolute mechanical accuracy in order to achieve high mechanical resolution and reproducibility. The accuracy of the visually controlled machine depends mainly on the resolution and stability of the visual sensor.

9.1.2. Practical Considerations Affecting the Use of Visually Controlled Machines

Since visually controlled machines may be used to fill the gap between fixed automation and manual operation, it is important that these machines blend easily with existing factory systems. They will often have to work in real time and might require operating cycle times of the order of 1–10 s. They should also be capable of being operated and maintained by normal factory staff. To satisfy these constraints, the designer can control the perceptual world of the machine. He can choose the sensor, tune it to the problem, and adjust certain external conditions such as the lighting or the projection of special patterns onto the object.

In order to take advantage of the inherent flexibility of a visually controlled machine, it is important that the reprogramming of the machine to perform a different task should be simple enough to be performed by someone unfamiliar with computers.

9.2. THE STRUCTURE OF A PRACTICAL VISUALLY CONTROLLED MACHINE

The type of visually controlled machine in which we are interested (Fig. 9.1) can be characterized by four main attributes. First, the machine is assumed to be working on a defined task within a limited real-world environment. A conceptual description of the task and the environment is given to the machine as *a priori* information. Second, the machine is equipped with sensors which enable it to gather from the environment the information it requires to perform the task. Third, by using both the *a priori* and the sensed information, the machine can construct (Klir and Valach, 1967) its own internal conception of the real world, i.e., its "world model" (Fikes *et al.*, 1972; Rowat and Rosenberg, 1972). The model is assumed to be dynamic and may be updated by new data from the sensors. Fouth, the main control unit can use the world model to interpret sensed information and as a result instruct the motor control to take actions in pursuance of the defined task.

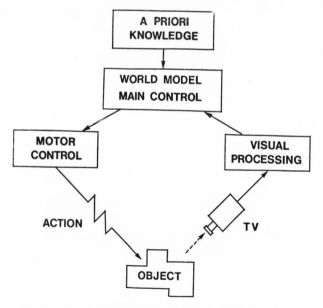

Fig. 9.1. Conceptual model of a visually controlled machine.

9.2.1. The Use of Visual Feedback

As well as using sensed information to determine a course of action, the sensors can also be used to monitor the progress of the action (Klir and Valach, 1967; Parks, 1973; Shirai and Inoue, 1973). If the action appears to be incorrect, this may be because the motor control is faulty, the mechanism is inaccurate, the machine's vision is distorted, the world model is incorrect, or the *a priori* information was false.

Some of these disorders can be detected and corrected by the computer itself, whereas others are beyond its capability. The machine can probably compensate for errors in the motor control or the mechanism by observing the results of the action and giving new instructions. This suggests that machines equipped with visual feedback may require less accurate mechanics. If the corrective actions do not succeed it is probably because the world model is no longer valid (perhaps the magnification of a zoom lens has changed). If the main control decides that it is the world model which is obsolete then it can enter a model updating phase similar to the model initialization phase (Section 9.3.3). False *a priori* information or distorted vision are likely to be hard to cope with, and in practice would probably result in calls for external help.

9.2.2. Hand–Eye Machines in University and Industry

Hand–eye machines of the type described above have been developed in various Artificial Intelligence laboratories. In many cases the aim of these projects has been the development of intelligent machines for their own sake, although the ideas have sometimes also been applied to practical problems in order to demonstrate the versatility of the system (Ambler *et al.*, 1973; Gill, 1972).

In industry, the situation is reversed. We are trying to solve real problems at the minimum cost, and we will only use machine intelligence where it offers a real advantage. Instead of trying to make machines solve complex problems, we try to make the problems and the equipment required as simple as possible. These differences in approach will be illustrated by the description of the visually controlled drilling machine.

9.3. THE APPLICATION OF VISUAL CONTROL TO DRILLING PRINTED CIRCUIT BOARDS

The drilling of printed circuit boards (PCBs) under visual control is a problem in which both the mechanics and the scene analysis are relatively simple. The solution to this problem involves the recognition of "blobs" and the evaluation of their spatial relationship. Many practical problems in navigation and target location as well as industrial tasks such as the placement of wheels onto hubs (Olsztyn, 1973) can be expressed in similar terms. The solution to the problem could, however, have genuine practical application.

In research laboratories and other establishments in which PCBs are made in very short runs of small numbers of boards, the boards are often drilled by hand using a single-spindle drilling machine such as that illustrated diagrammatically in Fig. 9.2a. A man views the board through the eyepiece, which contains a crosswire indicating the position of the drill Fig. 9.2b. To drill a hole, he moves the board until the point to be drilled coincides with the crosswire, and then actuates the drill. The experimental equipment built at Mullard Research Laboratories to automate this process and replace the man by a computer vision system, is illustrated diagrammatically in Fig. 9.3 and by the photographs in Fig. 9.4. The equipment is based on a standard Posalux numerically controlled PCB drilling machine, but with no numerical controller. The board is viewed, via a half-silvered mirror, by a TV camera which is interfaced to a Honeywell DDP 516 computer. The board can be moved over the table by means of stepping motors which are also computer-controlled. The function of the crosswire in the manual machine is accomplished by projecting a graticule (a Maltese cross) into the field of view of the TV camera. This also allows for automatic compensation of any drift in the TV system. The specification of the machine is summarized in Table 9.1.

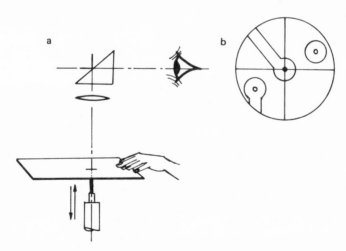

Fig. 9.2. The manual drilling of printed circuit boards. (a) Diagram of single-spindle drilling machine. (b) Eyepiece.

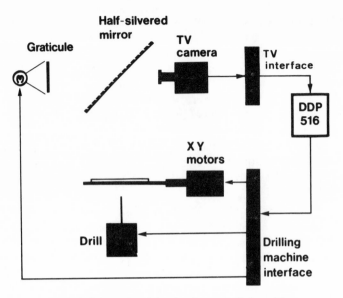

Fig. 9.3. Schematic of the visually controlled machine.

9.3.1. TV and Machine Control Interfaces

The TV videosignal can be sampled with a maximum resolution of 382 × 287 picture elements (pixels) over the field of view of the camera, and each pixel can be digitized to 5 bits (32 gray levels ranging from 0 for saturated white to 31 for black). A slow sampling system is used in which one point from each TV line is sampled during one field, giving data from one vertical column of points. During the next TV field the next adjacent vertical column is sampled and so on. The computer can specify the starting point and size of the scan as well as the pitch of the samples. To build up an array of 382 × 287 pixels would take 7.6 s, and this array would occupy more than 35K words of core memory (Packed 3 pixels/16 bit word). In practice only small areas are required by the program. The two sizes of scan used most frequently are 20 × 20 samples taken at every point ("high resolution") covering 2.5 × 2.5 mm on the board and taking 0.4 s, and 57 × 57 samples taken every second point covering 14 × 14 mm and taking 1.1 s.

The machine control interfaces allow mechanical actions to be specified from the program, provide interlocks between these actions, and enable the program to monitor the status of the machine. The available commands are:

1. Move the PCB by a specified number of steps in two orthogonal directions (X and Y).
2. Drill if not moving.
3. Switch the projected graticule ON/OFF.

The interfaces may also be switched to allow manual control of the machine.

9.3.2. Strategy

As mentioned in the introduction, the visually controlled drilling machine is intended to replace its manually operated equivalent. The human operator's task of drilling the board is approximately described by the flow charts of Fig. 9.5a. He first undergoes an "initialization" phase during which he establishes his "world model." He checks such things as the accuracy of the correspondence between the position of the drill and the crosswire, and the relationship between moving the board on the table and the resulting movement of the image of the board in his viewer. Once the initialization is complete, he systematically moves the board under the viewer, scanning the image for drilling points. As soon as he spots something that might be a drilling point, he stops moving the board and examines the candidate point more carefully. If he confirms it as a drilling point, he moves it to coincide with the crosswire, and drills a hole. At this point he can check whether the drilled hole is exactly in the center of the drilling point. If he notes a consistent error in the position of the

Table 9.1.   Specifications of MRL Visually Controlled Drilling Machine

Drilling machine	Posalux Copy for HV
Maximum traverse in X	250.00 mm
Maximum traverse in Y	350.00 mm
Smallest step in X and Y	0.020 mm
Maximum speed in X and Y	60 mm/s
TV camera	Philips LDH 150/160
Computer	Honeywell DDP 516
Store	16K words of 16 bits each
TV picture array	Up to 382×287 sampled points taking 7.6 s
Gray levels	32 (5 bits)
High resolution of scan of TV picture	20×20 sampled points taking 0.4 s
Equivalent pitch on board	0.125 mm
Low resolution scan of TV picture	57×57 sampled points taking 1.1 s
Equivalent pitch on board	0.250 mm
Performance	Approximately 12 holes/min

Fig. 9.4. Photographs of the visually controlled machine. (a) General view. (b) Close-up view.

hole, he may decide to adjust his "world model" accordingly. The above sequence of operations is repeated until all holes have been drilled.

The computer program in the DDP 516 follows a sequence analogous to the human operator's, as illustrated in Fig. 9.5b. It first enters the initialization phase during which the scale between table and image movement, the position of the projected graticule, and the relationship between the graticule and a drilled hole are determined. The main drilling phase is then entered during which the board is systematically searched for drilling points by scanning a sequence of

Fig. 9.4 (*continued*)

adjacent subareas (each of 57 X 57 pixels limited by the available space in the computer) of the board (Fig. 9.6). The TV interface samples the TV video and converts it into a digital signal of 5 bits per pixel (32 gray levels in which 0 is white and 31 is black). By this means, an image such as that illustrated in Fig. 9.7 is stored in the computer.

The stored image of each subarea is examined for patterns which correspond to "candidate" drilling points and their positions are recorded. A candidate is detected by a "blob operator" such as that described in Section 9.4.1.

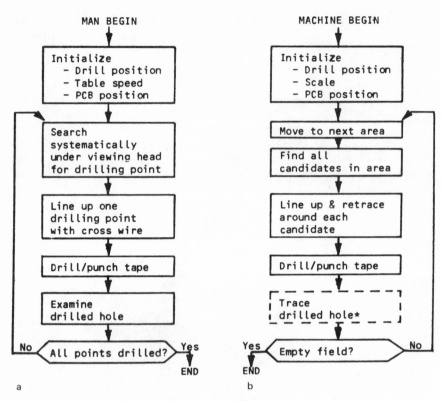

Fig. 9.5. Flow charts of PCB drilling: (a) by a human operator; (b) by the visually controlled machine.

Once the positions of all the candidates within a subarea have been determined, the table is moved to bring each candidate in turn to the center of the field of view of the TV camera. The region encompassing the candidate is then rescanned to obtain a higher resolution image. This is thresholded to a binary image which is examined by tracing the black–white boundary (Fig. 9.8) (Saraga and Wavish, 1972). If the shape of the boundary is consistent with that of a drilling point (i.e., it approximates to a circle), the table is moved to bring the center of the drilling point (the center of the circle) into coincidence with the assumed drill location, and a hole is drilled. This process continues until all subareas have been scanned and all drilling points have been drilled.

9.3.3. The World Model

We distinguish two categories of information within the world model (see Fig. 9.9), a set of "invariants" and a "state vector" (Lee and Wood, 1974). The

Invariants include the axioms about the real world specified by the program designer. In the case of the machine being described, these invariants include the assumption that the TV raster is parallel to the X displacement of the drilling machine table and that a linear relationship exists between movements in the real world (table movements Δ_{TB}) and corresponding displacements in the television picture (image movements Δ_{TV}). The actual value of the scale given by Δ_{TB}/Δ_{TV} is one of the elements of the state vector. In general the state vector includes those model parameters which might be changed when new information is received from the sensors. In the case of the drilling machine, state vector elements include the scale, the current location of the table, the drill, and the projected graticule in the field of view. Also contained within the state vector is a histogram of the gray levels in the picture currently being processed. If at any time the machine fails to calculate a value for any part of

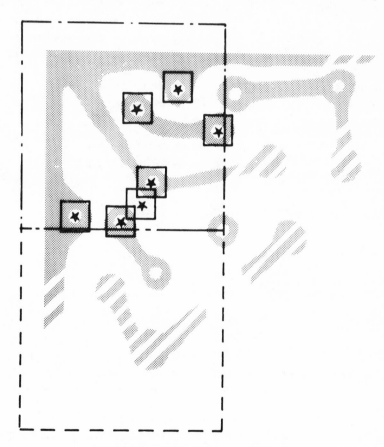

Fig. 9.6. Strategy for drilling-point location.

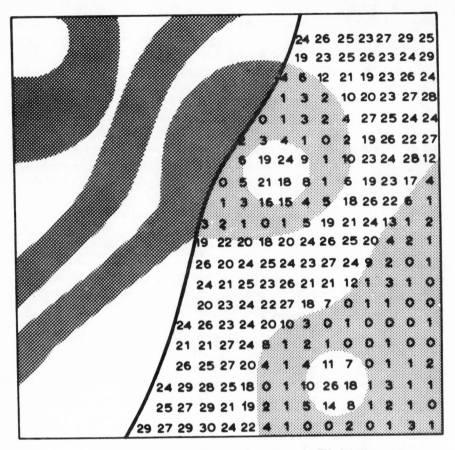

Fig. 9.7. Digital image of PCB extracted from the TV picture.

the state vector, it may either call for help or utilize the "default value" which is included within the *a priori* information given to the machine by the user when he sets it up.

9.3.4. Initialization

The object of the initialization procedure is to assign an initial numerical value to the elements of the state vector. The determination of these values can be either entirely automatic or in an interactive mode in which the user monitors each operation. If any automatic operation appears to the machine to have failed, the program assigns the default value to that element of the state vector.

The initialization procedure assumes that the board contains an "empty

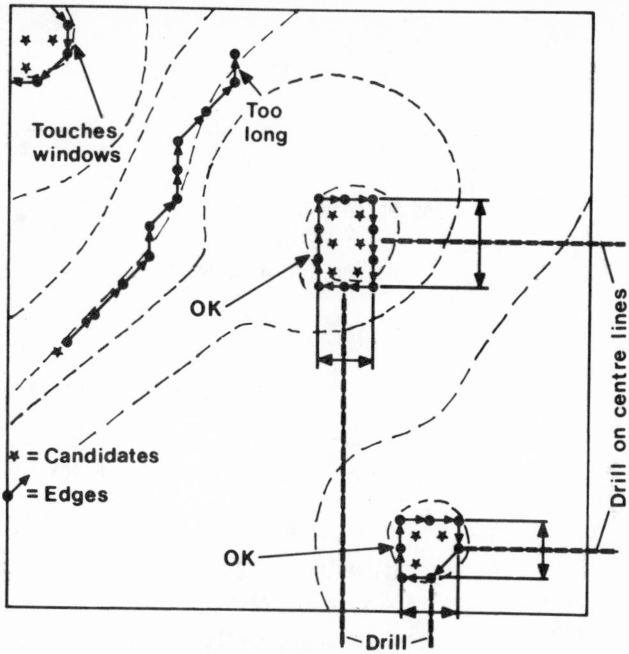

Fig. 9.8. Edge tracing of the picture of Fig. 9.7 around candidates.

region" without any pattern except a Maltese cross (which may either have been etched into the copper or stuck on to the board by the user). The user positions the board so that this cross is in the field of view of the TV camera and the program locates the center of the cross using a two-pass process. In the first pass the TV picture is coarsely sampled (every 16th point from the rectangular grid) and a simple operator (described in Section 9.4.2) is used to find "candidate" locations for the center of the cross. In the second pass the area around each candidate is rescanned at a higher resolution, and examined with the more elaborate operator described in Section 9.4.3. The center of the Maltese cross should be detected at one of the candidate locations. The board is then shifted a known distance in X and Y and the new position of the cross in the TV picture is determined. The "scale" in X and Y can now be calculated (Fig. 9.10).

The purpose of the projected graticule is to act as a secondary datum defining the position of the drill, the primary datum being the actual position of a drilled hole (Fig. 9.11). The graticule projector is mounted, independently from the TV camera, on the main frame of the drilling machine. Therefore, the relative positions, in the TV picture, of the graticule and the drill are unaffected by any mechanical movements or electronic drifts in the camera. Although by prior adjustment the center of the Maltese cross is arranged to correspond to the

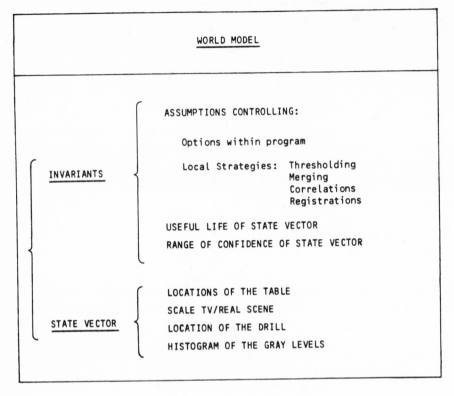

Fig. 9.9. Contents of world model.

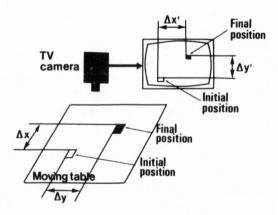

Fig. 9.10. Scale determination.

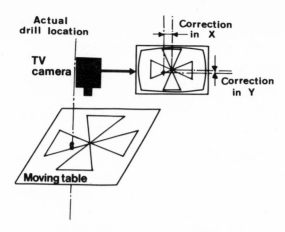

Fig. 9.11. Determination of the offset between the projected graticule and the drill.

drill position, mechanical drifts in the projector can introduce a displacement in their relative position. This offset is calculated by drilling a test hole in the board as described below.

In the next phase of the initialization procedure the Maltese cross graticule is switched on and appears in the TV picture. Its position XGRAT, YGRAT is determined by the same method as that used for the cross on the board. The projected graticule is then switched off, and the program relocates the cross on the board. It then searches for a small (20 X 20 pixel) empty area which it then centers at XGRAT, YGRAT. The test hole, which it is assumed will fall within the scanned area, is drilled and its position determined using the method employed to locate drilling points (Section 9.3.2). Since the position of the graticule has already been found, the offset between drill and graticule can now be calculated.

9.3.5. Movement Strategy and the Registration of Subareas

An important feature of the control program is its ability to explore the whole board without missing or reprocessing any regions. This is accomplished by starting in one corner of the board and by systematically processing successive subareas in a "boustrophedal" route as shown in Fig. 9.12. For each subarea being processed for drilling an area four times the size of the subarea is scanned, the other three subareas being concerned with the registration between one drilled subarea and the next. One of these three will be the next to be drilled and this subarea is also inspected for candidate drilling points whose locations are stored. When the board is shifted to process this next subarea for drilling, a new list of candidates is obtained whose positions could be checked against the list

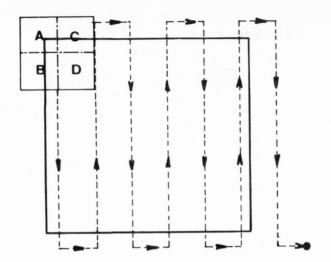

Fig. 9.12. Movement strategy.

obtained in the previous stop. Any misregistration which may have occurred could be computed and corrected.

9.4. PICTURE PROCESSING OPERATORS

9.4.1. A Simple Blob Detection Operator

During the execution of the program as described above, the computer has to recognize "blobs," defined as dark objects surrounded by a closed ring of picture elements which are lighter than the object itself. The shape of this ring is not defined, but its dimension must be smaller than D, where D is the size of the operator used to recognize the blob. (For the operator to have a center point, D is an odd number.) The operator (Fig. 9.13) is a square template composed of eight radiating limbs. Let $P(i,r)$ be the density of one picture element seen through the rth element of limb i of the template; and $P(0)$ be the density seen through the center of the template.

$$L(i) = \min \{P(i,r) \qquad \text{for } r = 1, \quad (D-1)/2\} + T \qquad (9.1)$$

where T is a given constant offset. Then, if the condition

$$P(0) > L(i) \qquad \text{for } i = 1,8 \qquad (9.2)$$

is satisfied, a blob is present in the region bounded by the outer elements of the operator and one of its elements is at the center of the operator. The contrast in

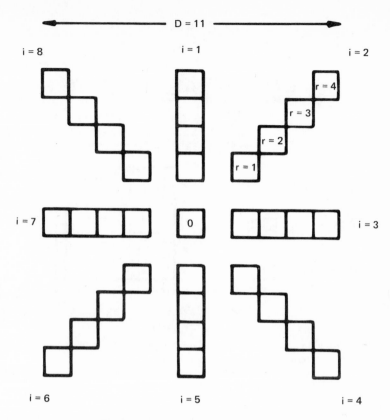

Fig. 9.13. Simple blob detection operator.

PCB images is usually good, and in such situations, this operator is fairly reliable. The value of T is not critical and is usually set to 5.

9.4.2. A Simple Local Operator for Locating a Maltese Cross

This operator is designed to detect the sequence of black and white areas which surround the center of the cross (Fig. 9.14a). The shape of the operator, which can be regarded as a variation of the blob detector described above, is shown in Fig. 9.14b. Let $P(i)$ be the density of the picture element seen through element i of the operator; and T be a given constant offset

$$L(i) = \max \ \{P(i) \qquad \text{for } i = 1,3,5,7\} + T \qquad (9.3)$$

Then if the condition

$$P(i) > L(i) \qquad \text{for } i = 2,4,6,8 \qquad (9.4)$$

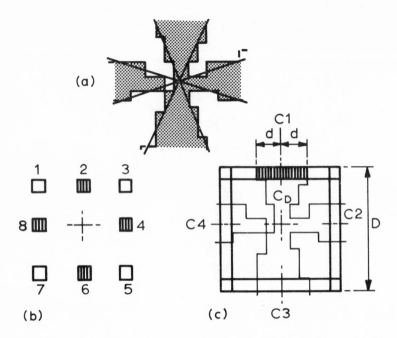

Fig. 9.14. Recognition of the Maltese cross: (a) binary picture; (b) simple operator; (c) more elaborate operator for locating the Maltese cross.

is satisfied, a candidate location for the cross is detected at the center of the operator. The value of T is not critical and is usually set to 5.

9.4.3. A More Elaborate Local Operator for Locating a Maltese Cross

This operator is designed to inspect the area surrounding the candidate location of the cross determined by the simple operator. It checks whether a cross is really present and, if so, determines its exact position. The operator is based on recognizing a series of increasingly long vertical and horizontal dark bars on a light background (Fig. 9.14c). First, a search path on a square of size D is defined, then for each side in turn a histogram of the density of the D pixels is computed. A binary pattern from this strip is obtained by thresholding at the density level defined by the first minimum level zero in the histogram. The binary pattern is "accepted" only if it contains a black bar the length of which is of the order of $D/2$ (within say 50% of this value). If the pattern is rejected a second threshold value (given by the next minimum in the histogram) is tried, and so on. If the pattern is accepted, its center C_1 is found and recorded. The procedure is repeated to find C_2, C_3, C_4 on the other sides of the square. A tentative center C_D for the cross is found by the intersection of line C_1C_3 with line C_2C_4. The center of the cluster of successive tentative centers C_D obtained for increasing values of D is taken as the center of the cross.

9.5. PERFORMANCE AND CONCLUSIONS

Although fairly complex at first sight, the program requires only the core size of a small minicomputer (16K words of 16 bits). The addition of more sophistication would require at least one level of overlay. The program is written in FORTRAN IV with single-precision integer operations and machine code device drivers. Although the speed of both the program and the scanning have not been optimized, the machine can drill holes at approximately 1 every 5 s. This time is approximately equally distributed between scanning, processing, drilling, and moving the board. The speed of the visually controlled machine is slightly slower than can be achieved by a man, and half the speed of a machine controlled by punched paper tape. The comparison with numerically controlled operation leads one toward an alternative use for the visually controlled machine. Instead of drilling the board, it could be used to prepare a punched tape for future numerical control, in which case the machine could work directly from the artwork. One can also consider extensions of this type of machine to other problems of automatic digitization.

In its present state, the machine can almost drill complete boards. The program is being improved to prevent occasional drilling points being missed. This is particularly important as the practical use of the machine in the laboratory workshop is now being considered.

As described, the machine treats each hole as an individual item. In practice, however, the relative position of holes in a PCB is often as important as the location of an individual hole at the center of a drilling point. For example, the pitch of the pattern of holes for a dual-in-line IC must be regular to ensure easy insertion of the device in the board. The drilling machine could be made intelligent enough to recognize such patterns of drilling points and to drill the holes accordingly. The recognition of patterns of holes could probably be achieved by comparison and matching with a stored pattern, using a technique similar to that employed in the registration of areas (Section 9.3.4).

Other features which could be added to the program include the recognition of other shapes as drilling points and an ability to check that the holes are actually drilled in the right place with respect to the copper pattern. If the machine discovered that the holes were misplaced, procedures for updating the contents of the state vector could be invoked.

ACKNOWLEDGMENTS

We would like to acknowledge the contributions of the other members of the team who have worked on this project, A. R. Turner-Smith and D. Paterson. We would also like to thank our Group Leader, J. A. Weaver, for his help and encouragement.

REFERENCES

Ambler, A. P., Barrow, H. G., Brown, C. M., Burstall, R. M., and Popplestone, R. J., 1973, A versatile computer-controlled assembly system, *Proc. 3rd Int. Joint Conf. Artificial Intelligence*, Stanford, Calif., p. 298–307.

Fikes, R. E., Hart, P. E., and Nilsson, N. J., 1972, Learning and executing generalized robot plans, *Artificial Intelligence* 3:251–288.

Gill, A., 1972, Visual feedback and related problems in computer-controlled hand–eye coordination, *Stanford Artificial Intelligence Memo AIM 178*.

Heginbotham, W. B., 1973, Reasons for robots, *Proc. 1st Conf. on Ind. Robot Techn.*, Nottingham, UK, p. R1/3–12.

Heginbotham, W. B., 1974, Robots – The trend to a visual knowledge of their environment, *Design Eng.* Sept.:280–283.

Hitachi Ltd., 1973, Japanese robots now have eyes, *Engineer* Oct. 4:21.

Klir, J., and Valach, M., 1967, *Cybernetic Modelling*, Iliffe Book, London, pp. 257–278.

Lee, M. H., and Wood, D. J., 1974, Sensory-motor control problems in a flexible industrial robot system, *Proc. 2nd Conf. Ind. Robot Technol*, Birmingham, UK, March 27–29, pp. C2/9–C2/20.

Olsztyn, J. T., Rossol, L., Dewar, R., and Lewis, N. R., 1973, An application of computer vision to a simulated assembly task, *Proc. 1st Int. Joint Conf. Pattern Recognition*, Washington, D.C., pp. 505–513.

Parks, J. R., 1973, Sensory devices for industrial manipulators and tools, *Proc. 1st CISM IFTOMM Symp.*, Udine, Italy, Sept. 5–8.

Radchenko, A. N., and Yerevich, E. I., 1973, On a definition of artificial intelligence, *Proc. 1st CISM-IFTOMM Symp.*, Undine, Italy, Sept. 5–8.

Rowat, P. F., and Rosenberg, R. S., 1972, Robot simulation sundries: Descriptions and plans, *Proc. Can. Computer Conf.*, Montreal, June 1–3, pp. 423301–423332.

Saraga, P., and Wavish, P., 1972, Edge tracing in binary arrays. In: Machine Perception of Patterns and Pictures. *Inst. Phys. (Lond.), Conf. Ser.* 13:294–302.

Shirai, Y., and Inoue, H., 1973. Guiding a robot by visual feedback in assembling tasks, *Pattern Recognition* 5:99–108.

10

Editorial Introduction

J. R. Parks describes some of the possible applications of pattern recognition to industrial inspection. This area is potentially very lucrative but is also very demanding, since the alternative solutions are already highly developed. At the moment, industrial inspection is usually performed by a human operator, who is slow and expensive but extremely versatile, even being able to detect flaws which were not realized as important during his training. This then is the standard which must be *surpassed* if pattern recognition techniques are to succeed. A machine has three advantages over a man:

1. It is probably much faster, possibly so fast that 100% inspection may be achieved in a situation which allows only sampling checks to be made by a human inspector.
2. It is more reliable than man, who is liable to suffer from fatigue and lack of concentration.
3. Although a machine is expensive to install, its operating costs are small compared to a person's salary.

Notice that the speed of a machine may often be exploited to offer something new: 100% quality-checking. Nevertheless, the standard which is to be bettered is remarkably high. Remember that industrial inspection, by human operator, has an evolutionary history of about 200 years! This has resulted in ingenious jigs being devised to make mechanical inspection faster and more reliable. However, jigs cannot be used to test apples for those indefinable qualities which makes them attractive to the housewife. Inspecting and grading fruit has progressed very little; the pattern recognition machine has a lower standard to better here, but there is a greater variability in the "product" to be inspected. Other fascinating examples, including inspection of patterned fabric, are discussed by Parks in this chapter. In many instances, the complexity of the inspection tasks can be reduced by very simple tricks, such as using slotted conveyors to fix the orientation of spindular objects, or using patterned and/or colored light to view the object during inspection. Parks emphasizes that there is no virtue in trying to solve a difficult problem entirely by pattern recognition, if by making a small adjustment to the basic pattern, a simpler overall solution can be achieved. Parks rightly sees pattern recognition as a tool, one tool among a host of other tools. We must use pattern recognition only when it is appropriate to do so; it is the overall solution that matters most, for it is this that will be used to judge the merit of pattern recognition to human society.

Parks (like Rutovitz *et al.* (Chapter 12), Amos (Chapter 11), Duff (Chapter 6), Brown and Popplestone (Chapter 8), Aleksander (Chapter 3), and Halé and Saraga (Chapter 9)) shows that "intelligent behavior" can be achieved by concatenating very simple procedures that have been carefully designed and selected. Halé and Saraga also emphasize the major lesson of this chapter, that there are often simple ways of greatly reducing the complexity of the pattern recognition methods. There is much that can be done, using existing pattern recognition techniques and a little ingenuity. This chapter shows that the greatest need in pattern recognition is for workers, with great enthusiasm, to dedicate themselves to areas like those described here. Only this will convince the layman that this subject has a significant contribution to human development; the subject has suffered for too long from "advanced theory," isolated from reality.

10

Industrial Sensory Devices

J. R. Parks

10.1. INTRODUCTION

This essay starts from the arguable proposition that the long history of pattern recognition and allied techniques, such as image analysis, has resulted in relatively little practical, i.e., commercial, application. The only significant commercial use is that of *character recognition* where, starting from the early systems for reading highly stylized character shapes printed (in magnetic ink) on bank checks, we now have devices capable of recognizing only lightly stylized handprinted characters, though at some cost.

The development of character readers has largely progressed through an intensely felt requirement, rather than (one might almost say, in spite of) theoretical (classical) pattern recognition theory. The relative success of character recognition began with the realization that, to make the solution to the problem (data entry to data processing systems) attainable, the problem itself had to be simplified. This was achieved initially by printing the characters in material which could be readily differentiated from the medium which carries it (thus easing the initial pattern detection problem) and by modifying the character shapes so that they are readily discriminated by available electronic techniques (though this reduced their visual separability).

Regrettably, there are not many "natural" patterns which are susceptible to manipulation in this way. Of course, such conditions as can be imposed on biomedical samples, for example, by staining techniques, should be used as far as is economically acceptable.

J. R. Parks · Computer System and Electronics Division, Department of Industry, Westminster, London

Potential areas for the use of pattern recognition (PR) abound. While many of them have been approached by intrepid research workers, and some of them even partially solved, very few, if any, with the notable exception of character recognition, have been solved in an economic fashion. Indeed, the problems to be faced in PR are so much greater than was initially anticipated that to have reached even an uneconomic solution to the objectives tackled speaks volumes for the researchers' ability and of the patience of the funding agencies.

It is conjectured that the reasons for this apparent failure are twofold and interdependent. One is the seduction of new entrants to the field by the beguiling familiarity and excellence of animal PR mechanisms. The other is the lack of glamour of apparently simple application, such as the inspection, counting, grading, and sorting of mechanical components, etc. Such mundane tasks are crying out for solution in manufacturing industry, being highly manpower-intensive and not performed at an adequate level. It is the object of this chapter to explore the possible use of PR methods in such industrial tasks, the motivation being to encourage the adaptation of PR methods to the improvement of production quality by the replacement of fallible (human) operators with less fallible mechanisms. The industrial situation is attractive as it is susceptible to a degree of manipulation to make the problems encountered more tractable. The potential savings to industry are comparable to the profit margin (i.e., about 7% of turnover).

10.2. POTENTIAL AREAS OF APPLICATION

In very broad terms, routine quantity production is of two basic forms:

1. Continuous, as, for example, in the oil, gas, glass, steel, and electricity supply industries
2. Segmented, as in the production of discrete items, such as electric light bulbs, nuts and bolts, sweets, motor cars

This distinction is significant. The continuous processes are essentially steady state in that the changes of associated (measurement) variables vary on a relatively long time scale. On the other hand, in the segmented process, production is in discrete, complete units and the process is thus characterized by a more-or-less continuous sequence of repeated transient events.

While it would be wrong to suggest that there are no problems in the control and monitoring of continuous processes, the development of instrumentation and control procedures is well in hand, and process control is a well-established engineering discipline.

On the other hand, instrumentation for dealing with the segmented production situation is still very dependent upon manually operated gauging, assembly, and inspection methods. The scale of the disparity between the two production situations can be seen by comparing the annual net production value per head in typical U.K. industries. In the oil industry this value was $37,000 per head in 1975. While the value in the mechanical and electrical engineering industries was about $7000 per head per annum. These values reflect the relative intensity of manpower in the two sectors. This situation is obvious to even the casual visitor to, say, an oil refinery and the best mechanical or electrical manufacturing units.

It is unreasonable to consider making the segmented production situations independant of "hands-on" operatives. This is possible in continuous production where the "components" for "assembly" come through metered pipes from large capacity feedstocks, and are "reacted" or mixed together in continuous-flow chambers, eventually to be collected in bulk storage. Production of some discrete objects such as light bulbs, ice cream, and bread, come close to this situation, but the production of the majority of discrete items, e.g., motor engines, telephones, and typewriters, require the use of manually fed, adjusted, and monitored material forming processes. Furthermore, human intervention is required in the "bundling" of components, handling and inspection, assembly, and often packaging.

Robotics is often put forward as the ultimate answer in this situation but regrettably is still rather a long way from being a shop-floor, cost-effective approach, as is discussed in Chapters 9 and 16. The pressure to reduce production costs through reduced manpower, reduced scrap and rework, and improved quality of product is both intense and immediate and cannot wait for the ultimate solution offered by robotic technology.

A large amount of mechanization is used in manufacturing industry, though this is mostly in the piece parts *forming* processes and only very little is used in *assembly*. The vast majority of this mechanization is "open-ended" in nature. Very little automatic control beyond simple sequencing is used; the majority of the control and monitoring functions are exercised through attendant operators. Assembly of components, though widely aided by power tools, jigs, etc., is, with few notable exceptions, very highly manpower-intensive. The manpower is not used primarily for its dexterity in assembly, as there is a wide range of mechanical feeding and assembly apparatus available and, to a large extent, unused; the chief attribute of the human assembler is her (his) ability to see and to feel, that is, to act as a sensory system par excellence.

This is the challenge awaiting attack: to develop artificial sensory systems, capable of rivaling that of human operators. Indeed this has been the challenge of pattern recognition since the early proposals for reading machines by Tauschek (1929) and Goldberg (1927). In industrial applications, we have a situation in which the objects involved are man-designed and can be manipulated to an extent limited only by cost-effectiveness of the total manufacturing process.

The following paragraphs attempt to indicate what initial steps might be followed by the PR engineer in defining an effective approach. It is the penchant of the engineer to take a set of given actual circumstances and technical methods and, through his ingenuity, to manipulate these resources so that new devices emerge. We can postulate, for the distant future, a mechanism which can look at a complete assembly such as a motor engine, or even a carburetor, and check it for completeness. Such objectives are a long way off in practical terms. However, we need not take the Irishman's point of view that if you want to get to Dublin you should not start from here. Rather, we have to take the point of view that our long-term aim is to reach Dublin and that there are many, many roads and many steps which we must take along those roads. Practically, we cannot, at this time, specify more than the first few, because we cannot see clearly. In the event, we may not finish up in Dublin at all but at some other, more or less desirable, place. This will depend on the result (successful or otherwise) of each deliberate step taken in our endeavor to travel purposefully.

10.3. SENSORY MODALITIES

We shall consider two types of patterns in what follows; these are *pictorial* and *continuous* patterns. Pictorial patterns are two-dimensional (spatial) distributions of energy, usually capable of being displayed as a pictorial display on a CRT; most of the methods discussed here use image (scene) scanning devices such as TV cameras. In principle, the use of tactile, sonar, or radar methods of scene scanning to provide pictorial patterns is also possible.

Continuous patterns are often single-dimensional variations as a function of time*; one-dimensional patterns are, of course, possible against bases other than time, e.g., displacement, wavelength, temperature.

10.3.1. Pictorial Patterns

In order to proceed in an orderly fashion, it is useful to consider pictorial systems according to the complexity of the image handled and to relate this to typical job requirements. This progression might be as follows:

1. Silhouette matching — determining whether the correct part is in the correct place
2. Silhouette dimensional checks, independent of the precise location of the object

---

*The EEG is not (see Chapter 15).

3. "Completeness" of silhouette, with independence of object position and orientation
4. Images in which distinct color or brightness contrasts exist
5. Continuous gray scale
6. Generalization from the limited two-dimensional situations of steps 1–5 into multiple-image systems for coping with the third dimension.

We may now consider these several classes in some detail, in order to explore the increase in capability needed at each step in the progression, taking the order suggested above. It is not the purpose of this chapter to promote or develop basically new image-analysis techniques, but rather to indicate that moderate adaptation of existing technologies can produce useful devices, which will then illuminate the need for improved techniques in a practical rather than theoretical context.

10.3.1.1. Simple Silhouette Matching

Figure 10.1 shows a schematic of a simple system for matching two silhouettes, or more strictly for matching an opaque object against a mask. Although the system shown is a primitive optical projector, alternative methods, using a TV camera and comparison of images in electronic or digital form, are possible and more realistic for practical use.

There are several severe practical limitations to this simple approach. The most significant is that in order to detect a "match" (indicated by reduced light transmission through the system) we must accurately locate the object for inspection. The system in Fig. 10.1 is a conventional *cross-correlator*, requiring exact correspondence of all points in both "images."

Precise alignment of objects by automatic means is clearly difficult even if highly specific jigs and clamps are used. Such a requirement is therefore expensive and a more economical approach would seem to be to accept the limited degree of alignment afforded by the use of primitive stops, channels, and simple guides which will locate objects approximately in position, and maybe also orientation. Manipulation of the electronic image can then be used to correct the position of the object image relative to the stored reference image.

Adaptive or automatic alignment methods have been developed, but unless all possible conditions are explored, such procedures are subject to *aliasing* when a reasonable but no perfect match is obtained (Fig. 10.2). It is obvious, of course, that human inspectors, employing a *shadowgraph* to project and compare, say, gear-tooth profiles, can use this approach. However, they have the benefit of intelligence and an understanding of how the world behaves, to aid them in the use of this simple tool. Fully automatic methods cannot yet call on either attribute, and better methods of image inspection are needed.

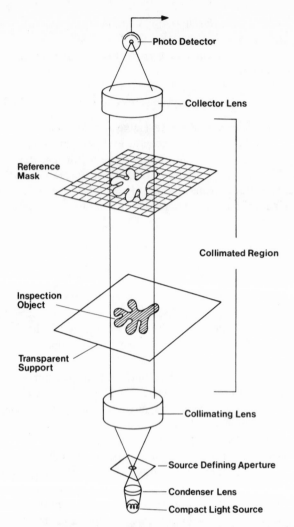

Fig. 10.1. A simple inspecting system using an optical cross-correlator.

Other shortcomings of the cross-correlation approach are:

1. The inability of such systems to indicate the various causes of any mismatch separately. Hence, it cannot discriminate different faults, such as under- or oversize, geometric distortion, misplacement, or absence of piercings.
2. As an extension of 1, it follows that different parts of an object cannot be allowed different tolerances, since all errors, from whatever cause, are aggregated.

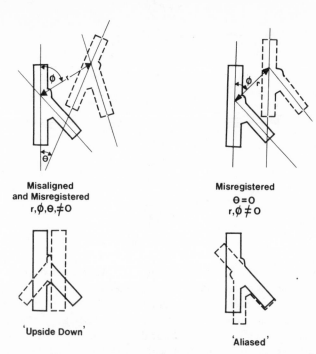

Misaligned
and Misregistered
r,φ,θ,≠0

Misregistered
θ = 0
r,φ ≠ 0

'Upside Down'

'Aliased'

Fig. 10.2. "Error" conditions to which simple correlator is susceptible.

10.3.1.2. Dimensional Inspection of Silhouettes

In order to allow the separation of various aspects of inspection, such as varying tolerance of dimensions, straightness and roundness, and position of features (holes, tags, etc.), it is essential to get away from the pure pictorial-handling methods, such as correlation, into a formalism, amounting to a *description* of objects to be inspected. Such a description must facilitate the separation of the various attributes of the objects which are to be considered in its inspection.

Consider the task of inspecting a particular class of objects: *spindular axially symmetric objects*. This general shape is characteristic of turned components, (Fig. 10.3). There is a considerable range of such objects, both from the morphological and dimensional points of view. The collection of objects in Fig. 10.3 can all be characterized from their silhouettes as a string of symmetrically disposed contiguous rectangles and trapezia. For inspection purposes the dimensions (length and width of each rectangle, or the length and major and minor widths of the trapezia), together with the center line (axis) of each component shape are adequate. This representation is equivalent to the conceptual form of the objects in the mind of their designer, when he is casting them as an engineering drawing.

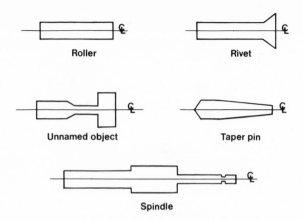

Fig. 10.3. A selection of spindular objects.

In practice objects for inspection do not, when seen through practical systems with finite resolution, appear to be so finely delineated as an engineering drawing. The image as seen by the "camera system" in Fig. 10.4 is shown in Fig. 10.5, though the quantizing effects are grotesquely exaggerated. The "camera system" consists of a linear (one-dimensional) photodiode array onto which a silhouette of the spindular object is projected. Motion of the object provides the second dimension of the scan, the object being transported approximately lengthwise on a translucent belt. Increments in the motion of this belt are indicated by a line-grating attached to the belt idler pulley. The linear photodiode array is at right angles to the direction of travel, so that the number of cells obscured corresponds to the diameter of the object at each sampled incremental position. Thus, an object is represented in the inspection equipment by a quantized image as shown in Fig. 10.5. Notice that this inevitably contains significant *quantization noise*.

In order to avoid the limitations and problems in matching objects against standard masks, we need to extract an "articulate" description of the object. This description must be symbolic, easily manipulated by suitable logic machine − probably a microprocessor − and efficient, though this does not necessarily mean minimization of redundancy. Convenience of articulation is more important in the representation of objects than the achievement of a highly compact coding. In order to avoid problems associated with irrelevant properties of the object presented for inspection, this description must be invariant (or nearly so) of the orientation and position of the object. Notice that by the use of silhouettes we also discard certain other properties of the object, such as surface color and finish, though these may not be totally irrelevant.

A number of approaches are possible and a few will be explored. However, most will later be discarded in the current consideration of "spindular" objects.

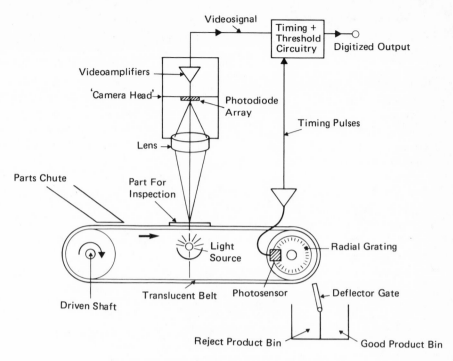

Fig. 10.4. Parts scanning system and camera timing.

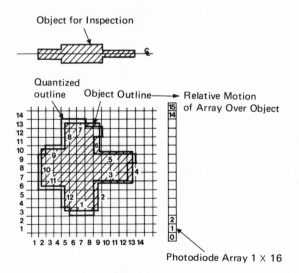

Fig. 10.5. Quantized image of object. (Note the effect of different size of quantizing interval on either axis.)

For example, the list of coordinates defining the position of black elements in Fig. 10.5 can be made *position-invariant*, if we first determine the center of density of the image and then refer the coordinate system to this point. As the image is defined on a discrete coordinate system using a matrix of picture elements (*pixels*), noninteger coordinates are meaningless, and such a shift of origin is not exact. By moving the coordinates (of both the specimen reference object and the object for inspection) we overcome problems associated with simple translation, but not rotation or inversion of the object (i.e., upside down). (Neither have we obtained a description of the object but have merely normalized its position.)

A slightly more sophisticated approach can be adopted for normalizing orientation, once we have found the center of the object. We radiate, at regular angular intervals, "rays" which will, at some distance from the center, cross the boundary of the image. Thus (see Fig. 10.6) we obtain a *position-invariant* description of the object as a function $r(\phi)$, where r is the distance of the image boundary in direction ϕ from the center of the image. Since ϕ is cyclic, changes in orientation of the object produce only a linear shift in the function along the

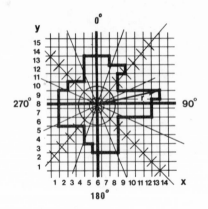

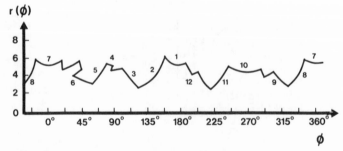

Fig. 10.6. Representation of object in Fig. 10.5 as a function $r(\phi)$. The center of density of the image is in cell (6, 8).

ϕ-axis. The inspection of an object now requires the comparison of two single-dimensional cyclic functions; in principle, the two functions should be compared for all values of ϕ. In practice, many actual objects will produce a function with well-defined and locatable discontinuities. Since we know that these discontinuities must be alignable if the two objects are closely similar, tests can be made to establish the best match of angular separation of the discontinuities in the function before proceeding to more detailed comparisons. We now have a full representation of the object image, which could, if we wished, be used to reconstruct the original image with a fidelity limited only by the resolution of the initial (xy) quantization of the image and the quantization of r and ϕ. To obtain a position and orientation-invariant representation by this method [using the mixed coordinate system, x,y and $r(\phi)$] and to obtain a given precision of measurement, the quantization will need to be finer than if a single system were used.

Although a normalized representation is obtained by the above "ray" method, the degree to which it can be articulated or manipulated is limited to translation along the ϕ-axis. Hence, we are no nearer to separating the errors (differences between the specimen and test objects) into those due, for example, to length rather than diameter variations.

In order to define the nature of errors, rather than just obtain a measure of the total error, a description is needed in terms related to the engineer/designer's concept of the properties of the object.

A slightly better approach, based upon analysis of the image boundary, allows some of the geometric properties of an object to be detected directly. We simply extract the (discrete) curvature of the boundary of the object for inspection. That function, relating distance s round the boundary (from some arbitrary starting point) and the slope ψ of the tangent to the perimeter, has a number of interesting properties. Figure 10.7 illustrates the difference between the continuous function ψ and its discretely quantized counterpart. Note that the functions are cyclic in s, the perimeter, and that ψ increases by 2π in each cycle. In either form, parts of the perimeter of constant curvature (either straight lines or circular arcs) are represented by straight regions in the function.

The (s,ψ) approach produces a single-valued function, whatever the complexity of the boundary described. It is reversible and therefore loses no information, and has a physical significance (i.e., relates in an obvious way to the geometry of the object boundary). For these reasons, the (s,ψ) approach is preferred to the $r(\phi)$ form, which is not single-valued, in general, and presents arithmetic-accuracy problems, due to the fact that it mixes (discrete) coordinate systems.

Thus, we have several techniques, of differing complexity, for use as the particular task in hand demands. The types of task to which they might be applied can illustrate their various strengths and limitations.

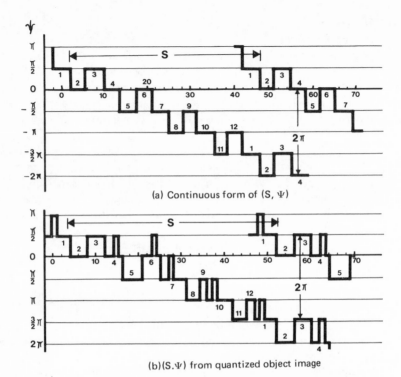

(a) Continuous form of (S, Ψ)

(b)(S.Ψ) from quantized object image

Fig. 10.7. (s,ψ) representation of object outline from both continuous and quantized object outlines.

Masking, or correlation, methods can be used in those situations where the objects for inspection are presented to the inspection device in known position and orientation. These conditions occur in some machine tools, for example, screw-thread profiles on die-making machines and head forms in a screw mill. However, the use of optical inspection methods on machine tools may be difficult in the presence of swarf, cutting fluid, and vibration. Alternative methods are suggested later. The "radial rays," $r(\phi)$, approach could be entirely adequate for detection of asymmetry, for example, in grading fruit on the basis of size and "regularity." Analysis of boundaries by the (s,ψ) approach can cope with more general shapes and can be developed to allow the identification of specific properties of the boundary. From discontinuities in (s,ψ) the morphological components of an object can be isolated and subjected to individual measurement or comparison.

These latter aspects of image analysis appear in many contexts other than the (s,ψ) case introduced here; the reader is referred to the articles in this volume by Brown and Popplestone and Halé and Saraga (Chapters 7, 8, and 9) for specific discussion.

However, while the methods discussed have some generality and intellectual appeal, we have wandered from the task with which we began: considering that of spindular objects. This type of component is produced by the most widely used machine tool, namely, the lathe. Many, but not all, objects produced on lathes and needing inspection have a length which is significantly greater than their diameter. Such objects (1) are relatively easily aligned (roughly) on a conveyor, say a belt with a groove in it, and (2) will exhibit axial symmetry along most of their length. We must consider methods which directly exploit these valuable, simplifying properties. A simple but novel procedure for inspecting such spindular objects, which are approximately oriented, has been developed at the National Physical Laboratory London.

Consider the unnamed object in Fig. 10.3 from the point of view of solid geometry. It is a collection of coaxial cylinders and a frustum, each individually specified by the lengths and diameters of the component solids. The whole ensemble is "assembled" in a prescribed order. (An exact reversal of this order also describes the same object.)

If this object is passed before the photodiode-strip camera of Fig. 10.4, we obtain a digital, electronic form representing the object (cf. Fig. 10.5). The image can be reduced to a simple numeric form by the hardware apparatus shown in Fig. 10.8. This produces the stream of number groups in Table 10.1, which are the photocell numbers in a continuous array of 512 cells (effectively) on 25 μm centers, corresponding to black/white or white/black transitions in the image.

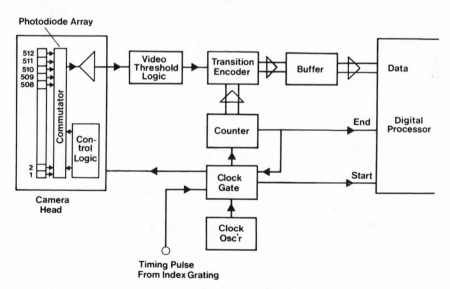

Fig. 10.8. Simple encoder for use with photodiode array "camera." Data input to processor is the cell number following a contrast transition.

Table 10.1. Raw Data and Derived Dimensions

Segment	Data: Cell numbers			Derived: Linear dimensions (arbitrary units)		
	Edge (L)	Edge (R)	Count (C)	Width	Center line	Length (N)
1	30 20 21 20 21	90 126 126 125 125	1 2 3 4 175	$\frac{1}{n}\sum_{n}(R-L)=\frac{1}{2n}\sum_{n}(R+L)$ 104.5	 72.5	 175
2	60 61 60 61	86 87 87 88	176 177 178 290	27.1	73.2	115
3	15 15 14 15 29	131 131 130 131 100	291 292 293 294 625	116.3	72.1	335
4 etc.	–	–	–	–	–	–

In practice, the scanned area in which the object will lie is somewhat wider than the object, in order that a limited amount of misalignment or mispositioning of the objects can be accommodated. This has the effect that the first and last cells in the array will normally not be in shadow. If they are, then a fault condition is detected in the apparatus or in the object being inspected. The test object must be rejected as equivocal and either discarded as waste (which it might not be), or returned for reinspection. If the first and last cells are both illuminated, then the sequence of cell numbers, corresponding to black/white or white/black transitions, is sufficient to describe the object. In general, there is an even number of such transitions in each scan, the odd-numbered ones indicating white/black transitions, and the evens black/white ones. The differences between the odd–even pairs indicates the width of each dark strip, and differences between even–odd pairs indicates the gap width between dark strips, should there be more than one. Objects having holes in them will produce a multiplicity of dark strips in each scan through the object, and so will other objects, like those in Fig. 10.3, if they are badly misaligned! We shall return to such problems when we have considered the simple situation in Fig. 10.5.

The information given by the diode-array camera will, in an idealized case, be pairs of numbers. One (successive) pair from each scan indicates the cell number corresponding to a transition from black to white and vice versa. These data, in digital form, are suitable for input to a digital processor. The difference between the elements of each pair of numbers indicates the diameter at one point on the object, while the mean of the two numbers corresponds to the position of the center line. Clearly, a step-transition, between one diameter (of the object) and another, will be indicated by an abrupt change in the measured diameter. Furthermore, any lack of straightness of the object will produce a gradual change in the position of the center line.

In practice, unless exact alignment of the object is obtained, various complications creep into this simple situation. In general, the first intersection of the leading edge of the image and the photodiode array will not produce a true diameter. This may be due to surface roughness of the object, or to variation of the photodiode's sensitivity, thus introducing equivocation in the quantizing. Typically, one or more "mutilated" diameters are obtained first. Although the measure of diameter obtained is equivocal, there will be no doubt that the image has reached the photocells. We may then begin measuring the length of the object by counting the number of scans which intercept the image. (Each scan of the photocells is a known linear distance from its predecessor.) Later scans will be free of the initial "edge effect" and will indicate the true diameter, moderated of course by the inevitable quantizing error of $\pm\frac{1}{2}$ of the effective diode pitch at each edge. Since the object-diameter will usually remain constant for many scans, the mean of the measured diameters within this range is an accurate measure of the true diameter of the object. This procedure is only allowable provided that the difference between successive object-diameters does not significantly exceed the anticipated maximum quantizing error on measured diameter (equivalent to ±1 diode pitch).

If a large change in apparent diameter is detected, it is likely that the transition to the next object diameter is occurring. There is a possibility, however, that, if the change is concentrated on one edge rather than symmetrically (indicated by an abrupt change in center line), the change is due to a piece of swarf or dirt attached to the object and should be ignored. If the change in diameter is symmetrical, then a segment of the image has been completely scanned. We can obtain accurate measurement of its diameter by averaging the measured diameters over its entire length. Moreover, the length of the segment is given by the number of scans counted up to this point. The procedure is repeated between successive changes of diameter until the image is completely scanned.

So far only constituent rectangles have been considered. The trapezium produced by the conical segments of the objects shown in Fig. 10.3 will give rise to a uniformly increasing but still symmetrical diameter. The point of transition, from cylindrical to conical form, may not be detected easily and accurately if the slope is slight (i.e., a gentle taper). To cope with this problem it is necessary to approximate the edges of both segments using linear curve-fitting methods. This is done for regions which are well clear of the suspected transition zone and its actual position is taken to be the computed point of intersection of the two linear edges. There are many more points of detail which must be incorporated in a practical machine, but these will not be discussed further here.

At the completion of a scan, we have a description of the image, in terms of continuous blocks of measured dimensions. These dimensions must now be checked for acceptability.

Thus, from an object composed of coaxial cylindrical forms, we obtain a description of the object as a series of segments, like those shown in Table 10.1.

Only the right-hand part of the table is needed for comparison with standard descriptions. Such a comparison is easily achieved by taking corresponding segment descriptions and determining the error on each dimension. These errors can then be individually tested for acceptability and the part accepted or rejected. Straightness of the object is indicated by variation in the center-line position. In practice, the object may not be accurately aligned axially. In this case, for a "straight" object, the center-line position will have a finite and constant slope. A variation in the slope of the center line then indicates that the object is bent. Significant misorientation will introduce errors into estimates of diameter and length. However, these can be corrected by simple trigonometry, having already derived the actual orientation of the object from the center-line positions. Measured diameters may then be corrected for such misalignment.

The description of the ideal object, against which samples are to be tested, can be obtained in two different ways. Most obviously, the component designer can specify the object in tabular form, similar to Table 10.1 but including permitted tolerances. These values are then encoded and input to the processor. More conveniently, the inspection machine can have two phases of operation as follows:

Phase 1: Carefully made (or selected) components, with dimensions corresponding to the upper and lower tolerance bands of acceptability, are fed to the inspection device. Each of, say, four objects will produce a set of measurements which can be analyzed simply, to extract the highest and lowest value of each relevant dimension. Thus, we can define an "envelope," within which each inspected object must fall if it is to be accepted.

Phase 2: These limit values are stored and production samples are accepted or rejected, according to whether or not they fall into or out of the stored "envelope."*

---

*This is very much like the procedure discussed in Section 4.5.4, there called recognizing "familiar" patterns.

In this way we have a self-setting or self-programming machine, needing no particular special ability (e.g., to handle and load paper tapes, etc.) on the shop floor. The detailed numeric specification of the object and the need for means to encode it have been replaced by the cost of producing (or selecting) components demonstrating the limits of acceptability. It is also worth observing that, in the second method, absolute calibration of the machine is not required, though stability is essential. This is set by the mechanical arrangement of the scanner and photodiode array alone and presents no particular problem (provided the scanner is not adjusted between the two phases). Absolute measurement can, of course, be achieved by accurately setting the optical-system magnification, but is not necessary for what amounts to comparative tests. The resolution of the scanner must nevertheless be better than the tolerance to be applied by an adequate factor (at least two times).

A prototype system operating as described above using a Texas Instruments microcomputer and a 512-photodiode linear array with a capability to measure four pieces per second has been produced by Gay's (Hampton), Ltd., Middlesex. Spindular objects with diameters down to 1 mm are measured with an accuracy of better than 5 μm. Similar instruments with capacity up to 10 mm in diameter are now being marketed by Thomas Mercer, Ltd., St Albans, Hertfordshire.

10.3.1.3. Checking for Completeness Independent of Object Location and Orientation

The system described in the above paragraphs was developed for accurate measurement of a restricted class of objects (i.e., elongated, spindular, turned). Although this is an important area, it is clearly not universal in its scope. Another class of inspection tasks, again of limited but useful generality, is that concerning essentially flat objects, for example, stampings, shallow pressings, and moldings, where inspection is required for completeness rather than size.

Let us consider the possibilities of inspecting plates, which are flat in the sense that they only have two stable states when arbitrarily laid on a table. For illustrative purposes, "FAC kit"* pieces provide a useful source of such parts which, though rather primitive when compared to production parts, serve for this discussion. A selection of various parts is shown in Fig. 10.9.

Metal-working processes based on punches and/or dies are not subject to significant change in shape during normal operation. Hence the parts produced cannot but be dimensionally correct. However, abnormal operation can occur, due to failure to feed metal properly, broken or chipped tools, or displaced tooling. All of these faults produce very marked effects in the product. Therefore, inspection methods are needed which can check a complex shape for the existence of specific properties, such as (overall) geometric shape and the

---

*FAC is a constructional toy of Swedish origin.

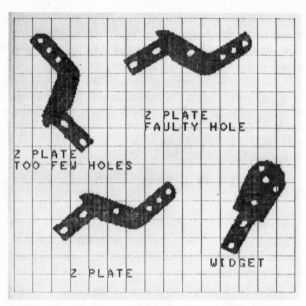

Fig. 10.9. Examples of punched components showing computer-labeled specimen and defective punchings. Definition of complete picture is 240 × 240 picture elements.

number and shapes of piercings. This may be achieved with relatively low precision, since if a piece is about correct, then there is a very high probability that it is exactly right; there is no fault likely to occur in the manufacturing tooling which can introduce small errors. Thus, this type of inspection is *existential* in nature rather than *metrological*, and different methods from those discussed above are appropriate.

Again a variety of techniques of varying sophistication and mathematical appeal are conceivable. We need not go through the discussion of cross-correlation methods again, although it is a natural starting point. Some success has been achieved by workers at Scientific Instrument Research Association, London, in the use of position-invariant transforms, such as Fourier analysis. The transformed image, or selected parts of it, is then more appropriate for matching by correlation methods. These techniques have been used in the inspection of printed packaging and labeling materials — circumstances which ensure the correct orientation of the material for inspection. Unfortunately, due to the integrative nature of both Fourier analysis and correlation processes, only aggregate errors are detected and not their specific distribution or nature. Therefore, very little diagnostic information can be obtained from these inspection processes.

A surprisingly simple process has been successfully demonstrated by Bell (1974) for sorting and inspecting pieces of FAC kit. The apparatus consists of a conventional closed-circuit TV camera, viewing pieces presented on an illuminated table and interfaced to a minicomputer. The analog video waveform is reduced to a two-state signal by thresholding in the interface.

Images are analyzed first to separate individual items in the field of view; each isolated object is then analyzed to extract a limited number of characteristic parameters:

1. Length of the outer perimeter
2. Area enclosed by the outer perimeter
3. Number of internal perimeters (holes)
4. Length of each internal perimeter
5. Area enclosed by each internal perimeter
6. Position of the geometric center of the area enclosed by the outer perimeter
7. Geometric center of each of the areas enclosed by internal perimeters

This information is easily extracted during a single scanning of an image by employing a number of logical "bugs," which can be "attached" to any black/white or white/black edge of the image. Such bugs are always required in pairs, in order to trace both branches of a boundary, from the point of first encounter of a contrast boundary. Any one perimeter may involve a number of such "bug-pairs," though the complete trace will result in all pairs being strung together, at the points where they meet on the perimeter. The pairs of bugs from separate perimeters will, of course, remain separate with no danger of confusion. Following boundaries in this way, through the use of tracking bugs, is easily accomplished in a digital computer. The discrimination of outer perimeters from inner perimeters (around holes in the object) can be achieved, either by remembering which perimeter contains the first bug-pair to be generated (though this method is fallible under some circumstances) or, in each bug-pair, by determining whether the total angle turned through is a clockwise or counter-clockwise revolution. (It should always be 2π in magnitude.) For this calculation it is necessary to adopt an arbitrary convention concerning which way a bug is facing. Inspection will show that if a bug may move forward or backward while keeping black (metal) on, say, its right hand while traversing a boundary, then the total angle turned through (observing sign conventions according to direction of movement) will be $+2\pi$ for an outer boundary and -2π for an inner.

From each perimeter and enclosed area a *shape factor*

$$F = (\text{perimeter})^2/\text{area}$$

can be computed. This factor has a minimum value of 4π for a circle, is 16 for a square, and generally increases for more complicated shapes. The value is not unique to any particular shape, but it is position- and orientation-independent and can be used to detect changes in shape, provided such changes do not produce the same value of F. This is an unlikely condition in practice, when we may expect the breakage of chipping of tools to produce a piercing which is more complicated than the original and will produce a markedly higher value of F. The distance of the centers of holes, relative to the center of the enclosed area, in most cases provides an adequate method for checking the positions of piercings.

An experimental system of this sort has been constructed using a small minicomputer capable of labeling and identifying defective aspects of pieces of the sort shown in Fig. 10.9. Processing speed is somewhat dependent upon the complexity of pieces, but all parts shown were each isolated and processed in a fraction, typically a third, of a second. When using a closed-circuit television camera, whose linearity is typically of the order of 1%, precise measurement of objects, to an accuracy of better than about 2% of the scanned field dimensions is not possible. However, this linearity is adequate for the type of sorting and inspecting task outlined above. Greater accuracy, even to metrological limits, as described for the spindular object, could be obtained using solid-state arrays whose geometry is precise.

In use, an existential inspection and sorting system would be "taught" the objects of interest, by demonstration in a setting-up procedure. In this procedure, certified good examples of the objects of interest are presented to the system, which analyzes each object and stores their descriptions for subsequent use.

In operation, the description of test objects are obtained and compared with each stored reference object until a good match is obtained. The comparison is arranged to test the grosser properties first, beginning with the total object area and outer perimeter, proceeding to the number of piercings and their individual areas and perimeters, and finally the distance of the centers of the piercings from the center of the object. Owing to the limitations in linearity and quantizing effects of a TV camera, a degree of mismatch has to be accepted.

Such systems can store several object descriptions and can also be used for segregating mixed components. A practical system might require rather higher capability than is obtained by the simple measures outlined. This could be introduced readily through more detailed examination of the perimeters to detect the number of, and distance between, sharp discontinuities. We could also determine the maximum chord or obtain moments of the object. These properties are all highly shape-dependent, though it will be noticed that, while these methods of description are only sufficient to certify with a high probability that a given object is correct, they cannot be used to reconstruct an image of the object.

10.3.1.4. Moving Away from Black/White Images

So far, only situations in which objects can be reduced to black/white images have been considered and inspection is restricted to boundaries defined by the outer edge of an object and piercings through it. However, there are more general inspection tasks in which the "scene" cannot be reduced to a convenient binary image form, but which exhibit only a limited number of contrast levels. Instances arise in the inspection of assemblies of steel and brass, as in sub-assemblies of riveted parts in automotive instruments. Further examples occur when inspecting printed-circuit cards (copper on a fiber board). For many such applications we need not treat the images as though they contain a continuous range of optical density, since the various materials are often capable of providing a distinct color contrast, as well as, or in place of, a brightness contrast. Again we have a situation which can be simplified by the use of a combination of lighting tricks. Selection of the color used will emphasize any color contrast between dissimilar materials, or a combination of frontal and rear lighting may be used to make the most of the particular situation. Such arrangements, yielding well-defined contrasts on "uniform" material, allow the image to be reduced to a coded form of two or more bits per picture element. These may be separate "mappings" of the same scene or may simply correspond to the crossing of different contrast thresholds. Figure 10.10 shows the separate images obtained from a printed-circuit card. These arise from the component lead holes and the copper strips. (The reader is referred to Chapter 9 for further discussion of PCBs.)

10.3.1.5. Into Gray Scale

Eventually, movement into full-range, gray level image analysis will be inevitable, though it is not yet an area where cost-effective technology is available at the industrial level. There are some application areas (e.g., nuclear particle detectors, aerial and space photographs, biomedicine, and fingerprints) where high expenditure would be justified. Space imagery and nuclear physics experiments continuously produce such a large volume of material for evaluation that some degree of automation in analysis is essential. (This assumes, of course, that the effort and cost of generating the material were worthwhile in the first place.)

It is perhaps worth observing that the results of much of the work in the fields specifically mentioned in the above paragraph has led to systems which can more truthfully be called *man-aids*, rather than *man-replacements*. Practical systems tend to provide methods for automatic *image sifting*, that is, systems for discarding the definitely noninteresting images. This may be expressed in the converse manner, as locating images which are of potential interest to a human evaluator, leaving the trained operator to analyze these (few) interesting events. Image measurement systems for fingerprint analysis have been developed, which can isolate major features but which depend upon human operators to correct and pick out the additional minutiae needed for effective classification and retrieval. The automatic system, given a jointly labeled fingerprint, then auto-

Fig. 10.10. Printed circuit board viewed simultaneously with reflected and transmitted light. Notice the small white circular regions corresponding to the component-lead holes, drilled in the board, and that in some cases these do not coincide exactly with those in the copper conductor (top center and top left). (Photograph supplied by P. Saraga, Mullard Research Laboratories.)

matically measures and encodes the relative position of annotated points on the prints and inputs such information to the classification/retrieval processes.

These comments should not be construed as being critical. It is a repeated theme in many attempts to automate inspection processes that complete automation is very difficult, expensive, and prone to error and hence failure. More modest approaches, which depend upon the intelligent behavior of an attendant supervisor can produce highly effective systems, which reduce the dependence on trained manpower and make better use of that which remains. (Chapter 12 develops this point at length.)

From the social point of view, the use of men to aid machines need not, in this context, lead to men being the sensory slaves of machines, as portrayed in Charlie Chaplin's *Modern Times* and demonstrated by Huxley's "clones." This time the machine can be the slave, enabling its human supervisor to exploit his innate intelligence and understanding of the real world, while the "idiot machine" performs the routine chores. Air traffic control is a case in point. Here the machine presents uniform data to the controllers, logs and interprets their instructions, analyzes a very complex system of moving objects, and draws the operator's attention to potentially hazardous situations. This process leaves the resolution of such problems to the controllers' far superior understanding of aircraft, fuel reserves, and local stochastic variations, such as short-term constraints due to breakdowns, temporary hazards, and emergency conditions. This latter-day symbiotic relationship between men and machine can, if it is properly planned, produce a result which is much more than the sum of the abilities of the individual parts alone. It also elevates man to the role of a professional, intelligent controller and not as a replaceable skilless slave.

Returning again to the main point of this essay, namely, the potential for the application, in the short-term, of PR methods in volume production, there is a variety of tasks which cannot be approached without entering the gray-scale-image area. These areas are exemplified by inspection of materials, for example, sheet metal, fabric, clothes, paper, powders, timber, plywood, and plastic films. Such application areas demand quality inspection rather than checks for dimension or the existence of parts. The information rate necessary in quality inspection of sheet material is often very high (continuous rates in excess of 10^7 picture elements per second are met when inspecting certain materials e.g., metal strip and paper), but the rate of defects is relatively low.

Two distinctly different attitudes to the required capability of the inspection apparatus are apparent. It is often sufficient merely to detect a defect, while other applications may require that detected defects be classified according to types. For example, the quality of a roll of paper or cloth from the visual point of view is adequately graded by the number of specks exceeding a given size (or possible range of sizes) per unit area. Conversely, steel plate is graded, among other properties, in terms of the number of gouges, pits, scale inclusions, blisters, etc. These must be separately classified and discriminated from other unimportant surface marks, such as drying and oil stains and scratches.

Work to date in this area has concentrated upon the problems of scanning material surfaces at high speed. Detecting defects and estimating their extent has been approached, and commercial systems are available with a limited but usable performance from firms such as Ferranti and Boyle Gauges. These machines are finding application in production monitoring of special steels (particularly stainless), plain plastic films, and paper. So far, the more general application to general steel, patterned paper, and cloth has been frustrated through the lack of systems which can adequately discriminate between defects and acceptable blemishes. The ability to classify, as well as to size "events" on the surface, is essential in many applications.

Work on the classification of detects is in hand, using classical pattern recognition methods (Norton-Wayne *et al.*, 1976), with little explicit use being made of the geometric properties of the defects. The reason for this is that systems have not yet been developed which operate upon inspection areas more extensive than a single point. Through the use of (digital) image storage in the equipment individual defects can be isolated and retained as images for analysis of shape and other characteristics at rates not directly determined by the basic strip-speed. This, of course, assumes that the density of "events" on the surface is low. The availability of two-dimensional, parallel, image-processing techniques could be critical in this area (see Chapters 2, 6, and 11).

Consideration of the inspection of cloth leads us into the field of *texture* in images and the problems of searching for defects in the form of local distortions of the texture. This is an area where some theoretical consideration has been given (cf. Rosenfeld, 1975), but little has been achieved yet, outside the area of aerial photography. However, cloth in particular has some exploitable properties, arising directly from the cause of the texture. In particular, the warp and weft of cloth are fairly accurately aligned, and also have a distinctly periodic structure. Unfortunately, traditional techniques for exploiting periodicity, e.g., autocorrelation and frequency analysis, are "integrative" by nature and will tend to obscure small variations of a nonperiodic nature so there are no ready-made solutions here. However, phase-locked methods might be developed to exploit the periodicity and to detect discontinuities in the otherwise regular structure. [In Chapter 13 Thomas proposes "filtering" the power spectrum and cepstrum analysis in this type of situation—*Ed.*]

These comments may have relevance to materials with strong regular structure: tweeds, plaids, etc., in addition to plain woven cloth. Extension to pseudorandom (i.e., long-term repeating) patterns, wallpaper, carpets, etc. is beyond our present capability—even to speculate.

10.3.1.6. The Third Dimension

In the foregoing discussions we have specifically ignored the fact that many objects, particularly assemblies of components, have more than the two stable rest states, characteristic of what we have considered as flat(ish) objects in an earlier section.

The attack upon the *general* world of three-dimensional objects lies in the realm of robot vision projects as described in Chapters 7, 8, and 16. However, the study of general-purpose robots is interesting, rather more for its intellectual challenge than for its approach to solutions of particular real problems. The objective of most robot research workers is to develop information-processing structures, which approach the abilities of man to conceptualize (model) his world and, through inspection of and interaction with the model, to perform *loosely defined tasks* in the real world. This is the domain of artificial (or machine) intelligence (AI or MI). Our objectives are merely to take note of the MI movement, in order to realize solutions to problems of limited but useful generality and practical value.

Again, we are permitted to impose our own constraints upon the part of the real, three-dimensional world which we are trying to conquer. If objects can be presented for inspection in the correct orientation, but not necessarily correctly positioned, then adaptation of the silhouette methods described earlier could conceivably be useful.

In attempting to provide visual means for directing a manipulator, workers at Nottingham University, England, have demonstrated the combined use of visual and tactile methods for examining solid objects (Page *et al.*, 1976). A visual sensor views the object from above on an illuminated table, while a tactile instrument, containing a matrix of vertical probe wires, is lowered onto the object to determine its vertical surface profile. The matrix of pendant wires is lowered gently onto the object and the height at which individual wires touch the object can readily be interpreted as a two-dimensional height profile.

A totally different approach to the visual inspection of three-dimensional solids has been demonstrated by Will and Pennington (1971). Though this work is believed to have been abandoned for the moment, it is a prime example of our freedom to manipulate a situation to our advantage.

In their experiment, Will and Pennington viewed a solid object onto which was projected the shadow of a periodic structure, such as a line-grating or lattice. When viewed from any direction except along the light path of the projected pattern, the differing attitudes of surfaces of the object cause the projected pattern to appear to be modified. Variations occur in the line pitch line direction and "orthogonality," in a manner which can be interpreted in terms of the ranges and attitudes of the surfaces relative to the projector and the observing camera. Figure 10.11 shows an example of a scene containing rectangular blocks, illuminated with a hexagonal mesh (3 sets of lines disposed at 120° intervals).

It will be seen that each plane-surface exhibits a different periodic pattern. These can be detected by a variety of techniques, for example, Fourier analysis and autocorrelation (in two dimensions), which can detect particular periodic structure. In this way, the range and orientation of each surface are determined, but the positions of the edges have to be deduced from the intersection of the surfaces of the object. Regrettably the technique becomes very complex, if objects containing other than plane surfaces are considered. Projecting a regular structure onto a well-figured object helps in the visual assessment of its topography but poses severe problems for automatic analysis on a routine basis.

10.3.2. Continuous Patterns

When considering the possible improvement of product quality by inspection, we are inevitably delaying the detection of malfunction of the manufacturing process until sometime after, and perhaps at some distance from, the point of manufacture. This is inevitable, as visual inspection is not generally

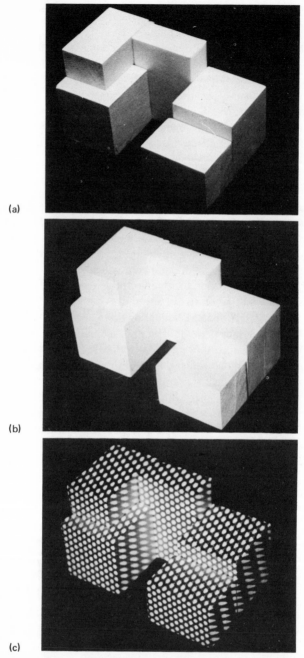

Fig. 10.11. (a) A "scene" carefully lit to emphasize the faces of the component cubes. (b) Frontal lighting giving low brightness differences between faces. (c) Frontal illumination with a mesh structure. Angular differences and ranges of separate faces are emphasized by apparent differences in the mesh pitch in the scene.

practicable within a machine tool, where cutting lubricants, swarf, and excessive vibration are often present. Postproduction inspection is therefore not ideal as a method of *process-monitoring*.

We therefore turn our attention to the possibility of monitoring the manufacturing process directly, rather than inspecting its product. This method is indirect, however. When a manufacturing process is seen to change, it is reasonable to assume that the part produced will also be subject to (unintentional) change and therefore is probably defective. The converse is not necessarily true; the product of a process in which no change has been detected is not necessarily similar to its companions. However, as a working principle, this should be adequate; and shortcomings could theoretically be made good by improving the monitoring of the process.

In continuous manufacturing processes, the monitoring is essentially the measurement of long-term variation in conditions, and is already well instrumented. However, discrete-product manufacture is rather different and falls into two main classes: (1) those processes which are violent and short-lived, such as: punching, broaching, forging, certain extrudings, and possibly spot welding; and (2) drawn-out processes such as turning, milling, and grinding. The approach to be suggested below is based on the premise that some indication of the energy absorbed in the metal-forming operation is obtainable and is indicative of the "trueness" of the process. This is not usually the case for the second class of production. The reasons will become apparent later.

In the "violent" processes, the forming of the product is completed quickly, through the absorption of a short-lived pulse of energy. Failure, due to any cause, to expend approximately the same amount of energy, at the same rate, will usually result in a poor product. Most violent processes accurately define the object dimensions, if completed correctly. This is obviously the case with a process such as metal-punching, when the energy expended in a punch stroke will change if a part breaks off the tooling. The energy will also change if the machine fails to feed metal properly, the feed metal runs out, waste material gets caught up in the tooling, or for a variety of other fault conditions.

Consider a process such as cold-forging screw-head forms from steel-wire stock. In simple terms the process has several stages:

1. Steel wire is fed through a clamping collet, to protrude into one-half of the head-form die.
2. The die closes violently, so that the protruding wire end flows to take up the form defined by the two halves of the die.
3. As the die opens, the wire is fed forward through the collet and is chopped off by a guillotine. This is done when the length required for the threaded screw (produced in a separate operation) is extended. The process then repeats, starting from stage 1.

The faults which can occur in this cycle include:

1. Incorrect amount of wire is fed forward.
2. Faulty wire, resulting in split heads.
3. Breakage or chipping of the dies (e.g., the "peg," which forms the cross-head recess is subjected to very severe loading and is therefore vulnerable).

The physical results of some of these faults are shown in Fig. 10.12. During the production of these faulty pieces, it is reasonable to anticipate that the energy used in their formation will be different in magnitude and distribution from that when the machine was working normally.

In order to monitor this energy waveform or signature, we have to attach some form of instrumentation to the machine. Ideally, we should like to measure the energy actually dissipated during the course of metal-forming. Unfortunately, this is difficult to measure directly. Energy input to the moving parts of the machine will be extensively moderated by the large inertial masses involved. Indeed, inertial energy storage is frequently employed to "smooth" the energy intake, over a relatively long period of time. Accordingly, on machines such as forges and punches, we turn to measurements of strain or vibration in, or close to, the tooling. This form of instrumentation is relatively cheap, particularly strain measurement, and does not require significant modification of machine tooling. Indeed, it can easily be attached to tool supports, mounting pillars, etc.

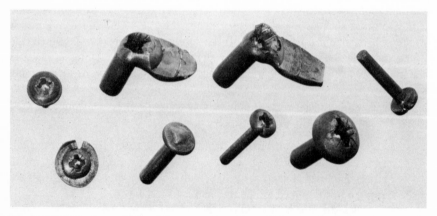

Fig. 10.12. Defective screw-head forms produced by cold forging. In the bottom row, the three leftmost items show the results of excessive metal feed and broken peg in tooling and defective stock material. (copyright, The Crown, United Kingdom.)

Signals obtained from an accelerometer, attached to a large multistage press tool, are shown in Fig. 10.13. This figure also shows the results of frequency analyses of the waveforms, from which it is obvious that marked changes occur under fault conditions. Continuous, or regular, monitoring of the signals therefore offers a means of tracking the behavior of certain machine tools. However, the calculation and comparison of spectra is likely to be too expensive in production situations and alternative, cheaper means of signal processing are needed; the time domain is probably more appropriate than frequency.

The attraction of this sort of technique is its generality. There is a large number of manufacturing processes which produce short-lived characteristic signatures. We have mentioned forging and punching, but there is an increasing

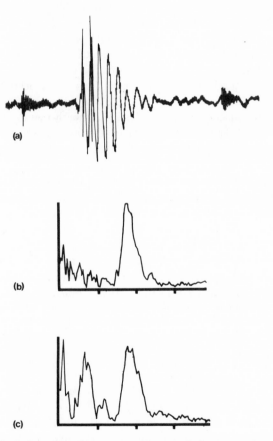

Fig. 10.13. (a) Typical accelerometer output when attached to a punch press. (b) Spectral analysis of the press in satisfactory condition. (c) Spectral analysis of press with defective tooling.

interest in new methods of high-rate metal-forming, involving large impulsive tool loadings. Monitoring of these processes is likely to provide useful, even essential, protection against machine and tooling damage under fault conditions. Other manufacturing processes are also thought to contain characteristic phenomena which could be monitored. For example: (1) metal die-casting, mass flow, or injection-pressure signature; (2) plastic molding, injection pressure signature; (3) spot welding, current-flow signature.

Little work has been done in this area yet, but clearly as signal-processing techniques become cheaper with the continued decline in the cost of electronic processing, and particularly microprocessors, this is an area offering very direct benefit in terms of production quality and efficiency.

It is anticipated that such systems would be used in medium-sized machine shops, in which there are a few tens of instrumentable machines. Each of these would have its own transducer, which need not be the same type for all machines. The transducer output voltages would be functions of time and have a bandwidth sufficient to enable the short-lived transients of the manufacturing process to be monitored. These signals would be routed to a central analyzing station which periodically examines the signal from each machine automatically. The analysis of these signals would make use of stored reference signals, obtained when the machine tool was initially (correctly) set up. It would remain the responsibility of the tool setter to certify that machines are correctly set and to personally monitor their performance while the central monitoring facility builds up a statistically significant sample of each process signature. These are obvious parallels with the method of training the visual inspection systems described earlier. Again, we avoid the problems of shop-floor personnel having to learn new skills to operate such systems. The ability of a machine to learn the range of variation expected of a single class is outlined in Chapter 4.

We can expect every process and each individual machine to produce its own characteristic signature. It is unlikely that two nominally identical machines doing the same job will produce identical signatures. Consequently, each machine will be individually represented in the monitoring system and it follows that one such system can simultaneously monitor a variety of machines.

10.4. APPLICATIONS

There is something of a paradox in approaching the general problems of automatic inspection. On the one hand, the declared and identified needs from users of such methods stem from the sheer difficulty of using human operators. Often the task is beyond their capability, for example, the close inspection of metal strip and paper, coated surfaces, and cloth. In these cases, the high speed at which the material is manufactured, prevents human inspection at the point of production and necessitates low-speed ("off-line") inspection. Alternatively the manufacture may "chance it" on the basis of sample inspection. Though this is a well-tried statistical technique, there is an increasing number of instances in which effective, 100% inspection of the product is necessary. These areas include some of the fundamental components in motor vehicles: braking system components, steering arms, etc. In such cases international or governmental regulations demand the highest level of integrity of the components achievable only by 100% inspection.

The paradox arises when we observe that those inspection jobs, for which we can foresee automatic devices, are as yet rather unsophisticated and can be done by human operators but *at a price* and with some doubt about consistency. Consistency and accuracy is the Achilles heel of human inspectors. The application of automatic assembly techniques requires product-quality levels of the order of 99.8% for cost-effective use of the assembly equipment. However, this is the familiar chicken-and-egg situation; automatic assembly cannot work well with the current quality levels (98–99%) and, because there is no large demand for better quality, there is no incentive to provide it. This argument is fortunately not quite circular. It can be broken, if the supplier of assembly equipment also sets about the problem of obtaining the necessary quality level.

The earlier sections of this essay attempt to indicate that there are some techniques of limited generality, which are beginning to emerge. It may be useful now to list other areas which need solution.

1. Plane-surface inspection: e.g., strip materials, steel, tin plate, paper, plastics
2. Textured plane-surface inspection: e.g., cloth, wallpaper, postage stamps, bank notes
3. Figured plane surface inspection: e.g., printed-circuit boards, integrated circuits
4. Component inspection/sorting: e.g., fasteners, instrument parts
5. "Integrity" inspection: e.g., quality of welds, completeness of assembly
6. Functional inspection: e.g., the correct positional relationship between the various lever-attachment points on car door locks (in whichever of several permitted states they are presented)
7. Shape conformity inspection: e.g., fruit and vegetable grading
8. Calibration inspection: e.g., correct indication by motor vehicle instrument assemblies
9. Impulse monitoring: e.g., stress waveform in tooling of violent metal forming processes

These are just a few of the more obvious areas in which automatic sensory systems, employing various degrees of pattern recognition capability could usefully be developed. Some applications will compete with more conventional gauging methods. Often, semiautomatic techniques already exist, while in other cases the only current method is human visual inspection. It is to be expected that PR techniques will become part of the increasing armory of methods available to the quality manager/engineer in the production industry. However, in order to do so, they will have to offer a cost-effective and convenient means of performing their intended tasks better than alternative methods.

10.5. CONSTRAINTS

It hardly needs to be said that a new technology, though welcome in principle, will be subject to aggressive examination, when it comes to judging the "value" of a device on a particular job. Since much of what PR-like techniques have to offer is direct replacement of human operatives, the cost constraints are fairly readily determined. It has to be recognized, however, that the value of replacing an operative in an existing factory plant is generally less than twice his wage rate, since no significant reduction in work-overheads is immediately apparent by a marginal reduction in manpower. A better saving, relative to manpower, can be expected in a new, rather than replacement, situation. Even here, the management will usually demand a return of investment within three years.

It is obvious that the serviceability of inspection equipment must be high, particularly if it is built into the production machines. This is critically important in introducing a new technology. In this case, high-speed digital technology will probably be used but will be unfamiliar to the normal factory maintenance staff. Dependence on service from the supplier is therefore inevitable and a high mean-time-between-failure will be a paramount factor in judging the cost-effectiveness of equipment.

The operation of inspection systems must be straightforward for conventional shop-floor staff. It is not reasonable to require extensive use of keyboards. Systems which are "self-training" will probably be most acceptable, as they require only familiar skills: selecting or making certified objects or conditions during the teaching period. In optical systems, some simple rules will be needed to ensure that the required resolution is obtained and maintained. In addition, adjustment to obtain a focused image will also be required, although all other functions will be preprogrammed into the machine's internal microprocessor. Any change in basic parameters from job to job would be achieved through the use of plug-in-memory modules (ROM) in the processor.

The benefits which could accrue through the improvement of manufacturing techniques are potentially enormous. According to British Institute of Management the "cost of quality" was, a few years ago, estimated to be £500M ($850M) per annum and accounts for between 4 and 20% of turnover, according to the industry.

As it is usually defined, "cost of quality" has three major components:

1. Prevention cost, design changes for quality, preventive measures, tero-technology
2. Appraisal cost, inspection of materials, product testing
3. Failure cost, scrap production, rework, rectification of failures in the field

Of these, the largest is the cost of failure, which accounts for more than half of the whole. This is really the cost of *nonquality*, since it is the consequence of inadequate control and inspection. Presumably, the overall manufacturing system is in some sort of economic balance, which finds this level of failure cost acceptable, having no low-cost methods available to reduce it. If the technology to reduce the basic defective rate can be introduced and is significantly less in cost than available methods, then the incidence of failure could be reduced, possibly dramatically. Existing methods, statistical quality control, visual grading, etc. have been pushed close to the limit (of cost-effectiveness) in highly competitive industry and are unlikely to result, in themselves, in a significant change in performance.

New approaches to quality assurance are necessary, principally to replace the sensory capability of human operators. Pattern recognition and allied techniques have much to offer here.

ACKNOWLEDGMENTS

The opinions expressed in this chapter are the responsibility of the author who gratefully acknowledges his Department's permission to publish them.

These opinions and ideas are, however, based upon innumerable meetings, drinkings, and dinings with production engineers and many engineers and scientists less directly linked to the size of the gross national product. A very direct debt to ex-colleagues in the Pattern Recognition group at the National Physical Laboratory, which it was my good fortune to lead for some years, is especially acknowledged.

REFERENCES

These references are by no means detailed and serve only to indicate areas of activity not otherwise covered in the companion essays in this volume.

Bell, D. A., 1974, Private communication.
Goldberg, E., 1927, Statistical Machine, US Pat. 1,839,389, April 1927.
Norton-Wayne, L., Hill, W. J., Finkelstein, L., and Popovici, V., 1976, Advanced signal processing in automatic inspection, *Proc. 8th IMEKO Cong.*, paper CTH/320.

Page, C. J., Pugh, A., and Heginbotham, W. B., 1976, New techniques for tactile imaging, *Radio Electron. Engr.* **146** (11): 519–526.

Rosenfeld, A., 1975, Picture processing, 1974, *Computing Surveys* **4**:133–155.

Tauschek, G., 1929, Reading Machine, US Pat. 2,026,329, May 1929.

Will, P. M., and Pennington, K. S., 1971. Grid coding a preprocessing technique for robot and machine vision, *Proc. 2nd Int. Joint Conf. Artificial Intelligence*, British Computer Society, London, pp. 66–70.

11

Editorial Introduction

This short chapter by L. A. Amos demonstrates how computer-based image analysis can assist in the interpretation of electron micrographs. Similar techniques have been proposed for the enhancement of industrial/medical X-rays and autoradiographs, satellite photographs of the earth, and thermographs (for locating tumors). In all these applications, the (initial) photographs are of indifferent quality, despite the great care and skill exercised in their preparation. In the cases of X-ray pictures and electron microscopy, there is an additional complication, due to the relatively poor focusing of the beam onto the subject. This causes the image brightness to be proportioned to the *integral* of the density across a fairly broad layer within the subject. Amos shows how this seemingly undesirable property can be put to good use in reconstructing a three-dimensional picture of the subject, from a series of two-dimensional images. Another important topic is the use of computer models to assist the experimenter to understand complex structure and how they can give rise to such electron microscope images. Computer modeling and simulation is an art that one would not normally include under the title of "Pattern Recognition." However, there is the possibility of iteratively modifying the model parameters, either fully automatically or with some human intervention, and such an adaptive facility would make most Pattern Recognition workers feel that this was within their sphere of interest.

The major contribution of this chapter is in the support it gives to the chapters by Ullmann (Chapter 2), Halé and Saraga (Chapters 7 and 9), and (to a lesser extent perhaps) Duff (Chapter 6). It typifies a whole new area of computerized image enhancement, which, if not strictly speaking under the aegis of "Pattern Recognition," certainly has a very significant contribution to our subject. Visual image analysis clearly has a most important role in the development of feature extraction systems for optical pattern recognition. It is only a short step from this chapter to that by Rutovitz *et al.* (Chapter 12) or even Parks (Chapter 10) on industrial inspection. In using Fourier transform methods,

Amos is venturing into areas which are often regarded as the domain of signal analysts (Thomas, Binnie *et al.*, Ainsworth and Green), although she is employing two- and three-dimensional transforms.

11

Image Analysis of
Macromolecular Structures*

L. A. Amos

11.1. INTRODUCTION

Straightforward electron microscopy has produced a wealth of information about subcellular biological structures. Up to a certain level of resolution, the interpretation of the images obtained is relatively straightforward. However, for a number of reasons which are discussed below, an objective interpretation of the fine details of biological structures requires relatively sophisticated techniques of image analysis.

Techniques for extracting reliable high-resolution information from electron micrographs have been developed over the past decade in this laboratory by a group led by Dr. A. Klug. The methods, which have much in common with the procedures followed by X-ray crystallographers, were originally developed for the purpose of investigating the structures of viruses and other molecular assemblies by electron microscopy, in conjunction with X-ray studies. The aim is to determine the arrangement of subunits within the subject and to recognize consistent features of the structure. Ways of assessing the relative reliability of images of different specimens and of averaging the informa-

*This chapter was originally published in *Journal of Microscopy*, Vol. 100, Pt. 2, March 1974, pp. 143–152. Permission to republish it, with minor revisions, was kindly supplied by Blackwell Scientific Publications Ltd., Oxford and the Royal Microscopical Society.

L. A. Amos   ·   M.R.C. Laboratory of Molecular Biology, Cambridge, England

tion from the better ones are important features of the methods. They are especially powerful when applied to periodically repeating structures (with either translational or rotational symmetry), but in principle can be extended to cope with any asymmetric particle. However, at present, because of the difficulty of data collection, their application is limited to the study of symmetrical macromolecular assemblies such as virus structures, large enzyme complexes, bacterial flagella, and subcellular microfilaments and microtubules. Some examples of such structures and the sort of information which can be obtained from their images are discussed below.

The contents of intact cells may be observed in thin sections of material which has been fixed, stained, and embedded in a supporting matrix. This treatment tends to limit the degree of resolution in the images to about 5 nm at best. Alternatively, the structures of interest may be extracted from the cells and spread directly on to a support film for observation in the electron microscope. The contrast in such specimens, whose intrinsic contrast is very low, is generally provided by negatively staining them, which means that the proteins are surrounded by an electron-dense stain such as uranyl acetate. Wherever the stain penetrates, it maps out the shape of the surfaces with which it comes in contact. Obviously the information obtained by this technique is limited by the extent to which the inner boundary of the stain conforms to the protein surface. There is, however, a further loss of resolution due to radiation damage. Under normal operating conditions, the effective resolution attainable has been found to be between 2 and 3 nm, which means that only the gross shape of individual protein molecules can be seen.

The first property of the image which may make its interpretation difficult by direct inspection, and require that more sophisticated methods of analysis be used, is that information from all levels of the structure appears superimposed. Unlike the light microscope which allows one to focus at different levels in turn, the electron microscope has a large depth of focus. In fact, the image obtained of a biological subject is generally interpreted as a projection of all the density within the structure upon a plane transverse to the electron beam. An experimental analysis of various images of thin catalase crystals (Erickson and Klug, 1971) has shown that this approximation is valid to the available resolution for thin negatively stained specimens such as those described here. The other main reason why image analysis techniques become necessary is that the signal-to-noise ratio is relatively low for the finer (and more interesting) details. In other words, individual molecules of 2–3 nm diameter cannot readily be observed in an unprocessed electron micrograph, because the specimen is imaged against a background of intensity variations from the supporting film, which is usually a film of carbon of 10–20 nm thick.

The purpose of image analysis in this case is to recognize features common to different images of supposedly identical objects. The results are most reliable when the individual objects are subunits of a symmetrical assembly in which the relative orientations of different members may be deduced from the symmetry of the assembly. Examples of such structures are two- or three-dimensional crystals, in which identical protein molecules are arranged on a regular plane lattice, or structures with helical or rotational symmetry. In such cases it is possible to filter out random variations in intensity while preserving the periodic ones.

Where the symmetry of a periodic structure is already known, the optical superposition method of Markham *et al.* (1963) can be used to average out random background noise. The method consists in superimposing a number of identical images either rotated relative to one another by an appropriate angle, in the case of a rotationally symmetrical structure, or translated by appropriate lattice vectors, in the case of a translationally periodic image. However, the method does not provide a reliable means of determining the symmetry of a particular structure. Such analyses have been attempted, particularly for rotationally symmetrical subjects, by comparing the images obtained by assuming different rotational symmetries. But the result depends upon a qualitative assessment of the final images and does not necessarily lead to the correct result.

In each of the procedures described below, two distinct steps are involved: the first is to discover and measure the strength of the regular periodicities and symmetry elements present in the subject; the second is to recombine the periodic information to reconstruct an improved two-dimensional or a three-dimensional image of the structure from those specimens whose symmetry has been shown to be well-preserved by the first step of the analysis. Normally only a few specimens from a large number of different micrographs pass the selection process.

11.2. TWO-DIMENSIONAL IMAGE FILTERING

Optical filtering provides a good illustration of the two stages. The first stage consists in obtaining the diffraction pattern of the image in an optical diffractometer, which reveals the regular periodicities present (see Figs. 11.1b and 11.5b). This technique was first applied to electron micrographs by Klug and Berger (1964). The second stage, resynthesis, is performed by selectively recombining the diffracted light rays. An opaque mask with appropriate holes cut in it is placed in the diffraction plane of the diffractometer in order to mask out all but the rays with periodicities which correspond to the symmetry determined in the analysis step (Klug and DeRosier, 1966).

Figure 11.1 shows an example of an image filtered in this way. In this case the subject, a "polyhead" of T_4 bacteriophage is a flattened tubular

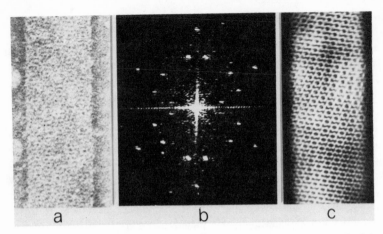

Fig. 11.1. (a) Electron micrograph of the tubular structures known as "polyheads," consisting of the major head protein of T_4 bacteriophage. ×160,000. (b) Optical diffraction pattern obtained from the image in (a). The outermost peaks correspond to spacings in the image of about 3.5 nm. (c) Filtered image of one side of the tube in (a). (From Yanagida *et al.*, 1972.)

assembly of protein molecules, so the image (11.1a) is of two superimposed layers. Because of this, and also because of the granularity of the carbon support film, one cannot see the individual molecules or their arrangement in the lattice. Figure 11.1b shows the diffraction pattern of the image in 11.1a. The intensity peaks lie on a pair of hexagonal lattices which are related by a vertical mirror line, one set of peaks arising by diffraction from periodic structures in the upper surface of the polyhead, the other from similar structures in the lower surface. Since the diffracted rays from the two sides are spatially separated, rays from one side may be filtered out along with the aperiodic noise and only the rays from the other side recombined. The result is a filtered image of a single layer, in which the hexagonal arrangement of the individual monomers can be clearly seen, as shown in Fig. 11.1c.

It is possible to perform equivalent operations computationally, that is, by digitizing the image and calculating its diffraction pattern, or Fourier transform, in a computer, which is useful in certain cases where the diffraction spots are close together or where the signal-to-noise ratio is particularly low (Amos and Klug, 1972).

11.3. ROTATIONAL IMAGE FILTERING

Optical diffraction cannot be used to analyze rotationally symmetrical subjects, since rotational components are not spatially separated in the diffraction pattern. Rotational symmetries are therefore most conveniently investigated in the computer, where the dominant symmetry can be determined in a quantitative way. The digitized image is analyzed in terms of cylindrical harmonic functions and the strength of each rotational component calculated to give a "rotational power spectrum" (Crowther and Amos, 1971a) analogous to the optical diffraction pattern of a translationally periodic image. If the power spectrum shows evidence for a single dominant symmetry, a rotationally filtered image is then synthesized using only the components consistent with that symmetry. It is possible to see from the power spectrum how much of the original image is included and how much discarded as noise.

The technique has been applied, for example, to images of the structure shown in Fig. 11.2, a small aggregate of the coat protein of tobacco mosaic virus, which consists of several rings of subunits stacked on top of one another. The problem was to determine the number of subunits in each ring. Slightly different power spectra were obtained depending upon the position defined as the center of symmetry. The curve drawn through the solid circles in Fig. 11.2b corresponds to the power spectrum obtained when the center was chosen to maximize the 17-fold component, while the triangles represent the spectrum obtained when either the 16- or 18-fold components were maximized (by chance

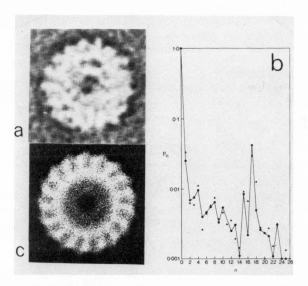

Fig. 11.2. (a) Electron microscope image of a short stack of "disks" of TMV protein seen end-on. ×1,000,000. (b) Computed power spectrum of the rotational components of the image in (a). The dots correspond to the spectrum obtained when the center was chosen to maximize the 17-fold component, while the triangles show the spectrum obtained when either the 16- or 18-fold components were maximized. (c) 17-fold rotationally filtered image of (a). (From Crowther and Amos, 1971a.)

the same position for the center maximized both components in this example). However, no matter which of the two points is used as the center of rotation, the strongest peak occurs at $n = 17$, indicating that there are 17 subunits in each ring (Crowther and Amos, 1971a). In this case the filtered image (Fig. 11.2c) does not show very interesting detail.

However, in the case of the structure shown in Fig. 11.3 the filtered image reveals the substructure of the rings which make up the protein coat of adenovirus (Crowther and Franklin, 1972). It is not clear from electron microscope images, such as the one shown in Fig. 11.3a, how many subunits there are in one of the small rings or "hexons," though the packing suggests the number is either three or six. The threefold nature of the whole group of nine is very well-preserved, as is shown by the strong peaks for the three-fold and higher harmonic components in the rotational "power spectrum" (Fig. 11.3b). Three-fold filtering of the image reduces the random speckling in the micrograph sufficiently to reveal the substructure of the hexons. The filtered image effectively includes three independent average images of hexons, each of which shows three distinct peaks, presumably corresponding to three subunits. Notice that the outer hexons appear to be less well preserved than the inner ones. This feature is noted in many images of closely packed arrays of particles, where the outer ones appear to support the inner ones.

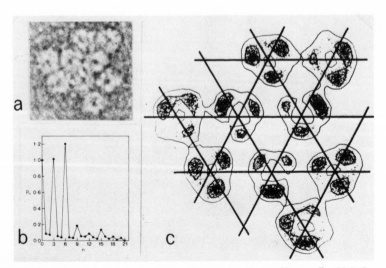

Fig. 11.3. (a) Electron microscope image of a group of nine "hexons" which forms the major part of one face of the icosahedrally symmetrical protein coat of adenovirus. ×550,000. (b) Rotational power spectrum for the image in (a). (c) A 3-fold filtered image of the structure in (a). ×1,380,000. (From Crowther and Franklin, 1972).

11.4. IMAGE SIMULATION OF THREE-DIMENSIONAL STRUCTURES

The structures of many subjects possessing helical or icosahedral symmetry have been worked out using simple model-building techniques (see e.g., Finch and Klug, 1966). The parameters of a model structure are adjusted until a set of simulated images is produced which agree in detail with different views in the electron microscope. The models obtained were also found to fit data from X-ray diffraction where these were available.

Figure 11.4 illustrates the results obtained by image simulation for bacterial flagella (Finch and Klug, 1972). The electron microscope images in Fig. 11.4a show typical examples of the three general types of appearance which

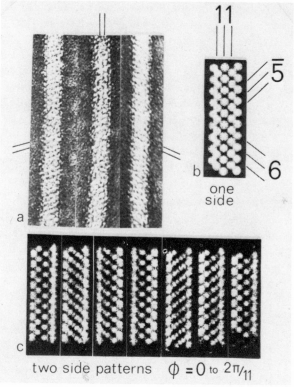

Fig. 11.4. (a) Electron microscope images of flagella from the bacterium *Salmonella typhimurium*. Sections of the central image appear longitudinally striated, while the two outer images have oblique striations, sloping in different directions in the two cases. (b) One side of a model helical lattice which can be used to explain the electron microscope images. (c) Different superposition patterns obtained from the two sides of the model helical lattice as it is rotated about its axis. (From Finch and Klug, 1972.)

are observed. The distinctive features are either longitudinal striations or obliquely transverse striations sloping in one direction or the other. A computer-drawn model structure which was found to agree well with the electron microscope images is shown in Fig. 11.4b, c. Figure 11.4b shows just one side of the model, which consists of spherical subunits arranged on a helical surface lattice. The complete helical lattice has 11 longitudinal rows of subunits, which are staggered laterally, so that the subunits also lie on two oblique families of helices, one with five members, the other with six members. The two oblique families are of equal pitch but opposite hand, so that, as the lattice is rotated about its axis, in certain orientations the near side of one family superimposes almost exactly on the far side of the other family and gives rise to an oblique striated appearance (Fig. 11.4c, second, third, fifth, and sixth images). In other views the longitudinal filaments on the two sides are superimposed, but the oblique helices are not (Fig. 11.4c outermost and central images). The model structure is capable, therefore, of approximately reproducing the three appearances seen in the micrographs. There is one obvious difference between the electron microscope and simulated images shown in Fig. 11.4; the transverse striations in the superposition patterns are at a steeper angle than those in the electron microscope images. From this we deduce that the bacterial flagella have flattened down on the grid and are no longer circular in cross section.

11.5. THREE-DIMENSIONAL IMAGE RECONSTRUCTION

However, to obtain a more objective interpretation of complex three-dimensional structures, it is desirable to use computer techniques of three-dimensional image reconstruction in which one combines a set of views from many different angles. There are many ways of solving the set of equations relating the three-dimensional density distribution with projections in various directions. The problem is simplified if it can be split up in some way into a series of smaller problems. One way is to use only views which are related by a common axis of rotation, so that the three-dimensional structure may be represented as a series of two-dimensional sections normal to the rotation axis. Each of the two-dimensional sections can then be determined separately from its corresponding one-dimensional projections, either by the simultaneous solution of a set of projection equations (Crowther *et al.*, 1970, 1972), or by an iterative procedure which modifies a grid of numbers representing the density distribution until it agrees with all the projection data (Gordon *et al.*, 1970; Gilbert, 1972).

However, it is the "Fourier" method, historically the first method to be proposed (DeRosier and Klug, 1968), which is found to be the most powerful in practice. The advantage of working in "Fourier transform space" is that the process of reconstruction can be separated into two steps just as in optical filtering. In the first step, transformation of the data into Fourier space allows one to analyze the symmetry components of each image independently. In the second step, reconstructing the image, Fourier space has the further advantage that any problem can easily be subdivided into a series of smaller problems, no matter which directions of view are included. This is because each view of the projected density, when two-dimensionally transformed, gives rise to a simple section passing through the center of the three-dimensional Fourier transform. Thus, the latter may be filled in progressively with planes from a number of views and a three-dimensional image synthesized by inverse transformation of the combined data. The number of different views required to reconstruct any three-dimensional object to a given resolution is the same for all methods but is particularly easy to determine using the Fourier method. Even if the views are unequally spaced it is possible to check that each region of transform space within a particular resolution limit is sufficiently filled with planes of data.

The required number of views is most easily obtained in practice by using subjects which have identical subunits arranged with known symmetry. In such cases each view of the whole structure includes many different views of the sub-units and correspondingly provides many symmetry-related planes in Fourier space. In particular, a single image of a helical structure provides a complete set of regularly spaced views of the asymmetric unit, and one image is sufficient for a three-dimensional reconstruction to a degree of resolution which depends on the helical parameters. In the case of the hemocyanin polymers shown lying side by side in Fig. 11.5a, individual hemocyanin particles are stacked end to end with relative rotations of 120°, so that the repeating unit along a polymer consists of three particles. This is deduced from the spacing of the layer lines in the diffraction pattern (Fig. 11.5b). Since each particle also has fivefold rotational symmetry, the image of each polymer contains 15 independent views of the asymmetric unit of the structure. This number is sufficient to reconstruct the particle to a resolution of about 5.5 nm, allowing us to determine the number and relative positions of the subunits which make up the particle. Figure 11.5c shows a solid model of the reconstructed density distribution, with the individual subunits (believed to be chemical dimers) outlined in black. A section through the three-dimensional map (Fig. 11.5d) shows that the particle is basically a hollow cylinder, capped at both ends.

Images of icosahedral particles, such as human wart virus (Fig. 11.6), provide a rather unequally spaced set of sections through Fourier space, so that at least two views are needed. In this example, six images were combined to ensure the reliability of the three-dimensional reconstructed image illustrated in Fig. 11.6a (Crowther and Amos, 1971b). The reconstructed density has been plotted section by section on to photographic plates which have been stacked together and is shown here with reversed contrast so that "protein" appears

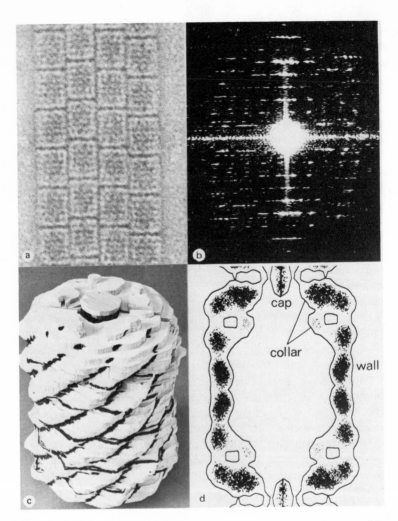

Fig. 11.5. (a) Electron micrograph of an array of hemocyanin polymers prepared from the blood of the snail *Helix pomatia*. ×220,000. (From Mellema and Klug, 1972.) (b) Optical diffraction pattern of the whole array in (a). (c) A three-dimensional model of a single hemocyanin particle reconstructed from one of the central polymers in (a). ×1,000,000. (d) A section through the reconstructed density distribution, which shows the hollow interior of the hemocyanin molecule.

white as in prints from the original micrographs. The morphological units represent rings of either five or six protein molecules, clustered around fivefold or local sixfold axes of symmetry. In Fig. 11.5b computed projections of this model for different angles of view are compared with individual electron microscope images of the virus. In each of the six pairs of images, the left-hand

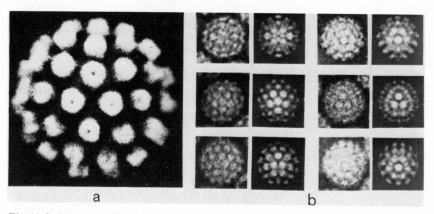

Fig. 11.6. (a) A three-dimensional reconstructed image of human wart virus (Crowther and Amos, 1971b). ×890,000. (b) A comparison of original electron microscope images of human wart virus with projections in different directions of the reconstructed density (the left-hand image in each pair being the electron microscope image, while the right-hand one is the corresponding computed projection). ×230,000.

member is from a micrograph, while the right-hand member shows the projection of the reconstruction. The selection of views shown covers the whole range of possible views, including those down 2-, 3- and 5-fold axes. The detailed agreement between members of each pair proves that the three-dimensional density distribution in the reconstruction is essentially correct.

In order to obtain sufficient information about a subject possessing no symmetry at all, one would obviously require a large number of different views. If these are to be obtained by tilting specimens in the electron microscope and imaging them from different angles, some improvements in technique are needed to reduce the damage caused by irradiating the specimen with the electron beam. Methods of minimizing the amount of irradiation for each image taken are at present under investigation in many laboratories and searches are being made for stains which will provide greater protection for the proteins they surround. It is possible, therefore, that in time it will be possible to see asymmetrical as well as symmetrical molecular structures in three dimensions using electron microscopy. Meanwhile, for symmetrical structures at least, quantitative analysis of subunit shape and size by image reconstruction is replacing the more qualitative approach of traditional electron microscopy.

REFERENCES

Amos, L. A., and Klug, A., 1972, Image filtering by computer, *Proc. 5th Eur. Cong. Electron Microscopy,* Manchester, pp. 580–581.

Crowther, R. A., and Amos, L. A., 1971a, Harmonic analysis of electron microscope images with rotational symmetry, *J. Mol. Biol.* **60**:123–130.

Crowther, R. A., and Amos, L. A., 1971b, Three-dimensional image reconstructions of some small spherical viruses, *Cold Spring Harbor Symp. Quant. Biol.* 36:489–494.

Crowther, R. A., and Franklin, R. M., 1972, The structure of the groups of nine hexons from adenovirus, *J. Mol. Biol.* 68:181–184.

Crowther, R. A., Amos, L. A., and Klug, A., 1972, Three-dimensional image reconstruction using functional expansions, *Proc. 5th Eur. Cong. Electron Microscopy*, Manchester, pp. 593–597.

Crowther, R. A., DeRosier, D. J., and Klug, A., 1970, The reconstruction of a three-dimensional structure from projections and its application to electron microscopy, *Proc. Roy. Soc. Lond.* A317:319–340.

DeRosier, D. J., and Klug, A., 1968, Reconstruction of three-dimensional structures from electron micrographs, *Nature* 217:130–134.

Erickson, H. P., and Klug, A., 1971, Measurement and compensation of defocusing and aberrations by Fourier processing of electron micrographs, *Phil. Trans. Roy. Soc. Lond.* B261:105–118.

Finch, J. T., and Klug, A., 1966, Arrangement of protein subunits and the distribution of nucleic acid in turnip yellow mosaic virus, *J. Mol. Biol.* 15:344–364.

Finch, J. T., and Klug, A., 1972, The helical surface lattice of bacterial flagella. In: *The Generation of Subcellular Structures*, First John Innes Symposium, Norwich (R. Markham and J. B. Bancroft, eds.), North-Holland Publ., Amsterdam, pp. 167–177.

Gilbert, P. F. C., 1972, Iterative methods for the three-dimensional reconstruction of an object from projections, *J. Theor. Biol.* 36:105–117.

Gordon, R., Bender, R., and Herman, G. T., 1970, Algebraic reconstruction techniques (ART) for three-dimensional electron microscopy and X-ray photography, *J. Theor. Biol.* 29:471–481.

Klug, A., and Berger, J. E., 1964, An optical method for the analysis of periodicities in electron micrographs, with some observations on the mechanism of negative staining, *J. Mol. Biol.* 10:565–569.

Klug, A., and DeRosier, D. J., 1966, Optical filtering of electron micrographs: Reconstruction of one-sided images, *Nature* 212:29–32.

Markham, R., Frey, S., and Hills, G. J., 1963, Methods for the enhancement of image detail and accentuation of structure in electron microscopy, *Virology* 20:88–102.

Mellema, J. E., and Klug, A., 1972, Quaternary structure of gastropod haemocyanin, *Nature* 239:146–150.

Yanagida, M., DeRosier, D. J., and Klug, A., 1972, The structure of tubular variants of the head of bacteriophage T4 (polyheads), *J. Mol. Biol.* 65:489–499.

12

Editorial Introduction

D. Rutovitz, D. K. Green, A. S. J. Farrow, and D. C. Mason describe the use of pattern recognition procedures in a cytogenetic laboratory. Their ultimate objective is to automate the inspection of chromosome preparations which, as they show, has very great potential benefits in genetic counseling, screening for genetic disease, etc. They do not attempt to solve their problem completely by machine, but realize that a human operator should be consulted to confirm doubtful decisions or arbitrate when there is equivocation between two rival hypotheses. This acceptance of the shortcomings of mechanized decision-making leads, almost certainly, to a more successful solution. (Their system is still being refined.) A similar philosophy, also based upon modest expectations, was very successful in the work of Halé and Saraga on the control of a drilling machine. It does not, in any way, belittle the contribution of either of these two chapters to acknowledge that they are simply based on good, sound, engineering judgment. The editor is convinced that unpretentious projects like these do much to establish a good name for this subject. Rutovitz and his colleagues also show that pattern recognition research should not primarily be concerned with questions like "Does procedure X converge in finite time?," but rather with questions relating the costs and speeds of proposed "solutions." Like Thomas (Chapter 13), Rutovitz *et al.* demonstrate that pattern recognition often requires advanced instrumentation methods. Like Bell (Chapter 5), Parks (Chapter 10), Brown and Popplestone (Chapter 8), and Halé and Saraga (Chapters 7 and 9), they show that "intelligent behavior" can be achieved by applying an appropriate sequence of simple operations. In this chapter, Rutovitz and his coauthors provide one of the most cogent arguments for continuing and even expanding pattern recognition research. What more worthwhile area for work could be envisaged?

12

Computer - Assisted Measurement in the Cytogenetic Laboratory

D. Rutovitz, D. K. Green, A. S. J. Farrow, and D. C. Mason

12.1. INTRODUCTION

Most living organisms are made up of cells in which there is a nucleus containing the genetic information for the duplication of the cell and of the organism, and a surround of varying dimensions and function known as the cytoplasm. This genetic blueprint is in the form of a code built into the structure of the thread-like bodies called chromosomes, which make up the major component of the nucleus. *Cytogenetics* is the study of the relation of nuclear, or more properly chromosome, constitution and variation to the transmission of information at cell division and in reproduction (Hamerton, 1971).

Control of inherited characteristics is arranged in independent groups within which there is a statistically linear linkage relationship—the closer together, the more likely characteristics are to be inherited together, the further apart, the less so. In the earlier years of this century it was shown that linkage

D. Rutovitz, D. K. Green, A. S. J. Farrow, and D. C. Mason • Medical Research Council, Clinical and Population Cytogenetics Unit, Western General Hospital, Edinburgh, Scotland

groups corresponded to physical structures which could be detected with the light microscope: the chromosomes. Figure 12.1 shows a stained human cell in metaphase, that is, the middle stage of cell division. The symmetrical halves constituting each chromosome are the chromatids; if division is allowed to proceed to its conclusion the chromatids separate at the joining region (the centromere) and migrate to opposite poles of the cell to form two daughter cells.

Human chromosomes occur naturally in 23 pairs in the female, 22 pairs and an "X" and "Y" chromosome in the male. These pairs can be classified by size and centromeric index [the ratio in which the centromere divides the chromosome (see Fig. 12.1)] into about 10 distinct groups. Recently introduced culture techniques (Paris Conference, 1972) can produce characteristic patterns of transverse bands ("G-bands") as shown in Fig. 12.2; these enable precise identification of each pair of chromosomes.

A number of important congenital abnormalities are caused by changes in chromosome constitution. Mongolism is the commonest of these and is due to the occurrence of an additional number 21 chromosome; next most frequent are variations in the sex chromosome complement which give rise to various disorders of growth and sexual development. These perturbations originate either at the time of fertilization of the egg or during the maturation of the germ cells. The final stage of the latter comprises a complicated process known as meiotic division which results in cells containing single rather than pairs of chromosomes of each type. An important part of this process is the random interchange of sections of genetic material between the chromatids of a pair (one from each parent) in the developing reproductive cells. During this phase of rearrangement and assembly, which is a powerful agency of genetic advance, all manner of mishaps can occur. A chromosome may combine with a nonmatching one resulting in a translocation, or parts of its short and long arms may be exchanged. This is illustrated in Fig. 12.1, where a new chromosome, in this case clearly recognizable as not belonging to any of the usual groupings, has been formed by inversion of long- and short-arm portions of a normal chromosome. Rearrangements also occur, however, during normal mitotic cell division, and most such abnormalities probably originate during the many replications of germ cells coming to maturity. If there has been no significant loss of genetic material, the resulting individual may be viable and unimpaired. Such individuals are highly prized by cytogeneticists, since the study of the inheritance of abnormal chromosomes is an important tool in mapping the gene-chromosome distribution. Unfortunately, much more commonly the result of such a happening is death of the embryo, or congenital abnormalities which may vary from slight to severe, but which almost always affect intelligence and fertility.

Life and environment interact with the process of cell division in the living organism to cause loss of chromosomes and breaks and rearrangements in somatic cells. Aging, chemicals, virus diseases, tumors, and especially ionizing radiation are among the more important agencies of such change.

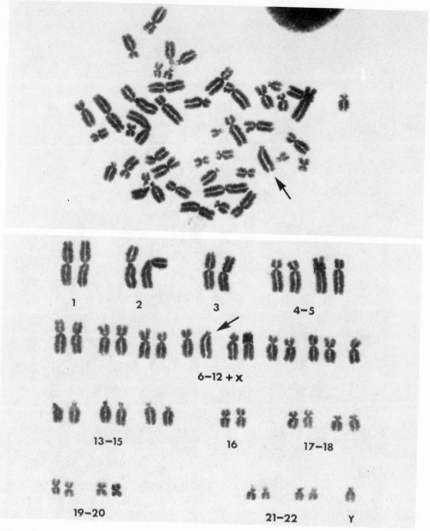

Fig. 12.1. A cultured human lymphocyte (white blood cell) in metaphase. The upper picture shows the cell as seen down the microscope, with, nearby, a portion of the nucleus of another cell which is not in division. The lower picture shows the 10 groups into which the chromosomes are arranged by size and arm-length ratio or centromeric index. When a cell is classified by visual inspection, the cytologist first counts the number of centric objects (objects with centromeres). For example, metacentric chromosomes, such as in group 16, are easily distinguished from the acrocentrics (centromere near the end) in groups 13–15. Chromosomes in the four smallest groups (13–15 downward) are identified, and the numbers in each group checked. The procedure is repeated for the larger chromosomes, groups 1–5. Finally the "C" group chromosomes, pairs 6–12 + X, and any structurally abnormal chromosomes are dealt with. The arrow shows a structurally abnormal chromosome formed probably by pericentric inversion, that is, by interchange of material between the short and long arms during cell division. The abnormal chromosome is detected in this case because its length and centromeric index differ from that of any normally found. Unusually shaped chromosomes can also be formed by translocation, that is, accidental transfer of material between arms of different chromosomes.

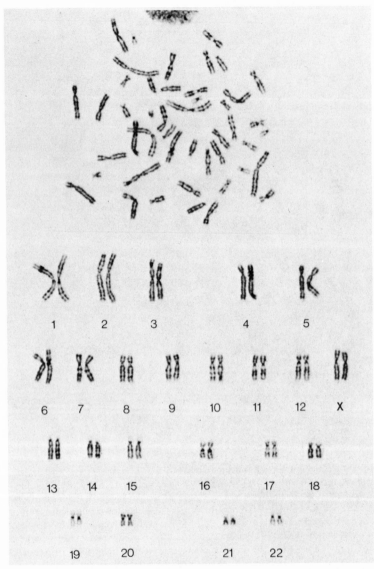

Fig. 12.2. A G-banded chromosome spread. The pattern of dark and lightly stained regions constitute a "signature" sufficient to identify each pair of chromosomes. Studying the banding patterns of translocations or other abnormal chromosomes often leads to precise location of the site of breakage and exchange.

12.2. THE LABORATORY WORKLOAD

One can distinguish three main areas of work. The first of these is called karyotyping: in many European laboratories this is done by direct visual inspection of the microscope image. A skilled technician can check to see if a cell has any abnormal chromosomes, and whether each of 10 normally recognized groups has its expected complement. This can be done in as little as 2 or 3 min, using a working protocol which has been developed and tested over a long period. The technician commences by identifying and counting the shorter chromosomes, distinguishing the acrocentrics, then metacentrics (Fig. 12.1), and so on. The procedure is repeated for the larger chromosomes, eventually eliminating all but the "C" group chromosomes, which are then counted. Sometimes a very rough matchstick-figure pencil sketch is drawn to aid the process, sometimes not. Another technique, very common in the United States, is for the cell to be photographed and a print prepared. The operator then cuts around the outline of each chromosome on the print and rearranges these physically to give a correct karyotype. In screening operations a surprisingly reliable job of detecting abnormalities can be done on the basis of as few as two cells per subject, although mosaic individuals, in whom a proportion of cells differ from the rest, may not be detected; these are fortunately rather rare. Banding analysis (Fig. 12.2) is rather more exacting and takes longer. Nevertheless, by comparing with standard banding charts, an operator can assess a cell in 5 to 6 min.

For some purposes, a quantitative rather than a qualitative karyotype is required; that is, one comprising the size (length, area, or integrated optical density) of each chromosome and its centromeric index, or the size of individual bands. Such information enables the detection of variation in the normal karyotype, as well as offering the possibility of finding less conspicuous abnormalities. It is particularly important in relation to banded preparations, as individuals often show considerable variation of band size; since inheritance of different size bands can sometimes be traced through a number of generations, one has in such bands a chromosome marker considerably more frequent in occurrence than structurally abnormal chromosomes.

In work on environmental damage, one must determine proportions of cells with additional, missing, aberrant, or broken chromosomes. These may be occasioned by, for example, radiation or virus attack, as illustrated in Fig. 12.3. One is usually concerned with low-dose-level effects: for example, nuclear powerplant workers or employees of a shipyard, working with lead and cadmium, show percentages of cells with breakage effects that are significantly different from controls in the same area. A typical finding would be an incidence of $\frac{1}{2}$% damaged cells, so that it would be necessary to inspect many hundreds or even thousands of cells from each individual. Clearly only a limited volume of such work can be carried out with conventional techniques of analysis.

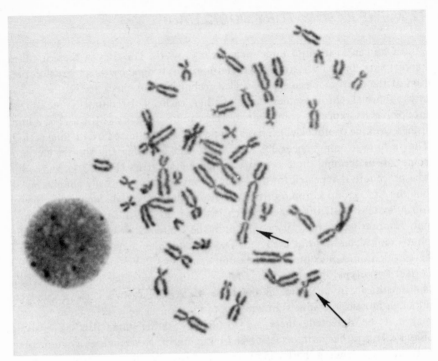

Fig. 12.3. Metaphase cell showing a dicentric chromosome (a chromosome with two centromeres) and a chromosome with a break in a chromatid. These are both examples of chromosome damage which can be induced during the cell cycle by a variety of agents (e.g., X-rays, chemicals, and viruses).

Before a cell can be analyzed, it has to be found. Preparations are normally made from peripheral blood samples which, when cultured and spread on a glass slide in a good laboratory, yield 10–100 cells/cm^2 suitable for analysis. Human operators are adept at locating suitable cells with a ×22 objective, then changing rapidly to a ×100 for the actual analysis. However, not all cultures grow well, and some material is notoriously difficult to handle, e.g., amniotic fluid samples. When good cells are sparse, locating them is a time-consuming task, and in certain cases, e.g., bone marrow from leukemic patients, it is a major part of the work.

There are also many simpler counting tasks to be considered: finding the mitotic index of a culture (the percentage of dividing cells among all cells), or the density of "film spotting" by radioactive tracer elements, and so forth.

12.3. AUTOMATIC KARYOTYPING

This comprises the classic pattern recognition sequence of digitization, field segmentation, feature extraction, and classification (Nobel Symposium, 1972; Ledley *et al.*, 1972; Mendelsohn *et al.*, 1966). Here, however, classification is a three-stage affair beginning with individual components, proceeding to entire cells, and then to a composite karyotype for an individual. This last stage consists of collating results from different cells. Our procedure is not fully automatic and relies on an operator for difficult or doubtful decisions.

12.3.1. Digitization

Many different systems are in use: we employ an image dissector tube. This is a high-resolution fully programmable nonintegrating image tube: in effect, a scanning photomultiplier. The signal-to-noise ratio depends on the electron output from the photocathode during the point measurement time. With a tungsten—halogen lamp (run at maximum continuous intensity and a spot size equivalent, in geometric optics, to a 0.125 μm diam disc on the slide) the average integration time per sample point is about 800 μsec for a signal-to-noise ratio of 64:1. To detect the geometry of chromosomes properly, it is necessary to scan at about 0.125 μm sample spacing. Since metaphase cell diameters range from 40–60 μm, we have to deal with up to 230,000 6-bit density values for each cell. Direct scanning would take about 3 min, and there is difficulty in storing so large a number of measurements. In consequence, to economize on both storage space and time we use a two-phase scanning procedure (Fig. 12.4), which commences with a coarse 0.5-μm grid giving about 14,000 points. As it proceeds, a table of "crossing frequencies" is generated (Butler *et al.*, 1962).

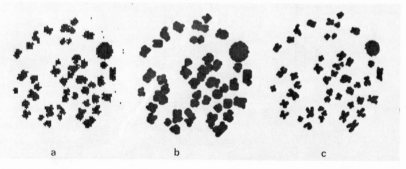

a   b   c

Fig. 12.4. (a) The regions obtained by segmenting a coarse $\frac{1}{2}$-μm raster scan of a metaphase spread. (b) Results of converting to a fine $\frac{1}{8}$-μm sampling grid and expanding isotropically by 5 pixels. (c) The chromosomes obtained by rescanning and resegmenting the fine objects, one by one.

This is a count, for each optical density (OD) value, of the number of transitions from below to above it. To reduce noise effects, we count upward transitions only when the minimum OD of the 4 points to the right of the current one is greater than the maximum of the 4 points to the left. To force the count to peak at about the optimum separation level for chromosomes, we include only those transitions which have been preceded in a 20-μm interval by a similar downward one (see Figs. 12.5 and 12.6). The OD giving the maximum crossing number is used as the threshold level for field segmentation (Butler *et al.*, 1962). This procedure was evolved after many disappointments with other approaches, the difficulty being that one must cope with preparations varying from well-spread, very lightly stained to dark, compact cells, often in the presence of extraneous nuclei which themselves may be dark or light and speckled.

The two-stage scan is essential to overcome the speed problems resulting from the use of a photodetector which does not integrate charge over a whole field scan. However, the technique also affords useful savings in store and processing time. Similar speed limitations apply to all except television systems, which are difficult to use because of the inherent mismatch between television scan speeds and computer-store cycle times. To facilitate the use of television, we have implemented a powerful set of field and resolution-limiting and data-compression facilities, with double-buffered rate adjustment for direct store input. This is presented diagramatically in Fig. 12.7, and we are presently evaluating its suitability for use in chromosome analysis (Farrow, 1974).

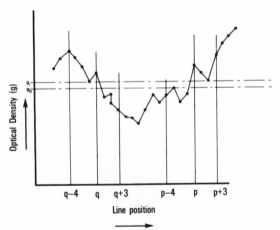

Fig. 12.5. Crossing number calculation. The minimum value g_1 of g at the four points p, $p + 1, p + 2, p + 3$ exceeds the maximum (g_0) of g at $p - 1, p - 2, p - 3, p - 4$. Since there has been a (smoothed) left-to-right upward transition through density values between g_0 and g_1, the crossing count associated with these values is increased by 1. A further restriction is sometimes made, namely, that there should have been a similar downward transition (as shown here at q) within a specified distance before p.

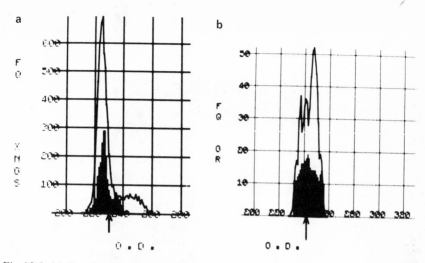

Fig. 12.6. (a) Crossing numbers associated with different OD levels in a metaphase spread. The bars show the number of smoothed upward transitions associated with each gray level in an 0.5-μm raster scan (at 4 times the actual scale). The polygonal line shows the frequency of each level. The large peak in the frequency graph corresponds to the background mean, the small peaks the chromosome means. These are not always clearly separate, as here. Crossing numbers, however, always peak strongly at a level between background and foreground. The threshold is taken at an offset from the peak value. (b) Mean gradient at different levels. The polygonal line shows the frequency $f(g)$ of OD level g in an 0.125-μm raster scan in the region of a single chromosome (Fig. 12.10). The bars show the mean gradient $\Sigma_{g=g_0}$ $(\delta g/\delta x)/f(g_0)$ associated with each level g_0. The peak value of this function gives a reasonable OD threshold.

12.3.2. Field Segmentation

A procedure for labeling, i.e., extracting the above-threshold connected regions of the picture is now invoked. By this time, the OD values for the coasely scanned field have been stored (partly in core, partly on disc). First a table of above-background "intervals" (line segments in the line-scan direction) is obtained (Rutovitz, 1968) (Fig. 12.8). This consists simply of the sequence of left- and right-hand interval endpoints, with a separate list giving the addresses of the beginnings of each new line. This table is now inspected for "adjacency" of intervals in one line with those in the following one. Each interval is assigned an index number, or pointer, which initially indicates its place-order in the table. When two intervals are found to be adjacent, the bases of their associated pointer chains are found by recursive look-up of index values, until a self-addressing one is encountered (Fig. 12.9). All elements of both pointer chains

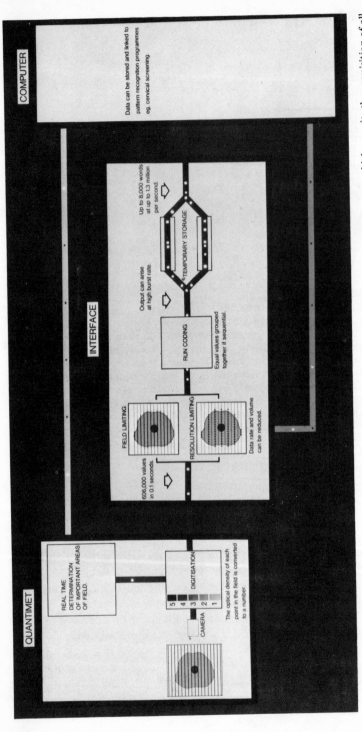

Fig. 12.7. The basic elements of an interface between a television scanner and a directly accessed computer store which permits the acquisition of all optical density data in fields of up to 880 × 688 sample points in a frame time of 90 ms. A field-limiting facility makes it possible to restrict the acquisition of data to an arbitrary interval in each line, that is, to regions of any shape which can be made up from single segments per line. The interface picks up the interval table from the computer memory as the scan proceeds. A resolution-limiting feature restricts the input to every nth spot and/or line, where $1 \leqslant n \leqslant 16$. Run coding: equal sequential optical density values are grouped into single 8-bit "vectors," whose components are the "run-length" (r-bits) and their common density value (d-bits); r and d are prescribable. Below-threshold points can be given 7-bit run-length whatever the (r, d) combination for other points. Two shift registers are used to smooth bursts of data acquisition activity. While one is being loaded with coded OD data, at speeds up to 4 MHz, the other is clocked into the computer memory at a rate controlled by the latter.

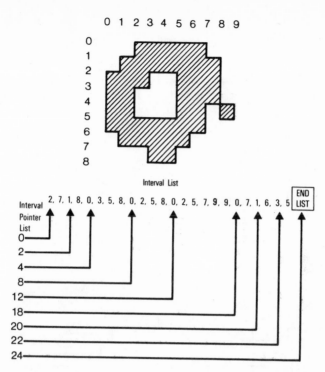

Fig. 12.8. A black–white pattern and its representation by a simple data structure.

are then reset to the minimum value of the two base elements. By this means all sequences of touching intervals, however complex their articulation, are eventually associated with that one of their members which occurs earliest in the table. This procedure is rapid, though somewhat expensive in store. However, since we now no longer require the table of density values, the space occupied by it can be reused for the interval lists.

On completion of the process, the like-numbered intervals constitute the maximal connected subregions of the picture. A data structure representing the domain (region occupied) of each such object is now generated: this consists of a short header list, a list of constituent intervals and a pointer list showing where each new line begins in the interval list (Fig. 12.8).

A procedure is then called, which converts from a 0.5 μm to an 0.125 μm coordinate system and expands by 0.67 μm in all directions. In effect we take unions of 5 pixel radius discs in the 0.125 μm coordinate plane; fortunately this involves no great computational burden, as it is done one interval at a time, using look-up tables for the disc coordinate calculation. The expansion is done for two reasons: (1) to make sure that no relevant material is omitted from the fine scan;

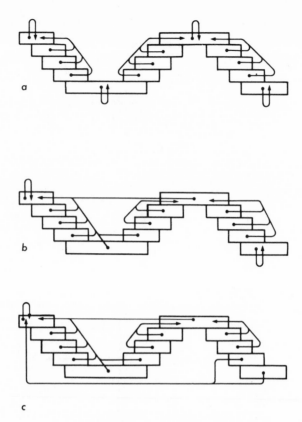

Fig. 12.9. Stages in the application of an algorithm for picking out connected parts of a picture set. "Intervals" (axial line segments) are initially furnished with self-addressing pointers. When, in the course of a top-down, left-to-right pass, two intervals are found to be adjacent, their pointers are followed recursively and reset to the minimum address encountered. (a) The state of the pointers before processing the bottom line. (b) After dealing with the left-hand interval in the bottom line, the hitherto distinct objects have merged. (c) All pointers encountered in chain-following have been reset, but not all chains are 1-long and this is still true after the bottom right-hand interval has been processed.

and (2) to provide a standard-sized background envelope for accurate threshold calculation (Fig. 12.10) (Mason *et al.*, 1975).

These objects are written to disc store as they are found and the main-store space is retrieved for their successors. At the end of the process, we shall have filed on the disc about 48 region descriptors (containing a total of about 20,000 points), identifying the areas of interest on the slide. These are now read back into core, one by one, rescanned and one of a number of different thresh-old calculations is carried out: for example, for each gray value g_0, we calculate the average rate of change

$$\sum_{g=g_0} \left| \frac{\delta g}{\delta j} \right| / \sum_{g=g_0} 1$$

[using $g(i, j + 1) - g(i, j - 1)$ as an estimator of $\delta g/\delta j$ at the point (i, j)]. In practice, we first find the

1. OD frequency distribution
2. OD frequency weighted by the gradient (in the line direction)
3. Quotient of the two preceding distributions

We have found that the smallest OD value which has half the maximum average rate of change (Fig. 12.6) yields a cutoff point somewhere in the base of the shoulders of the chromosome. This is taken as the new threshold value. Each component is segmented at its own new OD value (giving a reasonably effective local thresholding) and the new components are, once more, written to disc.

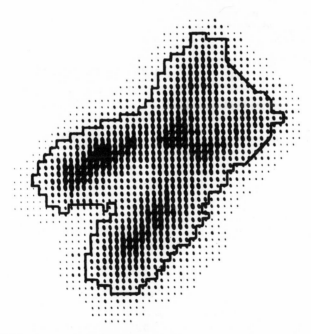

Fig. 12.10. The region occupied by a single chromosome has been found from a $\frac{1}{2}$-μm raster scan of a metaphase cell. It was then rescanned with an $\frac{1}{8}$-μm raster, each above-threshold picture point in the first scan having been replaced by a disc of radius 5 pixels. The expansion ensures the presence of some background points for local thresholding. The final chromosome outline is also shown; this is the contour for the OD value at which the average gradient is maximal.

This ends the initial field segmentation (Fig. 12.4). It is not always perfect, and indeed it cannot be as chromosomes may touch, overlap, or entwine (see Fig. 12.1). Nevertheless, most of the objects found will be single chromosomes: some will be composites and occasionally pieces of chromosomes, artificially split up by the process, will be found. One must also expect a small proportion of extraneous artifacts of one sort or another.

12.3.3. Modification of the Original Segmentation

The segmentation is now assessed by the machine. Decisions are based on simple likelihood rules, using statistics such as the area of the circumscribing rectangle, moment information, area:perimeter ratio, and so on (see also Section 12.3.4). If an object is classified as composite, a splitting procedure is called which erodes away the lower-density points from the boundary in successive layers. If, before a prescribed maximum number of layers have been removed, a newly exposed part of the edge meets another, from which it was originally separated by a significant distance, the labeling procedure is actuated. Points removed from the original composite are subsequently united with the nearest of the newly obtained picture segments. The latter are put through the entire parameter extraction and estimation procedure again and, providing neither appears to be a piece, the new segmentation is accepted. The splitting attempt may fail, due either to the layer-removal limit being exceeded or the derived objects being unsatisfactory. In this case, the object is shown to the operator (Fig. 12.11), who may be asked to decide whether it is a composite or not, or to position a splitting line.

If something is considered to be a piece (on the basis of statistical rule or operator decision), it is put into a holding list. If other pieces are subsequently found, this list is inspected by the program and, should two of its constituents be close enough together to justify their tentative combination, they will be joined and reassessed in the same way as the objects derived from composites.

12.3.4. Feature Extraction

Analysis of individual components begins with an investigation of the shape of their boundaries. To this end the convex hull (the minimum-size enclosing convex polygon) is found. This is done by a rapid procedure, requiring only a single pass through the list of line segments composing the object (Rutovitz, 1969). We look only at the left-hand end of the leftmost interval and the right-hand end of the rightmost interval in each line and apply, separately to

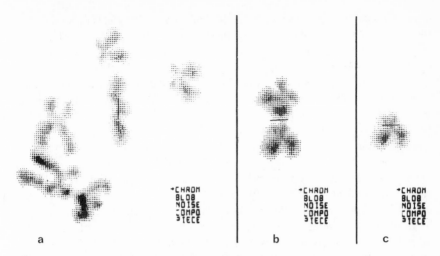

Fig. 12.11. (a) The machine has encountered a splayed acrocentric and has recognized it as a chromosome but could not find a centromere line. It has consequently referred it to the operator who has asked to see the surrounding objects as this helps in assessing doubtful cases. The operator must adjust the centromere line to what he considers an appropriate position. (b) A composite object consisting of touching metacentric and acrocentric chromosomes. The machine has wrongly identified it as a chromosome and found a centromere line. However, confidence is too low so the operator is called in. His responsibility is now to change the type class to "compo" by the use of a particular teletype key. (c) The machine has correctly located a centromere, but with low confidence. In this case the operator simply presses the space bar of the teletype to signal the machine to continue. Such confirmatory interactions can be done very speedily.

the left and right edges, the following algorithm:

> slope (0): $= \infty$
> > x_0: $=$ first line
> > y_0: $=$ right-hand end of first line
> > k: $= 0$
> > line: $=$ first line
> > *go to* 2
> 1: k: $= k - 1$
> 2: $(x_{(k+1)}, y_{(k+1)})$: $=$ (line, right-hand end of current line)
> > slope $(k + 1)$: $= (y_{k+1} - y_k)/(x_{k+1} - x_k)$
> > *if* slope $(k + 1) >$ slope (k), *go to* 1
> > k: $= k + 1$
> > line: $=$ line $+ 1$
> 3: *if* line $\neq$ last line $+ 1$, *go to* 2

This produces a vertex list $[(x_0, y_0), (x_1, y_1), \ldots, (x_m, y_m)]$ for the convex hull.

Once obtained, another pass through the line segment list yields information on the depth of the boundary of the original figure below each spanning chord (Fig. 12.12) of the convex surround, and the area delimited by the chord and the boundary. This constitutes a description of the general shape of the object, in terms of the depth and location of the principal concavities and the angular inclination and length of the spanning chords (Rutovitz, 1969). In addition, the area of the convex cover and the area of the object are found. We now make a first attempt at classifying the object as a single chromosome, composite or piece. If it can reasonably be regarded as a single chromosome, we also derive a possible location for the centromere line. This is a line through the centromere, dividing the chromosome into its long-arm and short-arm sections (Fig. 12.12). Here, and in all other routines for extraction of geometrical information, it is essential to remember that the procedures may fail and the results obtained may be incorrect. We therefore find, by experiment, the probability of success of a procedure as a function of any other information available at the time. For example, the likelihood of a successful prediction of the centromere line position (Ledley *et al.*, 1972; Gallus and Neurath, 1970) varies with the number,

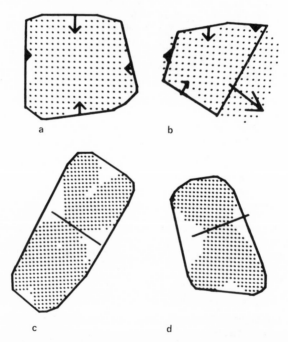

a   b

c   d

Fig. 12.12. (a) A chromosome and its convex hull. (b) The arrows show the extreme points in the principal concavities and residual convexities. (c) and (d) Chromosomes with centromere line computed from features of the convex hull.

depth and area of concavities, aspect ratio of the object, proportion of empty space in the convex hull, and so forth. Given sufficient confidence in a result, we accept it and proceed to the next stage. If confidence is too low, generally a backup procedure is called, the final backup being the human operator. More specifically in the case of the convexity calculation, if confidence is too low, we enter a further substantial processing phase. First a number of additional boundary parameters are found, then the OD values are accessed and first and second moments found. The principal axis of second-order moments (which may also be regarded as the line of least squares fit to the object treated as a density distribution) gives us a candidate for an axis of symmetry. A symmetry index of the parts of the object at either side of this line can be calculated using, for example, the expression

$$\frac{\sum |g(x_1) - g(x_2)|}{\sum g(x)}$$

where g is the OD value at a point x, and x_1 is the mirror image of x_2 with respect to the given line and x and x_1 range through the object (Hilditch and Rutovitz, 1969). In addition, we find the density profiles on either side of this line. The densities are summed in strips orthogonal to the axis of symmetry, and by finding the maxima and minima in the ordered succession of these density sums, we can estimate positions of the centromere, central parts of the arms, and so forth (Hilditch and Rutovitz, 1969; Hilditch, 1969) (see Fig. 12.13). We now have a wealth of measurements, which can be used for classification (once more dividing objects into chromosomes or other entities) for centromere location and for confidence estimates. If any of these parameters indicate strongly that an object is of one type (and none clearly indicates a different type), the indication is accepted. Otherwise, recourse is made to the operator for decision. Of course, multivariate analysis might yield a better decision strategy, but obtaining the data for it is a substantial undertaking, which we have not yet carried out.

12.4. CLASSIFICATION

The main classification problem is that of assigning chromosomes to groups, on the basis of their size and centromeric index. At present, we use area as a measure of size. With our system of mensuration, this has been found to be the most effective discriminant. Centromeric index is calculated as the ratio of the integrated optical density (IOD) of the short arms to that of the whole chromosome. (Variations in IOD which trouble us in relation to whole-chromosome size are less relevant for centromeric index, since we are taking the ratio of the IOD of two similarly stained and geometrically nearby parts of an object which will be subject to comparable errors.)

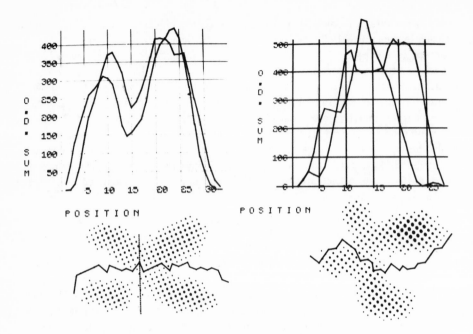

Fig. 12.13. Density profiles for two chromosomes. Optical density values are summed in strips perpendicular to the long axis of the chromosome. This results in a density profile from which the centromere position can be determined. Separate scores are kept for the two halves of the strips, resulting in separate profiles for the portions of chromosome on either side of the segment midpoint locus (wavy line). If the orientation has been correctly determined, the object should be symmetrical about this division line. The extent of the similarity of the two half-profiles is used as an acceptance criterion for the centromere determination.

12.4.1. Normalization

Absolute size has little significance in chromosome measurement, as this depends on the stage of the cell division cycle at which the cell was fixed. Length varies by a factor of 5:1, area less so, due to an apparent compensating broadening as the chromosome contracts. Consequently these measurements must be normalized in some way before being used as classification parameters. Conventionally, this is done by dividing the size of each chromosome by the adjusted total cell size. That is, the sum of the sizes of all chromosomes in the cell, adjusted to allow for any missing or additional ones. However, in the context of automatic analysis, one has to carry out the normalization, without prior knowledge of the completeness, normality, or otherwise of the cell. An initial attempt at normalization is made using unadjusted total size. Since extraneous matter, such as undivided nuclei, will have been rejected, the total will not be grossly different from the correct adjusted size. Even a cell containing an extra X chromosome, as in the case of an XXX female, differs by only 2% in total area from that of a normal cell. The chromosomes are now assigned to groups, on the basis of the provisional normalized size and centromeric index. At this stage, a more accurate normalization factor can be calculated. When a chromosome is assigned to a group, its size can be used, inversely, as an estimator of total cell size. Thus, if the average area of chromosomes in a group is G, in relation to a perfect cell of nominal total size C, assigning a chromosome of area A to this group makes AC/G an estimate of the adjusted cell size. By averaging these estimates of total size, for all the chromosomes assigned [in which we also use the probability of correct assignment (see below) as a weighting factor], we obtain a better estimate of adjusted cell size (Hilditch and Rutovitz, 1972). This procedure is iterated, and usually converges rapidly to within acceptable limits of variation.

12.4.2. Assignment of Chromosomes to Groups

How do we decide to which group a chromosome should be assigned? By examination of a large training sample, we have obtained a table of frequencies $f_G(i, j)$ for chromosomes of size i and centromeric index j belonging to group G. In order to overcome the problems of discrete sampling, we smooth this distribution slightly. The procedure is equivalent to replacing each occurrence of (i, j) in G by a normal distribution, centered at (i, j) with standard deviation equal to the root-mean-square measurement error (Paton, 1969). From this table, we calculate a maximum-likelihood matrix, which gives, for each position (i, j), the first, second, and third most likely group assignments, and their relative probabilities. This matrix is stored in a compact form and accessed by the program when making an assignment.

Initially, chromosomes are allocated to the most likely group. However, there are areas of ambiguity (see Fig. 12.14), which we believe a human operator resolves by making use of his prior knowledge of the expected karyotype. We have set up a procedure to do the same. This makes use of a figure-of-merit which is based on (1) the deviation from expected numbers in any group; (2) the confidence attached to the group allocations; and (3) the confidence attached to the measurements used in obtaining them.

This index is an increasing function of the "normality" of the karyotype. Allocating a chromosome to other than its most likely group will cause a term in the merit index to decrease, the more so if the measurement confidence and assignment probability are high. On the other hand, a chromosome with a low measurement confidence or assignment probability can be transferred to one of

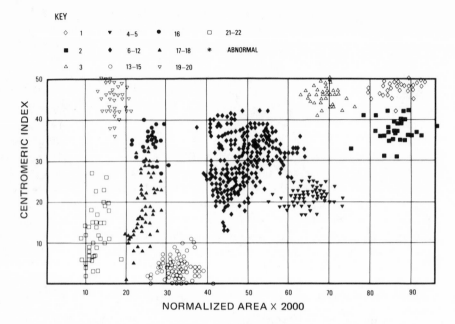

Fig. 12.14. Size vs. centromeric index table for chromosome classification. A table of 50 × 100 values gives sufficient resolution to classify most chromosomes correctly. Ambiguities can be dealt with by the method explained in the text.

its alternative group classifications, without significant decrement to this term of the index. If as a result of a series of intergroup transfers, the number of groups with too many or too few chromosomes is decreased, the merit index may show an overall improvement.

A specialized hill-climbing procedure is used to optimize the merit index by means of a series of trial intergroup transfers (Rutovitz, 1977).

12.4.3. The Composite Karyotype

A karyotype consists of a list of the numbers of chromosomes allocated to each group and of those unassignable to any group. We integrate the results from several different cells, first, by determining the mode in each group. If this is not the expected number, or if there is a bimodality, with a significant proportion of cells exhibiting different numbers of chromosomes in the same group, the individual is regarded as possibly abnormal. Single chromosomes, with apparently abnormal measurements, are ignored; but if a number of chromosomes from different cells cluster in the same part of the size/centromeric index plane, the program assumes that a genuine abnormality is present and reports it in the output.

12.4.4. Interaction

At various points in the karyotyping process, provision is made to seek the help of an operator (who requires training, but need not be a specialist cytogenetic technician). The way this is done has a substantial effect on the timing and economics of the whole operation. Some groups working in the field consider it necessary to ask an operator to check each step of the analysis and the final chromosome allocation, but we rely on our confidence estimates to avoid unnecessary human intervention. In most cases, the operator will only be called upon to confirm a correct, but borderline, machine decision: for example, that an object considered to be a chromosome, but with low confidence, really is one. The operator sees a display of the queried object and a list of the alternatives indicating the one preferred by the machine (Fig. 12.11). This list is an important detail, as it takes considerably longer to choose and specify an alternative than to confirm that what is shown is correct. The acceptance procedure consists of depressing a pedal or the space bar of a teletype; this is quickly done. Sometimes the operator needs to examine an object rather carefully before, perhaps, categorizing it differently from the machine's estimate or repositioning a centromere or splitting line. This takes longer but for the computer it is largely inactive time. Analysis and interaction therefore proceed asynchronously; if the machine is unable to continue with the analysis of an object until it has received the operator's response, it puts it aside in a queue of partially complete tasks and continues with the next one. There are awkward problems of organization here. Since store is always under pressure; one wants to avoid having to retain a displayed object in the mainstore along with one being analyzed. We are currently experimenting with a storage display for object portrayal, with only a small amount of material (for example a centromere line) shown actively, by a clock-driven independent program. It is not clear as yet whether there is always sufficient detail available in this form of presentation for reliable recognition. Different alphabetic characters are used to simulate varying density levels, and as shown in Fig. 12.11, in many cases this a very acceptable image reconstruction (see also Chapter 7). However, Fig. 12.11a indicates that this is not always so. In this case, it is the perception of the surrounding material, rather than the object itself, which helps the operator to make a decision. A system which, on operator demand, brings in and displays all surrounding objects (with consequent loss of store and disc time) might be satisfactory. Previously, we used a display directly linked to the scanning device, where reference to the original microscope image was possible. This undoubtedly makes interaction easier, but it lowers the efficiency of the duty cycle of the entire facility. In practice, it seems better to digitize asynchronously, allowing the front-end equipment to be optimally deployed, without being held up by the analysis phase. This implies, of course, that a series of images are written to and analyzed from the disc store, but good discs are cheaper than good scanners, and the approach does not cause much overall timing degradation.

12.5. COUNTING AND FINDING CELLS

We have so far considered the problems of analysis of a single microscope field. Since it takes even a skilled human operator minutes, rather than seconds, to assess the types of field discussed, there is at least the possibility that analysis by conventional computing equipment is time- and cost-competitive. However, when it comes to finding or counting easily recognized types of cells in relatively low-magnification fields (typically using a ×22 objective) the human eye is a formidable competitor; one either has to use an all-hardware system, or bring about a massive data compaction, before handing over to a conventional computer. As noted above, really fast scanning almost inevitably requires television equipment, but the speed mismatch must be overcome. A 625-line, 45 frame-per-second television system, resolving, say, 512 points per line, deals with a point in 70 ns. If we limit ourselves to, say, a 4-bit digitization and pack 4 density values to each 16-bit computer word, we are left with the problem of putting words into a typical 1 MHz store at rates of over 4 MHz. Existing solutions practiced include the rather elaborate scheme mentioned earlier: slowing down the camera; using interleaved independent stores; picking up, say, every 4th point on successive scans; and/or preprocessing the data. For purposes of metaphase cell selection and simple cell counting tasks, we have concentrated on the last of these possibilities, namely, prior processing and compaction of the data.

A point of importance in connection with television systems, often glossed over in manufacturers' literature, is that it is not possible to obtain accurate and high-resolution definition of different image fields, if these are presented to the camera on successive scans. The reason is that the charge pattern due to one image is not completely erased by a single read pass; it may typically take 3 or 4 frames before the camera settles down to the new charge pattern. Alternatively, a specially powerful erase pulse can be used to reduce the number of scans. Another point to be remembered is that, if a microscope stage is stepped from field to field and stopped for a television scan, it will be necessary to allow the vibrations of the stage to die down before accepting camera output. Thus, it is not uncommon to find that only every 10th frame can be effectively used for scanning different fields. We use Imanco's solution to this problem: namely, an arc lamp, synchronized with a Plumbicon camera, which strobes an image onto the photocathode while the stage is in continuous motion. This enables one to make use of alternate frames from a 44 frame-per-second Plumbicon camera to give an overall search rate of roughly 1 cm^2 of slide per minute. A good example of its use is given in the next section. Another possibility is the use of a photo-diode line array with a continuously moving stage.

Finding Metaphase Cells. A dividing cell may be characterized as an agglomeration, over an area of about 70 μm diam, of either (1) a discrete set of connected regions darker than the background, or (2) a set of above-background scan intervals of length less than 4 μm, separated from each other by not more than 20 μm. It is easy to detect suitable intervals and ends of connected two-dimensional objects (a delay line and comparator suffice), and the mean data rate of either phenomenon is not unmanageable (a maximum of about 10^5 intervals, or 10^4 connected regions, per square centimeter). On the other hand, the peak rate encountered when scanning within a metaphase cell is far too high for direct output to a computer. Both the data rate and the total quantity of information to be handled can be drastically reduced by notionally dividing each field into 16 X 12 squares of about 20 μm side, and accumulating counts of whichever feature is wanted, for each square. As a single scan line passes through only 16 such squares, 16 counters are enough. After each strip, amounting to 1/12 of the picture, has been scanned, the entire nonzero count pattern is transmitted to a PDP9 computer. From this information, a program is able to calculate the location of metaphase cells, and an index of quality, which takes into account the uniformity of count distribution in the cells symmetry and so forth. Occasionally, the scan will have to be held up to empty the computer storage buffers, although normally the machine is able to operate at the maximum scan speed (Green and Neurath, 1974; Green *et al.*, 1977).

12.6. ACCURATE MEASUREMENT

If accurate measurement, rather than rapid screening or counting is the goal, the justification for machine usage is the nature of the work rather than an economic advantage over conventional methodology. An operator's judgments of size, length, and density are useful if simple comparisons are wanted. However, if accurate quantitative or absolute measurement is required, instrumentation is essential. Even if the operator's speed of selection and identification of patterns is adequate, there is often considerable advantage in computer control of such instrumentation, both for subtlety of measurement and for ease of logging and retrieving data. A good example of this is offered by the problem of measuring the so-called C-bands of homologous chromosomes (Fig. 12.15). One of the recently developed techniques (Paris Conference, 1971) of preparing chromosome spreads results in a characteristic dark band, the C-band, in the centromere region of certain chromosomes. There is considerable variation in band size between different individuals, particularly on chromosome pairs 1, 9, and 16 (Fig. 12.15). These bands form useful markers for tracing the inheritance of chromosomes through successive generations, thereby offering new possibilities for genetic-mapping studies.

It is possible to group C-bands into size categories by eye, but this is an imprecise and unreliable procedure, since estimates have to be averaged over

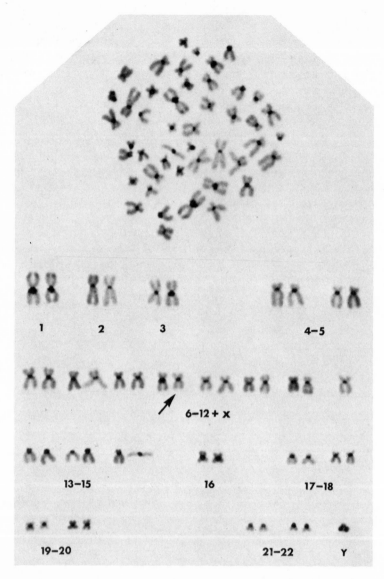

Fig. 12.15. The dark bands in the centromere region of these chromosomes are known as C-bands. In somatic cells, chromosomes are always found in pairs, one contributed by each parent. Although generally the members of a pair match each other exactly in size and shape (and have genes controlling the same characteristics at corresponding locations), there is sometimes a considerable size difference in the C-bands. An example occurs in pair 9, arrowed in the cell above. These differences in band size can enable the inheritance of a particular chromosome to be traced through several generations. Accurate measurement of C-bands is difficult, due to their small size and considerable stain-related variation.

many cells, due to variations in apparent band size from cell to cell. It is also quite difficult to measure the bands accurately with scanning equipment, because their dimensions (down to 0.5 μm X 0.5 μm) place them at the resolution limit of the optical microscope.

Another difficulty is that, due to the method of preparation, the heavy stain uptake in the band is matched by a light stain uptake in the rest of the chromosome. This makes the problem of segmentation more than usually difficult. We have found that, to obtain useful results, it is necessary not only to use special thresholding procedures, but also to optimize focus on the chromosome being measured, and to shift its image onto a known blemish-free portion of the photocathode (Mason *et al.*, 1975).

In this application, we rely entirely on an operator to select the chromosomes of interest and to assess and modify the action of the various procedures used. A suitable cell having been found, the operator positions a "steerable" ring on a displayed videoimage of the cell. Alternatively, the ring can be viewed down the microscope eyepiece: a half-silvered mirror is used to merge a CRT-generated picture with the microscope image. The area within the ring is then scanned and segmented at a threshold level corresponding to the mean gradient-weighted optical density (Section 12.3.3). The largest component within the field is selected by the program and shown to the operator (this eliminates any intrusive portions of other chromosomes). If the chromosome appears to be complete and detached from any other elements of the field, the operator signals the program to proceed to the next stage. Otherwise an operator-controlled trimming procedure is used (Fig. 12.16) to remove unwanted material.

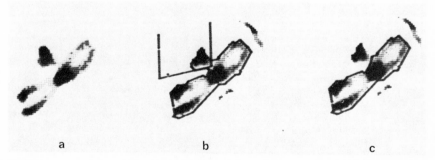

a   b   c

Fig. 12.16. A C-banded chromosome and its outline as shown to the operator. In this case a portion of another extraneous chromosome has intruded into the background envelope over which the scan is carried out (a). The operator uses console knobs to position the "trimmer" (b) and removes the unwanted material, as shown by the final chromosome boundary in (c).

The program now throws a border round the chromosome and carries out a number of scans over the expanded region at different focal planes; a final scan is then carried out at the plane of optimal focus which is found automatically. A special thresholding procedure is now used to try to define precisely the cutoff points for chromosome and band. Two optical density histograms are generated, one gradient weighted and one inverse-gradient weighted. The latter always displays two peaks, one corresponding to the background and one to the top of the chromosome. The chromosome threshold is chosen to be the OD value lying halfway between these two peaks (Section 12.3.2), a point in the steeply sloping density "shoulder" of the chromosome. The band cutoff level is taken as the median OD value of the section of the gradient-weighted distribution lying above the level of the top of the chromosome. The boundaries of the chromosome and band are now shown to the operator for approval. If necessary, either or both the objects found can be trimmed. Finally, the areas and integrated optical densities of band and chromosome are found.

In this case the measurements carried out are simple enough, but the thresholding and trimming procedure use the interactive capacity and flexibility of the computer-driven machine to good effect.

REFERENCES

Butler, J. W., Butler, M. K., and Stroud, A., 1962, Automatic classification of chromosomes. In: *Proc. Conf. Data Acquisition Processing Biol. Med.*, New York.

Farrow, A. S. J., 1974, TV scanner to computer in real time. In: *Oxford Conference on Computer Scanning*, Vol. 2 (P. G. Davey and B. M. Hawes), Nuclear Physics Laboratory, Oxford, pp. 407–422.

Gallus, G., and Neurath, P. W., 1970, Improved computer chromosome analysis incorporating pre-processing and boundary analysis, *Phys. Med. Biol.* 15:435–445.

Green, D. K., Bayley, R., and Rutovitz, D., 1977, A cytogeneticist's microscope, *Micro. Acta* 79:237–245.

Hamerton, John L., 1971, *Human Cytogenetics*, Vol. 1, *General Cytogenetics*; Vol. 2: *Clinical Cytogenetics*, Academic Press, London.

Hilditch, C. J., 1969, The principles of a software system for karyotype analysis. In: *Human Population Cytogenetics*, Edinburgh University Press, Edinburgh, p. 297.

Hilditch, C. J., and Rutovitz, D., 1969, Some algorithms for chromosome recognition. *Ann. NY Acad. Sci.* 157:339.

Hilditch, C. J., and Rutovitz, D., 1972, Normalization of chromosome measurements, *Computers Biol. Med.* 2:167–179.

Ledley, R. S., Lubs, H. A., and Ruddle, F. H., 1972, Introduction to automatic chromosome analysis, *Computers Biol. Med.* 2:107–128.

Mason, D., Lauder, I., Rutovitz, D., and Spowart, G., 1975, Measurement of C-bands in human chromosomes, *Computers Biol. Med.* 5:179–201.

Mendelsohn, M. L., Conway, T. J., Hungerford, D. A., Kolman, W. A., Perry, B. H., and Prewitt, J. M. S., 1966, Computer-oriented analysis of human chromosomes. I. Photometric estimation of DNA content, *Cytogenetics* 5:223–242.

Nobel Symposium 23, 1972, Sweden.

Paris Conference, 1971, Standardisation in human cytogenetics. Birth defects, *Original Article Ser.*, 8:7, 1972.

Paton, K., 1969, Automatic chromosome identification by the maximum-likelihood method, *Ann. Hum. Genet.*, 33:177.

Rutovitz, D., 1968, Data structures for operations on digital images. In: *Pattern Recognition*, Thompson Book, Washington, D.C.

Rutovitz, D., 1969, Centromere finding: Some shape descriptors for small chromosome outlines. In: *Machine Intelligence*, (B. Meltzer and D. Michie, eds.), Edinburgh University Press, Edinburgh, pp. 435–462.

Rutovitz, D., 1977, Chromosome classification and segmentation as exercises in knowing what to expect. *Machine Intelligence 8* (E. W. Elcock and D. Michie, eds.), Ellis Harwood, London, John Wiley, New York.

13

Editorial Introduction

In Chapter 13 D. W. Thomas discusses the problem of identifying vehicles from their acoustic signals. Despite some obvious military applications, this project is important for a number of reasons. For example, an automatic road-traffic census becomes a foreseeable possibility. Similar techniques could also be used in fault diagnosis of motor vehicles and domestic electrical equipment. Acoustic diagnosis techniques for making decisions about the condition of a motor bedded deep inside a larger machine have enormous potential benefits, since they are "noninvasive." Furthermore, expensive transducers are unnecessary and diagnostic methods could be developed for existing machines.

The identification of vehicles requires a hierarchical pattern recognition process. Since the choice of a suitable sampling interval is so important for correct recognition, the first step is concerned with the determination of the engine firing rate. From this, the appropriate sample length can be calculated. The necessity for this is evident when one realizes that the vehicle may be accelerating, cruising, ticking over, or braking, at any given instant. If the driver of the vehicle were cooperative, the method of analysis would be very different. Thomas emphasizes the general point very strongly: pattern recognition methods must always be selected to suit the problem they are intended to solve.

Another lesson which this chapter teaches is that advanced technology is often needed for pattern recognition machines. The aims of our subject may seem naive (e.g., light a LED when a Ford engine is heard), but the methods often require great technical sophistication (e.g., sample and store a 12-bit digital signal at a rate of 50 kHz, continuously for 5 s).

13

Vehicle Sounds and Recognition

D. W. Thomas

13.1. INTRODUCTION

The acoustic waveform that is of interest in the processing described in this chapter is obtained by recording the noise from a vehicle under normal operating conditions; it is a complex signal in which the characteristics of the engine, the exhaust system, the bodywork, and the environment are all intermingled. This is shown diagrammatically in Fig. 13.1 from which it can be seen that the task of classifying vehicle noises requires the selection from the total waveform of that part of the information relevant to the particular classification required.

Two tasks that are of interest are:

1. The "recognition" between different types of vehicles (e.g., rollers and wheeled, petrol and diesel, etc.).
2. The "identification" of a particular engine in a variety of environments, compared with other engines also in different bodyworks and environments.

Three words are used in this chapter in a defined manner and they will be delimited by quotation marks when they are so used. These words, then, take on the meaning given below:

Operation	Decision
"detection"	vehicle/no vehicle
"recognition"	light-wheeled/heavy-wheeled/tracked/rollers
"identification"	vehicle A/vehicle B

D. W. Thomas · Department of Electronics, University of Southampton, Southampton, England

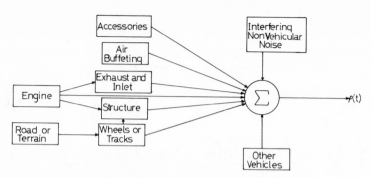

Fig. 13.1. Factors contributing to the acoustic signal $p(t)$.

It is prudent, before attempting to perform any pattern recognition task, to have some assurance that the task is in fact possible using the data given. This assurance could be obtained from human observers being able to perform the task, which should properly be assessed by controlled psychological experiments. In the absence of such evidence it was decided to accept subjective statements by people who work with industrial vehicles: at least the Ford 4D diesel engine can be identified when it is used as the prime mover in a number of different vehicles, even though some other engines cannot be so distinguished.

The analysis of acoustic signals is fraught with many conflicting and interdependent requirements. As a result, the order in which problems should be considered is in many cases a parallel one, and they have to be considered together. This chapter will discuss some of the more basic problems and how we view pattern recognition by considering feature extraction.

13.2. PLANNING AND PREPROCESSING

The importance of this stage cannot be overstated. Mistakes in this phase could, at worst, cause misleading results and therefore lead to false conclusions or, at best, would require the problem to be rerun. It is sometimes suggested that, for the solution of pattern recognition problems, it is not necessary to know anything about the data. This is a point of view which is not tenable in the present case and the following discussion should point out why this is not so.

This section is concerned with producing as accurate a digital representation of the original signal as possible and will do this by considering the system shown at the top of the next page.

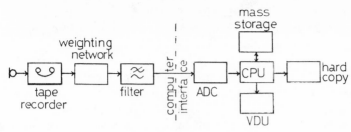

A Data Processing System

13.2.1. System Requirements

It can be seen from Fig. 13.1 that the acoustic signal $p(t)$ is a complex signal and that this complexity is the result of a number of systems that are likely to contain many resonant modes. This signal can have a large intensity range and could be present in the recording environment for a long time. For these reasons, it is necessary to consider the following three points with great care.

Dynamic Range. The recording and analysis system must be capable of handling the largest dynamic range needed. This may be restricted owing to limitations of the equipment. For most of the vehicle analysis to be considered, a dynamic range of 65 dB has been used. This corresponds to the difference in sound level between a suburban garden and a busy London street.

Frequency Range. The microphone, tape recorder, and any other equipment used must have a flat frequency response over the frequency range of interest. With some of the vehicle recordings, this has extended up to 10 kHz.

Length of Signal. The selection of a meaningful segment of signal is unfortunately complicated by a number of factors relating to the nonstationary nature of vehicle signals (Section 13.3.3), what information is required, and the inverse relationship between time and frequency. In one application, it may be necessary to store many minutes of data at a high sampling rate or to store a small number of samples taken over a relatively short time.

Each block of the processing system must then be constrained by the points given above. It would be as well, at this stage, to emphasize the necessity of keeping detailed, comprehensive, and complete records of recording sessions.

Recorder and Microphone. As we have already indicated, these must have adequate frequency and dynamic response, and the magnetic tape used should have low noise and be free from "dropouts" (which means keeping the tapes clean).

Antialiasing Filter. When a signal is to be *Fourier-analyzed* it is necessary that the signal contains no frequencies above the highest frequency of interest, f_H; otherwise, a confused spectrum will result. The antialiasing filter is set to remove unwanted frequencies. The filter should have a high rate of cutoff, possibly greater than 60 dB/octave. The minimum recommended sampling rate is $2.5 f_H$.

Analog to Digital Converter (ADC). The analog to digital converter should have an adequate number of quantization levels to cover the dynamic range of the input signals. As a guide, a 4-bit ADC (excluding the sign bit) has a 25 dB range and the range is increased by 6 dB/bit. Hence, an 11-bit ADC has a range of 67 dB. This system should be considered in conjunction with the discussion below on weighting functions.

ADC to Mass Storage Device. The mass storage unit must be able to receive and store data at a sufficiently high rate (assuming that the computer's main store is not large enough). The following calculation indicates the difficulties.

Let f_s be the sampling rate and f_H the highest frequency in the signal. Given that $f_H = 12$ kHz and the signal to be digitized is of 4-s duration, then

$$f_s = 2.5 f_H = 30{,}000 \text{ samples per second}$$

The volume of data to be transferred in real-time from the ADC to, say, a magnetic disc is then $4f_s$ (120,000 data words).

A 1.2M word disc would store 40 s of data at this rate, providing that the data transfers from the CPU to the disc are fast enough.

Visual Display Unit (VDU). The visual display unit is invaluable for program debugging and checking graphs before hard copies are needed. However, we consider it essential for monitoring the digitized signal *before* it is analyzed. Overloading of amplifiers and dropouts due to dirty tapes or other equipment malfunctions can cause havoc with the subsequent analysis.

Weighting Networks. This system is used basically to reduce domination of the spectrum by large-amplitude, low-frequency components. Compensation can be made for the weighting at a later stage in the analysis if necessary.

Prewhitening:

As the name implies the action of the filter is to "whiten" (flatten) the spectrum. The weighting network used extensively in vehicle sound measurements is the "A" weighting network (Hillquist, 1967) which has a high-pass characteristic, based upon the measured subjective response of the ear, and measurements made with this network are expressed in absolute decibels.

Weighting Functions and the ADC:

If the signal contains a large low-frequency component, then, since it is necessary to use the full dynamic range of the ADC, high-frequency information of low amplitude can be lost due to quantization effects. This low-frequency domination of the spectrum can be removed by use of an "A" weighting network.

Postdarkening:
> This means the removal of the effect of the prewhitening filter from the spectrum.

13.3. FEATURE EXTRACTION

Once the raw data (acoustic signals) have been preprocessed, digitized, and stored, they then form the data base upon which further transformations and measurements can be made. These transformations could involve sinusoidal basis functions, as with the Fourier transform, or nonsinusoidal, as with the Haar transform. The descriptive "features" of the signals are then derived in some way from the transformations or measurements of this data base and they reflect some identifying characteristic of the class to which a signal belongs.

The whole process makes use of the experimenter's prior knowledge of the signals and is iterative, in that modifications may be made to the process in the light of information gleaned from the analysis.

13.3.1. Dimensional Reduction

One of the objects of transforming the data is to reduce the dimensionality by selecting suitable features, so that the classification of the signals will be as simple as possible. The need for reducing the dimensionality can be appreciated if we consider the entropy of the feature-space in terms of Shannon's channel capacity, C:

$$C = B \cdot \log_2 |1 + (S/N)| \text{ bits/s}$$

where B = bandwidth of the signal, and S/N = signal to noise ratio.

At the first glance of a typical vehicle spectrum, it would seem that the spectrum is poorly structured compared to that of speech. If the vehicle signal being analyzed had a bandwidth of 6.6 kHz and a signal-to-noise ratio of 10 dB, the channel capacity is then of the order of 13,000 bits/s. The need to reduce the vehicle signal's dimensionality is put into perspective when one compares the human hearing capacity, which has been determined as 50 bits/s (Flanagan, 1965). This is assuming that we need to extract the same amount of information from the signal that the human apparently does.

Dimensional reduction is the operation of transforming the original signal, which contains a large amount of redundant information, to a new signal representation, containing a small amount of data with the information required.

13.3.2. Source Knowledge

As a result of computer limitations and the possible need to perform recognition in real-time there is a requirement to have a small amount of data with a high information content. Therefore, it would seem reasonable to argue that prior knowledge of the signal would be useful in detecting those features of the signal that achieve this requirement.

13.3.3. Vehicle Signals

The signals, or *signatures* as they are sometimes called, which are derived from vehicles, are *nonstationary*. By this we mean that their characteristics change with time. As a result of this nonstationariness, it is necessary to segment the signal in some way, and this means that the normal, continuous Fourier transform has to be modified by multiplying the signal by a "window" which then creates a modified transform. This procedure will now be described.

The double-sided power spectrum is defined as

$$F(f) = \lim_{T \to \infty} \frac{1}{T} \left| \int_{-\infty}^{\infty} p(t) \exp(-j2\pi ft) dt \right|^2 \qquad -\infty \leqslant f \leqslant \infty \qquad (13.1)$$

where $p(t)$ is the vehicle signal. Notice that the range of integration extends over all time.

This definition is modified for the short-term power-spectrum; the window multiplying the signal then has the effect of causing a convolution in the frequency domain of the window transform and the spectrum of $p(t)$.

The single-sided, short-term power spectrum is defined as

$$P(f) = \frac{2}{T} \left| \int_{0}^{T} \omega(t) p(t) \exp(-j2\pi ft) dt \right|^2 \qquad f \geqslant 0 \qquad (13.2)$$

where $\omega(t)$ is a symmetrical weighting function which tapers to zero at each end, so as to minimize the effects due to a noninteger number of "periods" of the signal.

It will be assumed for the rest of this chapter that it is $P(f)$ that is meant when discussing the power spectrum of signals and $\omega(t)$ is the Hanning window, defined as

$$\omega(t) = \begin{cases} \frac{1}{2} [1 - \cos(2\pi t/T)] & 0 \leqslant t < T \\ 0 & \text{elsewhere} \end{cases} \qquad (13.3)$$

13.3.4. Structural Systems

Referring to Fig. 13.1, we can see that the acoustic signal generated by a vehicle owes its form to a number of contributing physical systems. For the most part, these systems are excited directly or indirectly by the explosion and expulsion of gas from the cylinders. If the transfer functions of these systems are known it should be possible to generate invariant features characterizing a particular vehicle. Even if the transfer functions are not known in detail, as in general they will not be, it may still be possible to generate useful features over the operating range of the vehicle by using *a priori* knowledge.

We have said that the signal is nonstationary; as a result we must choose the length of signal in such a manner that, within the sampled time segment, the characteristics of the signal are relatively constant. How do we do this? We argue that over a small number of crankshaft rotations it is unlikely that the vehicle's characteristics will have changed markedly. Hence, we need to be able to select the signal segment in relation to the firing period of the engine.

13.3.5. Measurement and Selection of the Firing/Crankshaft Period

Periodicities in the signal are uncovered by making use of a technique termed *power cepstrum analysis* (Noll, 1967). The power cepstrum function is defined as the power spectrum of the logarithm of the power spectrum.

$$PC(\tau) = |\, F^{-1} \log P(f)|^2 \qquad (13.4)$$

where $P(f)$ is the short-term power spectrum and F^{-1} is the inverse Fourier transform.

The importance of $PC(\tau)$ lies in the ability of the logarithm to separate transfer functions that are multiplied together in the frequency domain. If we consider a simplified case of Fig. 13.1,

$$\begin{array}{cc} n(t)\ x(t) & p(t) = \int_{-\infty}^{t} n(\tau)x(t-\tau)\,d\tau = n(t)*x(t) \\ \boxed{\text{ENGINE}} \!\!\rightarrow\!\! \boxed{\text{EXHAUST}} \!\!\rightarrow & \\ N(f)\ X(f) & F\{p(t)\} = N(f)X(f) \end{array}$$

where $N(f)$ is the engine-source spectrum of $n(t)$, $X(f)$ is the spectrum of the impulse response, $x(t)$, of the exhaust system, and $P(f)$ is the short-term power spectrum of the acoustic signal $p(t)$, then

$$P(f) = |N(f)|^2\, |X(f)|^2$$

Taking the logarithm of this function gives

$$\log P(f) = \log |N(f)|^2 + \log |X(f)|^2$$

The effects of the two systems can now be separated, provided they have sufficiently different characteristics.

Figure 13.4 demonstrates the effect of separating the transfer functions. The ripples in the log-power spectrum, which in future will be referred to as the *spectrum*, are a function of frequency, not time, and are called *quefrency* components. The quefrency of a ripple is measured as the inverse of its "frequency period" and hence has the dimensions of time.

The spectrum of Fig. 13.4a is then seen to contain high quefrency components, "short wavelength" ripples, corresponding to the transfer functions of the gas pulses from the engine, superimposed on low quefrency components, "long wavelength" ripples, which are due to the resonant modes of the exhaust system.

Figure 13.2 shows the power cepstrum obtained for a vehicle (a moving Fordson tractor) with a clear peak at a quefrency of 95 ms, which is the point corresponding to the firing period, and with additional peaks showing *rahmonics* which indicate further periodicities in the signal. By comparison, the power cepstrum shown in Fig. 13.3 was obtained from rain falling through trees, and as we would hope, no indication of periodicities is obtained. This suggests that the cepstrum can provide an effective method of detecting the presence of a vehicle in noise ("detection") based simply on the principle of finding a peak in the cepstrum. Provided that the position of the peak on the time (quefrency) axis can be determined, low-frequency periodicities present in a vehicle signal can also be obtained. It is generally easy for a human subject to locate a peak or to decide that no peak is present. Unfortunately, as with so many visual discriminations, there are difficulties when attempting to automate the task. The main

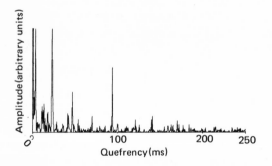

Fig. 13.2. Power cepstrum: moving Fordsen tractor. The peak at 95 ms indicates a firing frequency of 10.5 Hz.

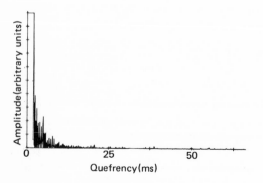

Fig. 13.3. Power cepstrum: heavy rain falling through trees.

problem is that, when identifying a peak, we have to ensure that we are, in fact, looking at the major amplitude variations (the "obvious peaks") and not at fortuitous variations in the noise level. This requires some kind of thresholding operation to be performed, and this in turn requires the threshold to be set at a level that is based on an incomplete knowledge of the expected range of vehicle signals. Even having removed the noise and isolated the peaks, the problem is still not solved, because there remains the decision as to which of the peaks corresponds to the firing or crankshaft period. The rules developed, both for the choice of threshold level and for the identification of the relevant peak, tend to be pragmatic rather than scientific, and an entirely satisfactory system has yet to be derived.

A major difficulty with thresholding is that, although the spectrum is normalized, the numerical values of the cepstral peaks can vary for different sound segments over a very wide range. The implications of this problem will be discussed later. This effect cannot be removed by a normalization of the power cepstrum because, as can be seen in Fig. 13.3, the area under the curve is small when examining a segment with no vehicle present, so that normalization would simply have the effect of amplifying the very variations that we are trying to eliminate.

A further point that has to be observed in interpreting the power cepstrum is that we may wish to conclude that no vehicle is present, on the basis of the absence of peaks in the "working part" of the power cepstrum. It is clear that large values of the cepstral function (Fig. 13.4) always appear at low quefrency values whether a vehicle is present or not (see Figs. 13.2 and 13.3). These quefrency peaks would correspond to relatively high frequencies, and they can therefore be discounted as indicating impossibly high firing rates. This is yet another thresholding operation, bringing with it the usual danger of placing the threshold, which is vehicle-dependent, in the wrong place and so losing information.

In a practical situation, we are likely to be faced with a signal consisting of an engine sound contaminated with environmental noise, such as rain, birdcalls, aircraft, nearby road sounds, etc. It is therefore important to establish the limits beyond which detection is no longer possible. For our tests, signals were constructed by mixing engine sounds and white noise in controlled amounts. The engine was mounted on a test-bed in a soundproof enclosure, with the exhaust pipe leading through the wall to the outside of the building, and the recording was made close to the exhaust outlet. Figure 13.4a,b shows the spectrum and power cepstrum obtained from this engine signal. The noise was obtained from a white-noise generator, giving the spectrum and power cepstrum shown in Fig. 13.4c, d. Composite signals were produced with power ratios (engine signal to noise) varying from 0 dBA to −20 dBA. The results of the power cepstrum analysis are shown in Fig. 13.5, where the amplitude scale,

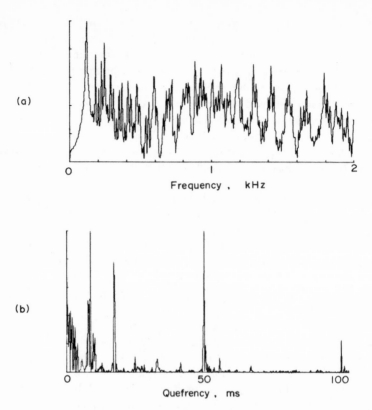

(a)

O   l   2

Frequency , kHz

(b)

O   50   100

Quefrency , ms

Fig. 13.4. Engine to noise tests. (a),(b) Spectrum and power cepstrum from the engine sound. (c),(d) Spectrum and power cepstrum from the white noise source. Amplitude in arbitrary units.

which is in arbitrary units, is changed between Fig. 13.5d, e by a factor of about 10. This shows that the cepstrum technique gives a visually detectable peak, even when the engine sound is −12 dBA relative to the noise. It should be noted that spectral averaging, which is a technique commonly employed for detecting signals in noise, is difficult to implement in the vehicle recognition problem as the engine may not be in a steady state. Furthermore, the analysis must be carried out over a relatively short time interval.

The change of amplitude scale in Fig. 13.5, while showing that a discernible peak is still present, also emphasizes once more the difficulties of thresholding. If the threshold is set to a constant fraction of the maximum, which is generally satisfactory when dealing with a straightforward vehicle sound, it will also detect peaks in a noise sound. To distinguish this case, we observe that the number of peaks (see, for example, Fig. 13.5e) will be large when there is no engine present,

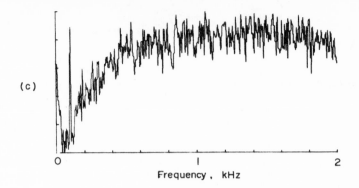

(c)

Frequency , kHz

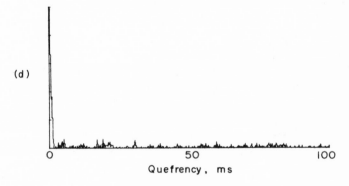

(d)

Quefrency , ms

Fig. 13.4 (*continued*).

or when the engine sound is deeply embedded in noise. This can be used as the basis for rejecting a signal that does not give a usable indication of periodicity.

Signal Segmentation. Using techniques for "peak picking" with a frequency threshold above which crankshaft rotation speeds cannot exceed, a peak is found in the power cepstrum that gives the crankshaft rotation rate.

The window width used to segment the signal is then selected by reference to this crankshaft rotation rate. The width is selected to be some multiple of this rotation rate such that a reasonable peak may be generated. The whole process of analysis is repeated for the next time segment so that we have an adaptive window, adjusting itself to variations in the signal's fundamental periodicity.

The stage has been reached where it is now possible to consider looking for features of the signals being analyzed.

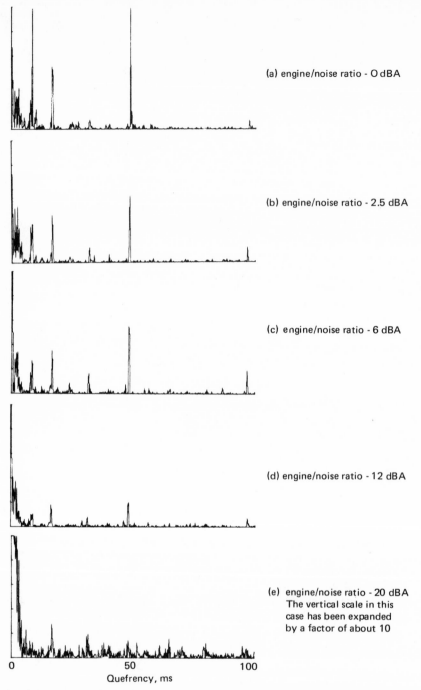

(a) engine/noise ratio - 0 dBA

(b) engine/noise ratio - 2.5 dBA

(c) engine/noise ratio - 6 dBA

(d) engine/noise ratio - 12 dBA

(e) engine/noise ratio - 20 dBA
The vertical scale in this
case has been expanded
by a factor of about 10

Quefrency, ms

Fig. 13.5. Power cepstrum. Derived from an engine sound embedded progressively deeper
in white noise.

13.4. MOMENT FEATURE SPACE

Since we are analyzing a signal that has resulted from exciting a resonant structure with a periodic or quasi-periodic forcing function, we might reasonably expect information about both the forcing function and the structure to be obtainable, by examining the shape of the frequency distribution of the energy in the signal.

It is common practice in statistics to describe a distribution in terms of "dispersion indices." These are measures of the spread (dispersion) of the distribution about some central value. As far as this chapter is concerned, the central value is the mean of the distribution of interest, which in this case is the short-term power spectrum. These indices (*moments*) give information about the shape or dispersion of the short-term power spectrum about its mean frequency. The first move, therefore, in analyzing the vehicle sounds was to obtain a short-term power spectrum of a segment of the signal and to describe the shape of this spectrum by measuring the central moments of the resulting envelope.

The nth moment $M(n)$ of the short-term power spectrum $P(f)$ is defined by

$$M(n) = \int_0^\infty P(f) \cdot f^n \, df \tag{13.5}$$

The *mean angular frequency* is given by

$$\bar{f} = \frac{M(1)}{M(0)} \tag{13.6}$$

The *central moments* are obtained from

$$U(n) = \int_0^\infty P(f) \cdot (f - \bar{f})^n \, df \tag{13.7}$$

The *normalized central moments* are then given by

$$NU(n) = \frac{U(n)}{M(0)} \tag{13.8}$$

The second moment, $U(2)$, known as the *variance*, is a measure of *compactness*; the third moment, $U(3)$, is related to *skewness* and gives a measure of the symmetry; while the fourth moment, $U(4)$, related to the *kurtosis*, is a measure of the "peakiness" of the distribution.

In multivariate analysis, as opposed to bivariate analysis, the difficulties are greatly increased when trying to disentangle the interrelationships among the variables and of interpreting the results of the analysis. This is especially true when the distributions are multimodal. (See Chapter 4.) In an attempt at reducing the number of central moment variables, combinations of the first five central moments were plotted as bivariate scatter diagrams (Thomas, 1971).

From experience gained in studying these scatter diagrams it was found that clustering and separation of the clusters were most pronounced in the $NU(2)/NU(3)$ plane.

It is of interest to note the relationship between the frequency domain moments and the time domain (Hjorth, 1970), the first three of which are given by, for example,

$$M(0) = \int_0^\infty P(f)\, df = \frac{1}{T} \int_0^T [p(t)]^2\, dt$$

$$M(2) = \int_0^\infty (2\pi f)^2 P(f)\, df = \frac{1}{T} \int_0^T \left(\frac{dp}{dt}\right)^2 dt$$

$$M(4) = \int_0^\infty (2\pi f)^4 P(f)\, df = \frac{1}{T} \int_0^T \left(\frac{d^2 p}{dt^2}\right)^2 dt$$

where

$$P(f) = \frac{2}{T} \left| \int_{-\infty}^\infty p(t)\, \exp(-j2\pi f t)\, dt \right|^2 \qquad f \geq 0$$

(See also Chapter 15.) If $P(f)$ is double-sided, $-\infty \leq f \leq \infty$, then all *odd* moments are zero and $M(n) = U(n)$.

Figure 13.6 shows the distribution of segments of vehicle sound in the $NU(2)/NU(3)$ plane under steady running conditions, the vehicle being stationary with the engine ticking over. The moment values have here been further normalized so that they all lie between 0 and 10. It will be seen that an "identification" task can be readily performed on the basis of these two measurements. The Ford and Perkins engines are clearly distinguishable from each other, but for each engine two of the vehicles are very difficult to separate. In the case of the Ford engine, this similarity supports the subjective evidence that the engine has a similar sound when in different vehicles.

These results appeared encouraging, and similar analysis was applied to segments of the vehicle recordings in which bursts of acceleration were present. The results are shown in Fig. 13.7, which shows that the previous, fairly tight, clusters have now been dispersed and the boundaries overlap. Under these conditions simple classification can clearly no longer be performed and a comparison between Figs. 13.6 and 13.7 reveals some interesting features. For the Ford engine, the points tend to move in the direction of lower values of both $NU(2)$

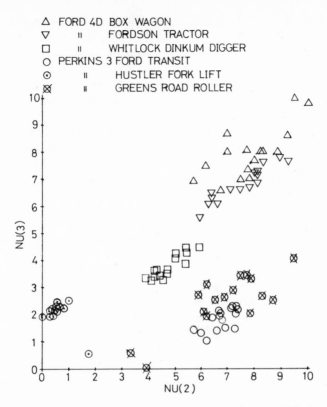

Fig. 13.6. Normalized moment distribution. Derived from stationary vehicle signals (engines ticking over).

and *NU*(3), while for the Perkins engine, the points tend to move in the direction of higher values of both *NU*(2) and *NU*(3). A third engine, the Petter, for which results are not shown here, seems to have the characteristic that the values of *NU*(2) and *NU*(3) are much the same for tick-over and during acceleration. Figure 13.8 includes moments of other sounds where some confusion of boundaries occur. This figure suggests that if a third dimension of frequency, that is, the fundamental frequencies of the signals, were plotted, then the clusters would be more separable as rain would be assigned to one end, while the box wagon would be at the other end. Further weight is given to this from Fig. 13.8, where the previous results are replotted, together with various interfering noises: rain, speech, and a jet aircraft. The last, when flying overhead at the start of the track shown, has moment measures that are very different from those of the vehicles previously analyzed, so that it can be readily identified;

after it has landed and is taxiing, the point in moment space is confused with some of the land vehicles. At all times, however, the "firing period" is very different from those of the land vehicles. Rain, moreover, which appears among the vehicle areas of the moment space, could also be separated by a "firing rate" measurement, since it should not have any distinguishable fundamental frequency.

The confusion of vehicles in moment space possibly arises as a result of the underlying structure of the power spectrum being dominated by the crankshaft harmonic structure. Moments give undue weighting, by multiplying the spectrum by the factor $(f - \bar{f})^n$, to the ends of the spectrum, and it seems reasonable that confusion does occur when some of the low-frequency peaks are 35–40 dB above the underlying structure. The removal of this harmonic structure is discussed in Section 13.6.

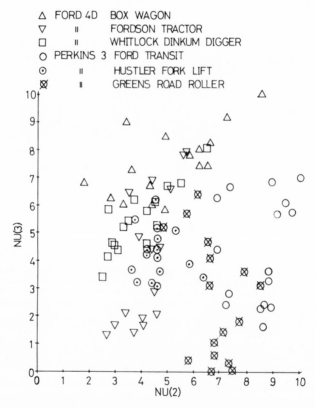

Fig. 13.7. Normalized moment distribution. Derived from vehicle signals (including bursts of acceleration).

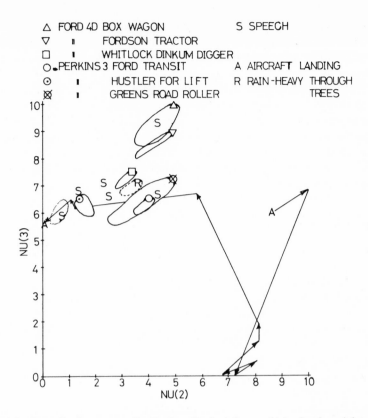

Fig. 13.8. Normalized moment distribution. Derived from vehicle signals and interfering signals.

13.5. NONSINUSOIDAL TRANSFORMS

Since the publication of Harmuth's book (1969), a great deal of interest has been shown in rectangular basis functions. In particular the Walsh transform has received particular attention. The Haar transform on the other hand, though generally ignored, has the useful property that, within the window width (the length of the segmented signal), the basis functions have increasing resolutions which can be used to detect the presence of bursts of energy (Thomas, 1973).

Automatic analysis schemes using the conventional techniques of Fourier analysis, do not cope adequately with identifying the presence of burstlike characteristics within an ongoing signal. In frequency analysis, the problem is that the integration over a given time interval (the window width) introduces ambiguities; as a result of this averaging effect, energy bursts cannot be identified with any degree of precision. A time window, selected to resolve certain basic characteristics from a signal, may not be optimal in detecting bursts of energy; the window width may be such as to smooth out the effect of the bursts in the spectrum. It is therefore necessary to be able to detect the occurrence of a burst within the previously selected window width. This is even more important when short-term spectra are being calculated automatically, using an adaptive method (Thomas, 1971). In this method, the window width is constantly being adjusted to the characteristics of the incoming data. Given the restrictions outlined above, it becomes apparent that what is needed is the ability to scan along a signal segment using various resolutions until a burst is found. The Haar transform supplies this scanning ability.

13.5.1. Energy Bursts

Bursts have been defined as "the excursion of a quantity which exceeds a selected multiple of the long-time average of this quantity" (Cochran, 1964). For the purpose of this chapter a burst will be redefined as being "a high-amplitude, narrow-band signal of short duration, where the 'attack' and 'decay' in both domains is abrupt." The quantification of this definition tends to be problem-oriented, but the nature of the bursts for the purpose of this discussion can be seen in Fig. 13.11. We are, in effect, dealing with the concept of a "cell" in the frequency—time domain. An interesting paper on "acoustical quanta" was put forward as long ago as 1946 by Gabor, but unfortunately it has not been pursued and published, as far as this author is aware.

The Haar transform is shown in Fig. 13.9 for a resolution of 2^3. Figure 13.10 gives the grouped functions and two corresponding notations with the row vectors of each group, collapsed together, shown to the left of the figure.

13.5.2. The Haar Transform Used in Burst Detection

A typical series of bursts distributed over frequency and time are shown in the sonagram of Fig. 13.11 and it is the detection of the presence of such bursts that can perform the operation of "recognition."

A 300-ms segment of signal, taken from a recording of a moving vehicle containing bursts of energy, is shown in Fig. 13.12, where a major-intensity burst is seen at a time of 150 ms, with two or three minor bursts elsewhere in the segment. The function $G(k)$, the ordinate of Fig. 13.12c, is equal to $S^2(k)$ (reduced to unit area, since absolute values are of no particular interest for the present application) where $S(k)$ is given by

$$S(k) = 1/D \sum_{d=0}^{D-1} X(d)\phi(k,d) \tag{13.9}$$

where

$k = 1,2, \ldots, 2^n$

n = maximum number of groups

$D = 2^n$

$S(1)$ is the d.c. term and is not assigned to a group

When dealing with time functions, the normal Fourier series uses the notions of frequency f and time t. Analogous dimensions for the Haar function are

Sequency:

S = one-half the average number of *zero crossings* per unit time.

Average period of oscillation:

$\tau = 1/s$; this gives the average separation in time of the zero crossings, multiplied by 2.

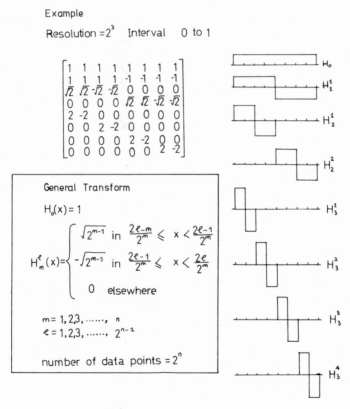

Example

Resolution $= 2^3$ Interval   0 to 1

$$\begin{bmatrix} 1 & 1 & 1 & 1 & 1 & 1 & 1 & 1 \\ 1 & 1 & 1 & 1 & -1 & -1 & -1 & -1 \\ \sqrt{2} & \sqrt{2} & -\sqrt{2} & -\sqrt{2} & 0 & 0 & 0 & 0 \\ 0 & 0 & 0 & 0 & \sqrt{2} & \sqrt{2} & -\sqrt{2} & \sqrt{2} \\ 2 & -2 & 0 & 0 & 0 & 0 & 0 & 0 \\ 0 & 0 & 2 & -2 & 0 & 0 & 0 & 0 \\ 0 & 0 & 0 & 0 & 2 & -2 & 0 & 0 \\ 0 & 0 & 0 & 0 & 0 & 0 & 2 & -2 \end{bmatrix}$$

General Transform

$H_o(x) = 1$

$$H_m^{\ell}(x) = \begin{cases} \sqrt{2^{m-1}} & \text{in } \frac{2\ell-m}{2^m} \leq x < \frac{2\ell-1}{2^m} \\ -\sqrt{2^{m-1}} & \text{in } \frac{2\ell-1}{2^m} \leq x < \frac{2\ell}{2^m} \\ 0 & \text{elsewhere} \end{cases}$$

$m = 1, 2, 3, \ldots, n$

$\ell = 1, 2, 3, \ldots, 2^{n-1}$

number of data points $= 2^n$

H_0

H_1^1

H_2^1

H_2^2

H_3^1

H_3^2

H_3^3

H_3^4

Fig. 13.9. The Haar transform, $H_m^l(x)$, matrix and basis functions of the orthonormal Haar transform for a resolution of 2^3.

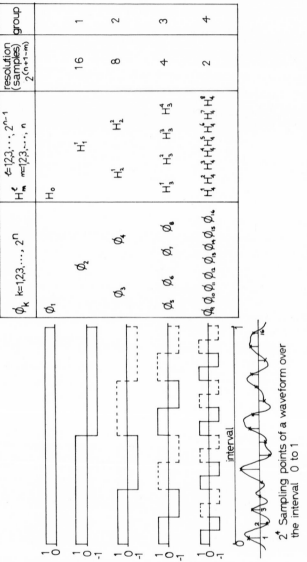

Fig. 13.10. Grouped orthogonal Haar functions: two notations ϕ_k and H^l_m, showing each group superimposed for 2^4 data points. ϕ_k is an alternative notation found in the literature.

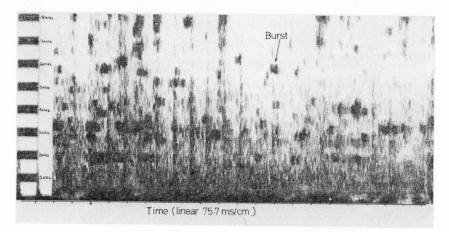

Fig. 13.11. Sonogram from a moving vehicle that generates characteristic squeaks associated with the burst shown.

The Haar transform of Fig. 13.12a is shown in Fig. 13.12c, where the largest peak, which is time-position-sensitive, in the middle of group 10, locates the position of the burst at $t = 150$ ms. (These groups are also related to the frequency-content of the signal.) Providing the peaks in this group can be located and that they do not occur in other types of vehicle sound to any extent, then we have available an analysis technique that will "recognize" the presence of this particular type of feature in a signal. An analysis of a typical heavy, moving, wheeled vehicle is shown in Fig. 13.13, where group 10 indicates that there are no bursts in the frequency area of interest. Using thresholding and peak-picking techniques within the sequency spectrum, bursts of energy can then be located automatically and segmented out of the signal for further analysis. Figure 13.14a shows the burst segmented out of the signal shown in Fig. 13.12a ($t = 150$ ms), together with its corresponding spectrum.

This rather brief discussion on the Haar transform has indicated two applications which may be usefully applied together for feature detection or separately, as the detection of the presence of a burst for "recognition" and/or as a data reduction method for further feature detection as shown in Fig. 13.14. If a comparison is made between Figs. 13.12b and 13.14b it is clear that the resonances associated with the burst are more clearly defined in Fig. 13.14b.

13.6. HOMOMOPHIC FILTERING

Homomorphic filtering (Oppenheim, 1968), is a general term applied to transforms where linear filtering techniques are applicable, and in this section

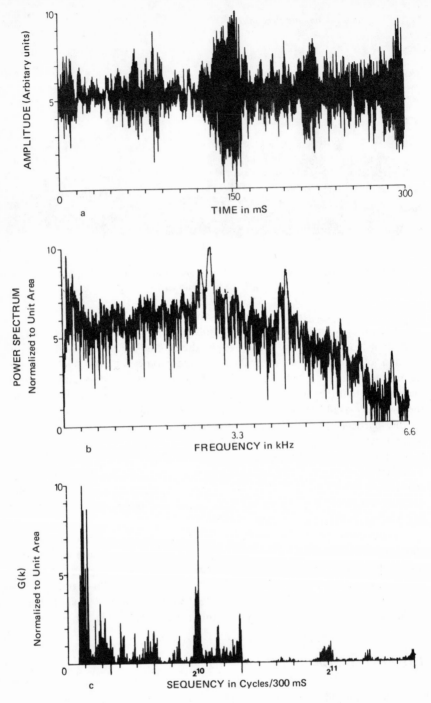

Fig. 13.12. (a) Waveform. (b) Power spectrum. (c) Sequency spectrum from a moving vehicle sound containing energy bursts.

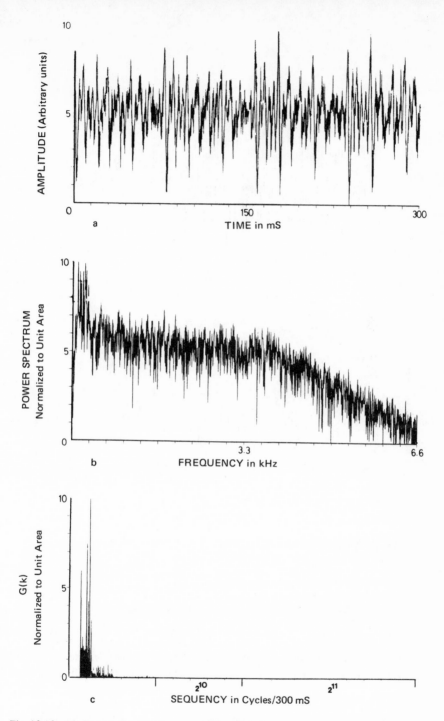

Fig. 13.13. (a) Waveform. (b) Power spectrum. (c) Sequency spectrum from a wheeled moving vehicle.

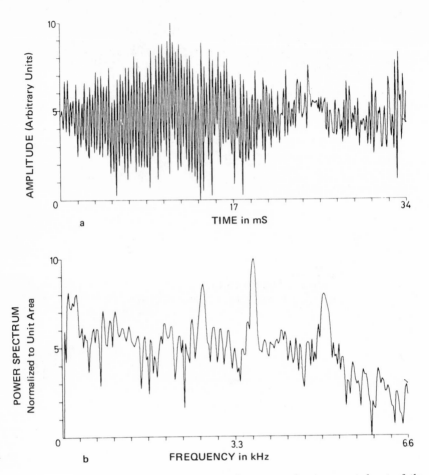

Fig. 13.14. (a) Waveform. (b) Power spectrum of an energy burst segmented out of the waveform of Fig. 13.12.

we will consider a method of creating a modified power spectrum by "filtering" the cepstrum (Bright and Thomas, 1975).

The power cepstrum was defined in Section 13.3.5 as

$$PC(\tau) = |\,\mathrm{F}^{-1}\,\{\log P(f)\}|^2 \qquad\qquad (13.10)$$

We were not interested in phase information. If it is required to transform back to the long $P(f)$ domain, it is necessary to use the cepstrum, defined as

$$C(\tau) = \mathrm{F}^{-1}\,[\log P(f)] \qquad -\infty < f < \infty \qquad (13.11)$$

In the simplified case of an engine and an exhaust system, it is possible to filter out that portion of $C(\tau)$ corresponding to $\log |N(f)|^2$ so that we may

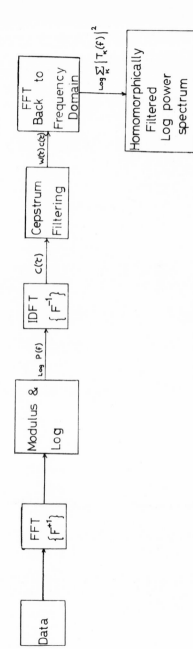

Fig. 13.15. Analysis system for producing homomorphed spectra.

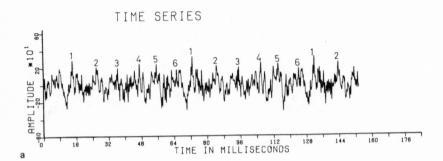

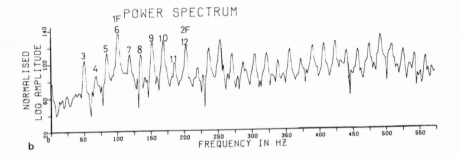

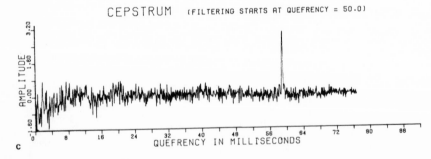

Fig. 13.16. Analysis stages in the homomorphic filtering of an exhaust noise signal obtained from a 6-cylinder test-bed engine. (a) Waveform. (b) Power spectrum. (c),(d) Cepstra for two filtering values. (e) Homomorphed spectrum.

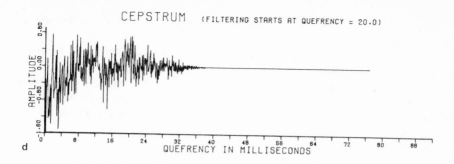

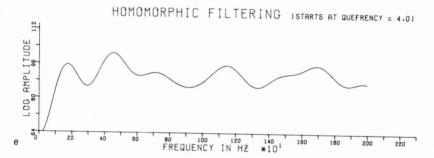

Fig. 13.16. (continued).

reconstruct the power spectrum, then

$$\log P(f) = \log |X(f)|^2 \qquad (13.12)$$

and we are left with a spectrum that is now directly related to the exhaust system.

In the normal case, involving the total vehicle sound, we have included sources not connected to the engine (cf. Fig. 13.1), so that we should write,

$$\log P(f) = \log |N(f)|^2 + \log \sum_k |T_k(f)|^2$$

where $T_k(f)$ are transfer functions of other systems. These other systems tend to obscure some of the detail in $C(\tau)$. Nevertheless, it is still possible to filter $C(\tau)$ and produce a transfer function that can be manipulated to produce useful features.

The procedure for generating homomorphically filtered log-power-spectra, adopted for the present discussion, is shown diagrammatically in Fig. 13.15. The generation of a homomorphically filtered log-power-spectrum is best appreciated by considering a sequence of typical graphs taken at various stages throughout the process outlined in Fig. 13.15.

Figure 13.16a—e shows the waveform of an engine sound; its log-power-spectrum (with a well-defined harmonic structure, showing the peaks associated with the crankshaft rotation period and peaks associated with the period between cylinders firing); two cepstra at different filtering quefrencies and finally a homomorphed power spectrum. It can be seen that Fig. 13.16e, which has been extended to 2 kHz, is a "filtered" version of Fig. 13.16b and that we are left with the "underlying structure,"

$$\log \sum_k |T_k(f)|^2$$

after $\log |N(f)|^2$ has been removed. This underlying structure is relatively stable in comparison with $\log |N(f)|^2$ in many cases and can form a data base whose features are now relatively invariant within a given class. Although in this type of analysis, the engine output harmonic structure has been removed, the underlying structure is less obscured. This does not imply that $|N(f)|^2$ contains no useful information. In fact, the reverse is true as will be shown in the following section.

Crankshaft Rotation Period and Cylinder Firing Period. The time series of the Rolls-Royce engine shown in Fig. 13.16a is of interest, since two phenomena are seen that can be associated with $N(f)$. The first is that the period between peaks of the same number is the same and gives the crankshaft rotation period. The second phenomenon is that the intervals between the (successive) smaller peaks are different. These are associated with cylinder firings. These two effects are shown more clearly in the harmonic structure $\log |N(f)|^2$ seen in Fig. 13.16b, where the harmonic spacing, (peak 6—peak 12) is determined by the crankshaft rotation period. The reason for this is probably that exhaust manifold lengths are frequently different, hence there are different time delays associated with gas pulses reaching a common point in the exhaust pipe. The constant factor is from one cylinder firing to the same cylinder firing again, and this time delay is the same for all cylinders. Another factor that comes out of this consideration is that, dependent on the various lengths of an exhaust manifold, an enhancement of some harmonics may be expected. Fig. 13.16b shows enhancement of the 3rd, 6th, 9th, . . . harmonics for a six-cylinder engine.

13.7. CONCLUSIONS

We have briefly considered the first data-processing stage of a pattern recognition problem by looking at some methods of extracting features from acoustic signals associated with vehicles. There are many problems in data processing which may be reduced by appreciating the bounds of the data itself when applied to the particular form of data analysis methods being undertaken. We believe that no matter how sophisticated the mathematical processes being used the solution to a particular problem depends on the reliability of the original data.

ACKNOWLEDGMENT

Part of this work was carried out with the support of the Procurement Executive, Ministry of Defence (U.K.).

REFERENCES

Bright, K. O., and Thomas, D. W., 1975, Vehicle acoustics and homomorphic filters, *Proceedings of the Digital Signal Processing Conference,* Institute di Ricerca sulle Oude Electromagnetiche del C.N.R. Italy.

Cochran, W. T., 1964, Burst measurement in the frequency domain, *Proc. IEEE* 54:6, 830.

Flanagan, J. L., 1965, *Speech Analysis Synthesis and Perception,* Springer-Verlag, Berlin.

Gabor, D., 1946, Theory of communication, *J. IEE* 93:429.

Harmuth, H. F., 1969, *Transmission of Information by Orthogonal Functions,* Springer-Verlag, Berlin.

Hillquist, R. K., 1967, Objective and subjective measurement of truck noise, *Sound Vibration,* 8 April.

Hjorth, B., 1970, E.E.G. analysis based on time domain properties, *Electroencephalogr. Clin. Neurophysiol.* 29:306.

Noll, A. M., 1967, Cepstrum pitch determination, *J. ASA* 41:293.

Oppenheim, A. V., 1968, Non-linear filtering of multiplied and convolved signals, *IEEE Proc.* 56:1264.

Thomas, D. W., 1971, Vehicle Sound Recognition, A Pilot Study, Ph.D. Thesis, University of Southampton, England.

Thomas, D. W., 1973, Burst Detection Using the Haar Spectrum, Colloquium, Theory and Applications of Walsh and other Non-sinusoidal functions, Hatfield Polytechnic, England.

14

Editorial Introduction

In Chapter 14, W. A. Ainsworth and P. D. Green trace the development of speech recognition from the early attempts to produce a phonetic typewriter to today's more ambitious work on question-and-answer dialogues between men and machines. Speech recognition has long been regarded by many experienced research workers as being one of the most difficult problems in pattern recognition. The reason for this is clear from this chapter; continuous speech is not simply a sequence of isolated words, with short interword gaps. Phonemes are not immutable "atoms," but can be modified by their context. To analyze speech properly, it is necessary to study its generation by the human body. The same general message appears many times elsewhere in this book; we should always consider the nature of the data source and look for clues to assist our building a recognizer.

It is only to be expected that there are many points of comparison between this chapter and those by Thomas (Chapter 13, on the analysis of vehicle sounds) and Binnie *et al.* (Chapter 15, on EEGs), since they are all concerned with time-varying signals. While Thomas keeps his work firmly in the area of statistical pattern recognition, the other two chapters also move into syntactic pattern recognition. Binnie *et al.* do so only briefly, but Ainsworth and Green show that, for speech recognition, this move is much more pronounced and is inevitable, owing to the changing emphasis of this subject. They also show that speech recognition is one aspect of artificial intelligence. For this reason, this chapter has close parallels with those by:

1. Brown and Popplestone (Chapter 8, on visual scene analysis)
2. Bell (Chapter 5, on the manipulation of decision trees)
3. Newman (Chapter 16, on syntactic pattern recognition)

The concepts of syntactic analysis, parsing, and tree traversers all feature significantly in these chapters. Another important point is made here, as it is elsewhere: pattern recognition frequently places severe demands on computer hardware. This may even require the development of new architectures. This is also evident from the chapters by:

1. Ullmann (Chapter 2, on optical processing techniques)
2. Duff (Chapter 6, on picture processing hardware)
3. Aleksander (Chapter 3, on new methods of processing using networks of memory elements)
4. Batchelor (Chapter 4, on classification hardware)

What applications can we foresee for speech recognition? Although these are numerous, only two will be mentioned, since they alone can justify all of the research in this field:

1. Voice-operated typewriters, computer terminals, ticket machines, and typesetting consoles. Here there are vast potential economic benefits for a successful solution.

2. Aids for bedridden, paralyzed, or other disabled people. Imagine being able to talk to a "tame" computer and asking it to

"Telephone Aunt Mary"

or

"Make me a cup of tea"

Imagine how much more tolerable life would be if *you* were paralyzed but had such a machine at your constant command!

14

Current Problems in Automatic Speech Recognition

W. A. Ainsworth and P. D. Green

Automatic speech recognition may be defined as any process which decodes the acoustic signal produced by the human voice into a sequence of linguistic units which contain the message that the speaker wishes to convey. At one extreme this includes the "phonetic typewriter," a hypothetical device which types any words spoken into it, and at the other, "speech understanding systems" which extract the intended meaning from the sounds and carry out some appropriate action such as replying to a question or controlling a robot. During the last two decades the emphasis in research in automatic speech recognition has gradually shifted from the former type of device to the latter.

Glossary of Linguistic Terms *

Acoustic-phonetics: the relation between the physical structure of an utterance and its transcription

---

*Abbreviations: ARCS, Automatic recognition of continuous speech; *ATL,* Analog threshold logic; *CAPRU,* Currently accepted partially recognized utterance; *KS,* Knowledge source; *MPCS,* Machine perception of continuous speech; *PSUD,* Partial symbolic utterance description; *ROVER,* Recognition overlord; *SEF,* Single equivalent formant; *VDMS,* Voice-controlled data management system.

W. A. Ainsworth • Department of Communication, University of Keele, Keele, Staffordshire, England **P. D. Green** • Department of Computing, North Staffordshire Polytechnic, Stafford, England

Coarticulation: the influence of sound segments on the physical structure of neighboring sound segments

Content words: words which carry the meaning in a sentence

Formants: the main resonances of the vocal tract

Function words: words which indicate the structure of a sentence

Intonation: the pitch contour of a sentence

Lexicon: a data structure containing information about each word in the vocabulary

Orthography: conventional spelling

Parse: grammatical structure analysis

Phoneme: smallest unit of speech which has meaning in a given language

Phonetic transcription: translation of an utterance into a sequence of phonetic symbols

Phonological rules: the systematic ways in which the realization of words or phonemes may change with their environment

Pragmatics: additional cues or constraints affecting the perception of an utterance which depend on the current context or situation

Prosodics: features such as rhythm and intonation which relate to a whole utterance

Rhythm: the stress pattern of an utterance

Semantics: the meaning of an utterance

Stress: the emphasis on a syllable or word

Syntax: the grammatical rules of a language

Vocabulary: the set of allowed words

14.1. PRELIMINARIES

14.1.1. Speech Production

In order to understand the problems involved in automatic speech recognition it is perhaps best to begin with phonetics, which has mostly been concerned with the classification of speech sounds and the generation of speech.

As the sounds of speech do not correspond exactly to the letters of the alphabet, it is useful to have a set of symbols in order to refer to these sounds. The most convenient labels for these sounds are the *phonemic* symbols shown in Table 14.1. This table also shows the phonemic classes and examples of words in which the phonemes are used. It is important to note that the phoneme is not a physical entity, but is defined linguistically such that if one phoneme is substituted for another in a word, the meaning of the word may change.

The mechanism of speech production is as follows. By expanding the rib cage and lowering the diaphragm, air is drawn into the lungs (Fig. 14.1). The

Table 1. The Phonemes of English

Phoneme class	IPA Symbol	Interpretation
Stops	b	*b*at
	d	*d*og
	g	*g*olf
	p	*p*op
	t	*t*en
	k	*k*id
Nasals	m	*m*ug
	n	*n*ine
	ŋ	si*ng*
Glides	w	*w*olf
	r	*r*at
	l	*l*ot
	j	*y*ou
Fricatives	h	*h*ot
	f	*f*og
	θ	*th*ick
	s	*s*ad
	ʃ	*sh*e
	v	*v*oice
	ð	*th*e
	z	*z*oo
	ʒ	a*z*ure
Vowels	i	m*e*
	ɪ	b*i*d
	ε	r*e*d
	æ	l*a*d
	ɑ	h*a*rd
	ɐ	r*o*d
	ɔ	f*o*rd
	ʊ	c*ou*ld
	u	f*oo*d
	ʌ	c*u*p
	ɝ	h*ear*d
	ə	th*e*

lungs are then contracted, and the pressure difference between air in the lungs and atmospheric pressure provides the energy for the speech wave. This pressure difference causes the air to flow up the windpipe. At the top it encounters the larynx which contains the vocal cords. The pressure of air causes a slit in these cords, called the glottis, to open. As the airflow builds up the local pressure

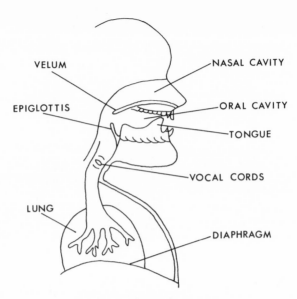

Fig. 14.1. The vocal apparatus.

drops, by the Bernoulli effect, and the tension in the cords causes the glottis to close. The pressure then builds up again, and the cycle repeats. This process, known as *phonation*, creates pulses of air which excite the resonances of the oral and nasal tracts. These resonances depend, of course, on the size and shape of these tracts. The shape of the oral tract can be changed rapidly by moving the articulators: the tongue, teeth, jaw, and lips. This forms the basis of the method by which vowels are encoded in speech.

In phonetics the vowels are classed according to the shape of the mouth which produced them. A vowel is said to be *close* if the oral tract is narrow, *open* if it is wide, *back* if the tongue hump is at the back of the mouth, and *front* if it is at the front of the mouth. The shapes of the tracts for producing some of the vowels are shown in Fig. 14.2. The glides are produced by a similar process, but with the articulators moving.

If the oral tract is closed by either the lips or the tongue, and the velum is opened, the pulses of air from the larynx excite the nasal tract and the sound is radiated from the nose. In this way the nasals are produced, /m/ with the lips closed, /n/ with the tongue on the roof of the mouth, and /ŋ/ with the velum closing the oral cavity.

Some of the consonants, the fricatives, are not generated by phonation but by a different process. A narrow constriction is produced at some point in the vocal tract. This causes turbulent flow of air passing through it, which sounds like a hiss. Any subsequent resonances in the path are excited by this random noise instead of by the periodic pulses generated by the vocal cords.

The sound /s/ is produced by making a norrow construction between the tongue and the teeth, whereas /f/ is produced with the teeth against the lips. An /h/ is generated with the glottis constricted but not vibrating. As this excitation is generated in the throat it is modified by the resonances of the oral tract.

There is another set of sounds, the voiced fricatives, which are excited by both pulses from the larynx and by noise excitation from a constriction. Each voiceless fricative /f, θ, s, ʃ/ has a voiced counterpart /v, ð z, ʒ/, which is generated with the same articulator configuration but with mixed excitation.

Finally, there are the plosive, or stop, consonants. These are produced by closing the vocal tract completely at some point then, when the pressure has built up, releasing it suddenly. With the voiced stops /b, d, g/ voicing begins at or before the moment of release, but with the voiceless stops /p, t, k/ it begins some time later.

14.1.2. Speech Analysis

Speech thus consists of a stream of sound with the positions and motions of the articulators which produced it encoded in the frequencies of vibration of the ripples of the wave. The waveform can be decomposed into its constituent frequencies by Fourier analysis. (Also see Chapters 13 and 15.) This is accomp-

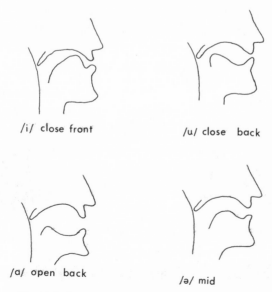

/i/ close front   /u/ close back

/ɑ/ open back   /ə/ mid

Fig. 14.2. Vocal tract shapes for the vowels /i, u, ɑ, ə/.

lished most conveniently by means of an instrument called the sound spectro-graph or sonagraph (Potter *et al.*, 1947). This consists of a disc of magnetic recording material on to which the speech to be analyzed is recorded. As the disc rotates the speech is replayed, over and over again, through a narrow bandpass filter. The intensity of the output of this filter is written on sensitized paper by a stylus. The center frequency of the analyzing filter is gradually swept through the audio range, and at the same time the stylus is swept across the paper. The result is a three-dimensional analysis of the signal with time along the *x*-axis, frequency along the *y*-axis, and intensity represented as the blackness of the trace. A typical sonagram is shown in Fig. 14.3.

Sonagrams of some of the vowels are shown in Fig. 14.4. The resonances of the vocal tract can clearly be seen as the black horizontal bars. These are called *formants*. The frequencies of the formants are different for the different vowels. The vowel /i/ has a first formant (F1) at about 250 Hz and a second formant (F2) at about 2500 Hz, while /u/ has a similar F1 but a F2 of only about 800 Hz. The vowel /ɑ/ has a high F1 and a low F2 which seem to fuse into a single broad formant at about 1000 Hz. The middle vowel /ə/ is produced with a vocal tract of fairly uniform cross section so the frequencies of its formants are simply related to its length, and are odd harmonics of the first formant. For a male speaker the frequencies of F1, F2, and F3 are approximately 500, 1500, and 2500 Hz, respectively. The relationships between the frequencies of the formants and the shapes of the vocal tract have been derived by Fant (1960).

The sonagrams of the nasals are similar to the vowels in appearance (Fig. 14.5) except that, because the nasal tract is longer than the oral tract, there

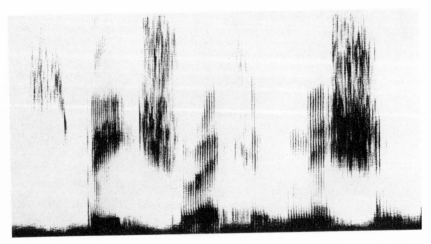

Fig. 14.3. Sonagram of the phrase "speech recognition."

Fig. 14.4. Sonagram of the vowels /i, u, ɑ, ə/.

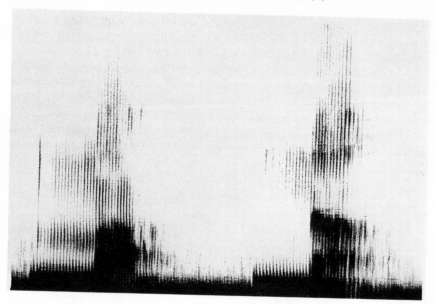

Fig. 14.5. Sonagram of the syllables /mə, nə/.

is always a low frequency F1. The higher formants occur at different frequencies depending upon the point of closure of the oral tract.

The glides /w, r, l, j/ are produced with the articulators moving, so their sonagrams (Fig. 14.6) show formant frequency transitions. A /w/ sound is generated with the lips initially close together but opening. This motion is

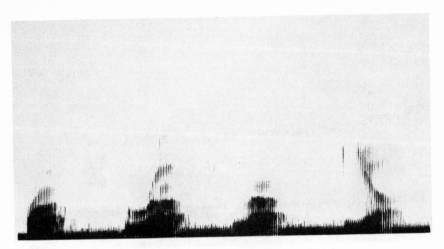

Fig. 14.6. Sonagram of the syllables /wə, rə, lə, jə/.

reflected in the sonagram by rising first and second formants. For a /j/ sound the articulators are initially in a position appropriate for an /i/, which has a low F1 and a high F2. As the /j/ is enunciated F1 rises and F2 falls.

All of the sounds considered so far are voiced. This is shown in the sonagrams by the vertical striations. Each pulse of air from the larynx produces one of the vertical stripes. With the voiceless fricatives (Fig. 14.7) there are no larynx vibrations, so there is no regular striped pattern. Instead there is fricative excitation which gives rise to a random pattern. The various fricative sounds consist of random noise which covers different bands of frequency. As might be

Fig. 14.7. Sonagram of the syllables /sə, ʃə, fə, θə/.

expected, the voiced fricatives produce similar patterns to the voiceless ones except that, for the voiced fricatives, the voicing can be seen at low frequencies (Fig. 14.8).

Sonagrams of stop consonants are characterized by a period of silence immediately prior to the release (Fig. 14.9). The different stops are produced by closure at different points in the vocal tract. This is shown in sonagrams by different formant transitions for the voiced stops /b, d, g/ and by the fricative energy being concentrated at different frequencies for the voiceless stops /p, t, k/.

Fig. 14.8. Sonagram of the syllables /zə, ʒə, və, ðə/.

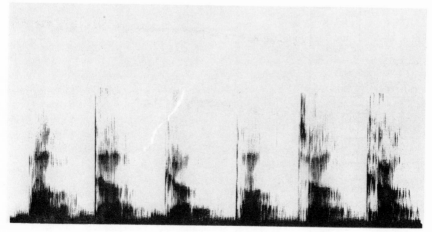

Fig. 14.9. Sonagram of the syllables /bə, də, gə, pə, tə, kə/.

14.2. ISOLATED WORD RECOGNIZERS

14.2.1. Early Speech Recognizers

The importance of F1 and F2 in distinguishing speech sounds from each other follows directly from speech analysis. By synthesizing sounds which have formants similar to those found in natural speech (Lawrence, 1953), it has been confirmed that formants are useful in perception. It is not surprising, therefore, that the first electronic speech recognizers were based on frequency analysis.

One of the first investigations was by Davis, Biddulph, and Balashek (1952). Their apparatus was intended to recognize spoken digits. The principle is shown in Fig. 14.10. The speech wave was first split into two bands by passing it through a 1000-Hz high-pass and an 800-Hz low-pass filter; then the number of zero crossings per second in each band was counted, giving approximate measures of the frequencies of the first and second formants. The first formant range (200–800 Hz) was quantized into six 100-Hz segments and the second formant range (500–2500 Hz) into five 500-Hz steps. A matrix with 30 elements representing the F1–F2 plane was thus produced. For a given spoken digit the time for which the F1–F2 trajectory occupied each element was determined.

A reference pattern was established for each digit. When a new digit was spoken, the pattern produced was cross-correlated with each of the reference

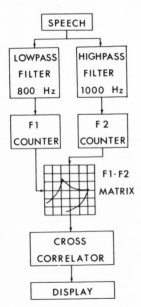

Fig. 14.10. Schematic diagram of the digit recognizer of Davis, Biddulph, and Balashek (1952).

patterns. This gave an approximate measure of the likelihood that a particular digit had been spoken. The most likely digit was thus chosen.

Provided that the reference patterns were adjusted for a particular speaker, the machine was quite successful and recognized the spoken digit correctly about 98% of the time. When a new speaker used the machine with no adjustments, however, the recognition score could be as low as 50%.

In a later development of a similar system, Dudley and Balaskék (1958) performed a spectrum analysis of the speech with a bank of bandpass filters, each 300 Hz wide. The output of the filter bank was cross-correlated with stored patterns, and the best match was selected as the spoken digit. This scheme also produced good results with the speaker who generated the stored patterns, but was less successful for the other speakers.

In order to produce a more general purpose speech recognizer, which would recognize phonemes rather than words and so be able to deal with larger vocabularies, Wiren and Stubbs (1956) built a device which was based on the *distinctive feature* hypothesis of Jacobson *et al.* (1952). In this scheme a phoneme consists of a bundle of distinctive features each of which may take the value 1 or 0 depending on whether the feature is present or absent. The features are qualities such as voiced/unvoiced, vowel/consonant, interrupted/continuent, etc. A number of distinctive feature schemes have been proposed since the original.

The principle of the device of Wiren and Stubbs (1956) is shown in Fig. 14.11. First, the voiced sounds were separated from the unvoiced sounds. Then the unvoiced sounds were separated into the unvoiced stops and the fricatives. Finally, the strong and the weak fricatives were separated. In the "voiced" branch, the turbulent sounds (voiced stops and voiced fricatives) were separated from the vowel and vowel-like (nasals and glides) sounds. This principle of binary classification according to linguistic features can be repeated until a single phoneme is isolated. Fairly good results were reported. With vowels in short words pronounced by 21 speakers an accuracy of 94% was obtained. The system, however, was not completely implemented.

Fry and Denes (1958) made a rather different attempt to use linguistic information. In English not every phoneme can follow every other phoneme, so once one phoneme has been successfully recognized the range of phonemes which may follow it is narrowed. The system of Fry and Denes took advantage of this fact. It consisted of a spectrum analyzer, a spectral pattern matching system, and a store containing the probability of any phoneme following another in spoken English. The phoneme recognition set comprised four vowels and nine consonants. The performance of the machine was not particularly good, but the use made of the diagram probabilities improved word recognition accuracy from 24% to 44%.

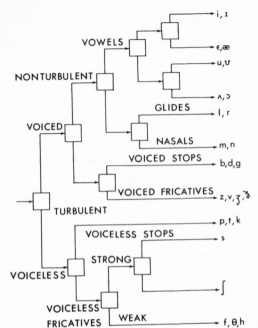

Fig. 14.11. The binary classification scheme of Wiren and Stubbs (1956).

14.2.2. Other Forms of Preprocessing

As it seemed impossible to achieve accuracies of near 100% for a large vocabulary and a wide range of speakers, it was thought for a while that the fault lay in the way in which the information was being extracted from the acoustic waveform. Earlier, Licklider and Pollack (1948) had shown that speech which had been amplified and clipped (Fig. 14.12) was fairly intelligible, and speech which had been differentiated, amplified, and clipped was almost completely intelligible. This being the case, it was argued that most of the information required for speech recognition was contained in the time intervals between the zero crossings of the speech waveform.

Sakai and Doshita (1962) reported a speech recognizer which was based partly on this kind of analysis. Zero-crossing measurements were combined with measures of the variations of energy in various frequency regions. It was claimed that the device correctly recognized 90% of the vowels and 70% of the consonants, although not all phonemes were allowed as input. Other speech recognition devices which employed zero-crossing time-interval analysis have been described by Bezdel and Chandler (1965), Purton (1968), Lavington (1968) and Green (1971). Each of these showed promise, but none achieved good recognition scores except by severe restriction of vocabulary and speakers.

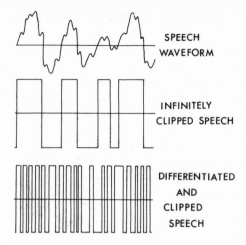

Fig. 14.12. Infinitely clipped speech.

Another type of preprocessing was introduced by Teacher *et al.* (1967). Their recognition system was based on the premise that a single, dominant formant was the main feature responsible for the recognition of any particular phoneme. They termed this the *single equivalent formant* (SEF). This is shown for the vowels in Fig. 14.13. For the back vowels, the SEF is identical with F1 and for the front vowels it is near to F2, while for the middle vowels it lies between F1 and F2. The frequency of the SEF can be determined by picking the dominant peak in the spectrum after preemphasis of +6 to 9 dB/octave. Teacher *et al.* measured three parameters of the SEF (frequency, amplitude, and manner of excitation) and used these as the data for a computer program which recognized a number of words. With a vocabulary of 10 digits an accuracy of 90% was obtained with the 10 speakers whose voices had been used to train the machine.

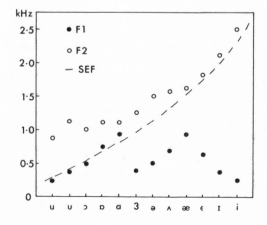

Fig. 14.13. The single-equivalent formant (SEF) of Teacher *et al.* (1965).

An interesting development was reported by Reddy (1967). In a computerized speech recognition system Fourier analysis often accounts for a large proportion of the processing time, so that in many systems a cruder form of frequency analysis, such as zero-crossing analysis, is employed. Such forms of analysis, however, lack the precision required for some of the discriminations required in speech recognition. In Reddy's system, zero-crossing analysis was used to segment the speech signal into portions corresponding approximately to phoneme units, then Fourier analysis was employed to make the necessary discriminations in the centers of the voiced regions. This was one of the first instances in which an attempt was made to select the kind of analysis which was appropriate to the input signal. (The reader is referred to Chapter 13, where this theme is expanded with reference to vehicle sounds.)

14.2.3. Some of the Difficulties

In spite of these ingenious methods of extracting the relevant parameters from speech signals, there seems to be an upper limit of 90—95% for a vocabulary of about 10 words spoken by about 10 speakers. For larger vocabularies or a larger number of speakers, this limit is considerably reduced.

One of the reasons why the systems fail to recognize the speech of a large number of speakers is the considerable variation in the acoustic signals produced by different speakers. To begin with, there are the differences in accent which often take the form of the substitution of one vowel or diphthong for another. It would be unreasonable to expect a machine to recognize that /baθ/ spoken in a southern English (U.K.) accent was the same as /bæθ/ spoken by a Northerner, unless this had been specifically programmed. The number of variations, however, is not too large, and rules to account for most of them could probably be devised.

Another source of variation derives from the different physical sizes of the vocal tracts of different speakers. The length of the vocal tract determines the mean frequencies of the formants. Peterson and Barney (1952) measured the fundamental and formant frequencies of the vowels uttered by a large number of men, women, and children. They found that the formant frequencies of childrens' vowels were about 30% higher than the vowels of men, and their fundamentals were about an octave higher.

Forgie and Forgie (1959) built a speech recognizer which took advantage of the correlation between formant and fundamental frequency. They used a 35-channel filter bank to analyze the speech wave. The spectrum was entered into a computer, and F1, F2, and the fundamental frequency were determined. From these measurements the vowels in isolated words such as /b/-vowel-/t/ were recognized. It was reported that for 21 male and female speakers each producing 10 vowels, and with no adjustment for speaker, the accuracy was 93%.

It is probable that the human speech recognition system employs similar mechanisms which automatically compensates for speaker variations. Fujisaki and Kawashima (1968) showed that by manipulating the pitch of synthesized vowel sounds, Japanese listeners could be made to categorize some of the sounds differently. Broadbent and Ladefoged (1960) had previously shown that by shifting the second formant frequency of a carrier phrase, listeners could be induced to hear different vowels in a test word. In an experiment by Ainsworth (1973) it was found that, for English listeners, the formant frequencies of the preceding vowels in a phrase had a greater influence on the perception of a test vowel than did the pitch of the phrase.

Gerstman (1968) showed that if all the vowels analyzed by Peterson and Barney were normalized by scaling the formants by means of the highest and lowest values of F1 and F2 of the vowels of each speaker, then a vowel recognizer could be devised which misclassified only those vowels which were ambiguous to human listeners.

Another source of variability between speakers, and between different utterances of the same phrase by the same speaker on different occasions, is the time course of the utterance. This varies to such an extent that it has been suggested that it should be ignored in speech recognition, and merely the *sequence* of acoustic events utilized. Others have suggested *time normalization*. The simplicity of the first course is attractive, but it ignores the fact that certain phonetic distinctions are made on the basis of the relative durations of events (Ainsworth, 1972).

The latter course of action is probably more realistic, but it is difficult to employ. In fast speech, the durations of all phonemes are not uniformly reduced, so a straight forward time normalization algorithm may lead to error. Also, the spectrum of a speech sound may depend on the speed of the utterance. Lindblom and Studdert-Kennedy (1967) have shown that the vowel in a syllable such as /wɪw/ will be correctly perceived even though F1 and F2 never reach the values they would assume in the isolated vowel /ɪ/. Thus, time normalization must take into account such factors as the rate at which the formants are changing.

Another factor which must be allowed for in a successful speech recognizer is *coarticulation* (Ohman, 1966). The articulators move into the positions in which they will be required as soon as possible. Thus the acoustic form of the current phoneme may be much changed by the fact that some articulator which is not involved in its production, is moving to its position for the phoneme which occurs next.

Finally, it has been found that in normal conversational speech many phonemes are missed out, and a few others are inserted for ease of pronunciation. In the sequence "six seven," for example, the /s/ at the end of "six" is hardly ever pronounced (unless the speaker makes a special effort to do so). Similarly, in "who are you" a /w/ is normally inserted between "who" and "are."

14.2.4. Analog Feature Recognition System

Many of the single speaker, small vocabulary recognizers were quite sucessful but failed when a greater vocabulary or more speakers were introduced. It may be that the best plan is to work in the other direction, to design a more complex system which contains mechanisms for coping with the difficulties. Then, if the task to be performed is not too hard, reduce the complexity of the recognizer. There is one speech recognition system, however, the EMI-Threshold VIP-100, which incorporates some of the best features of the single-word recognizers, and which is available commercially, rather than as a laboratory model (Ackroyd, 1974). For this reason, not all the technical details have been published, but a block diagram of its structure is shown in Fig. 14.14.

The bandpass filters are sufficiently broad so that a time resolution of 10 msec is attained. It is claimed by Nelson *et al.* (1967) that this is necessary in order to follow some of the rapidly changing features of speech. The rectifiers in the preprocessing circuits have a dynamic range of about 60 dB, so that the energy of the loudest and quietest sounds can be faithfully measured. This range is then reduced by logarithmic compression to about 20 dB, giving the useful result that differences in level between channels correspond to ratios in the original sound. The system is thus fairly independent of overall loudness.

In many systems the formants were found as the positions of the peaks in the spectrum, and were easily missed. In this system, the regions of increasing or decreasing slope are detected instead. These can be found more reliably and are less dependent on the exact frequency of the formants.

The *analog threshold logic units* (ATL)* were designed to extract the features useful for recognition. Excitatory or inhibitory connections can be made, and the output represents the probability that a particular feature was present in the input. The actual connections were decided by the designers on the basis of their knowledge of speech, and a great deal of experimentation.

The outputs of the ATLs are connected to a small computer. Each word or phrase is stored in the computer as a sequence of extracted features. The present model has a vocabulary of 32 words or phrases. The system must be adjusted for each individual speaker, but this can be done rapidly using the computer to store a number of examples of each member of the vocabulary.

It is claimed that for a vocabulary of phonetically *dissimilar* words a recognition rate of 100% is readily obtainable, but if the vocabulary contains *similar* words some errors may occur.

---

*The same topic, under different names, is discussed in Chapters 3 and 4.

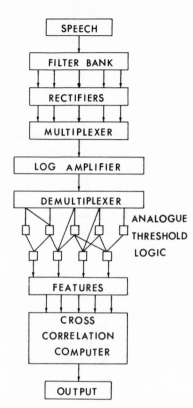

Fig. 14.14. Schematic diagram of the VIP-100 word recognizer.

14.3. MACHINE PERCEPTION OF CONTINUOUS SPEECH (MPCS)

14.3.1. Introduction

In the production of natural, continuous speech (rather than isolated words), the variations mentioned above become more acute, pronunciation is less careful, speaker differences are underlined, speaking rate less constant, co-articulation effects exist between words as well as within them. In addition, new problems arise for the speech recognizer: the importance of a word in the message will affect its stress and intonation and hence its acoustic realization. Although speech is perceived as a sequence of separated words, there is often little acoustic evidence of word boundaries.

In attempting to understand or recognize continuous speech, men and machines are presented with an input message which is imprecise, and in which not all the information necessary for decoding is completely or unambiguously encoded. In such a situation, traditional pattern recognition methods become inappropriate: rather than passively identifying segments of speech as phonemes, words, etc., one has to interpret the signal actively with the aid of knowledge of the standing and transient properties of the speech world.

There is ample evidence that human listeners make use of an extensive repertoire of linguistic cues and constraints in speech perception. In the spectrogram-labeling experiments of Klatt and Stevens (1973), experts achieved a 75% success rate in partial phonetic transcription using acoustic evidence alone, but this figure rose to 96% when they were permitted to use knowledge of the vocabulary, syntax, and semantics of the utterance. Pollack and Pickett (1964) have demonstrated that spoken sentence fragments are not more than 90% intelligible to human listeners if their syntactic and semantic structure is not apparent. The listeners intelligent application of linguistic knowledge means that portions of speech are sometimes redundant: one can often guess the remainder of a sentence after hearing the first few words. If machines are to approach human performance in speech recognition, they must develop at least some of this linguistic expertise.

14.3.2. Knowledge Sources

The *a priori* information available to a listener can be thought of as embodied in a number of distinguishable knowledge sources (KSs). Current research in MPCS is focused on the problems of how best to implement the KSs and how to control the interactions between them. The identifiable KSs are as follows.

1. *Acoustic-Phonetics* — The correlations between perceived phonetic sequences and physical parameters extracted from the speech signal. This usually involves some form of segmentation and labeling of the speech signal, though strictly the labels refer to acoustic events rather than phonemes. The techniques used in acoustic-phonetic analyses have developed from those employed in isolated word recognition (described above), and will not be discussed in the following sections. In MPCS systems, the acoustic-phonetic module is usually treated as a preprocessor or "front end" interfacing the incoming speech with the rest of the system. Since phonetic segmentation and labeling cannot be expected to be reliable, probabilistic information reflecting the confidence of the preprocessor about its decisions may be retained. Alternatively an "error set," a statistical compilation of the types of mistakes introduced by the pre-processor, can be made available.

2. *Vocabulary* — It is obviously useful to have some kind of dictionary giving the phonetic constitution of every word in the language. Dictionary entries may consist of "ideal" phomeic transcriptions, or of the "quasi-phonemic" labels assigned by acoustic-phonetics. Multiple entries are needed for pronunciation variations which cannot be predicted by rule.

3. *Phonological rules* — Express the systematic ways in which the realization of words or phonemes may change with their environment. This knowledge can help to reconcile the results of acoustic-phonetic labeling with the "normal" word forms stored as dictionary entries. Examples of phenomena for which context-dependent rules can be formulated are vowel substitution ("spectrogram" → /spektrəgræm/) and deletion ("boundary" → /baʊndrɪ) and a number of consonant cluster corruptions ("did you" → /dɪdʒu/, "mostly" → /məʊslɪ/). Oshika *et al.* (1974) give a useful introduction to this work.

4. *Syntax* — The grammar of the language, expressed in a suitable form such as a set of syntactic rewrite rules, specifying how word sequences can be built up to form legal sentences.

5. *Semantics* — Ways in which words or phrases may be linked so as to suggest conceptual structures which are meaningful to the listener.

6. *Pragmatics* — Additional cues or constraints which are dependent on the current context or situation. For instance, in an ongoing dialogue it may be possible to predict words or syntactic constructions which are likely to occur in the next utterance, or to impose additional semantic constraints.

7. *Prosodics* — In addition to phonetic information, the natural speech waveform contains cues which relate directly to the syntax and semantics of the utterance. These prosodic features — stress, rhythm, intonation — are used, for instance, to distinguish between

"They are (flying planes)"   vs   "They (are flying) planes"

or

"Light housekeeper"   vs   "lighthouse keeper"

and to indicate whether utterances like "this coat for five pounds" are to be interpreted as statements or as questions. In addition, stress patterns can be used to indicate the parts of the utterance where the most reliable acoustic-phonetic information is likely to be found. Lea *et al.* (1974) have investigated the extraction of prosodic features from energy and voicing parameters and fundamental frequency contours, and their use in an MPCS system (described later).

In general, no one KS can be relied upon to function perfectly (they are essentially heuristic rather than algorithmic), but suitably used in combination they should enable the system to build a grammatical, meaningful, and appropriate translation of the utterance. Since any deduction made from information supplied by a KS may be shown by later processing to be (probably) in error, it is necessary to keep track of alternative possibilities in a suitable data structure. Analogous (though perhaps less difficult) problem situations are encountered in other areas of machine intelligence research such as semantic information processing (Winograd, 1972) and robot vision (Winston, 1972). (See also Chapters 3, 7, 8, 10, and 16 for discussions of this topic.)

The following sections outline some of the major MPCS research projects reported in the period 1972–1974. For more detail, the reader is directed to the collection of papers from the 1974 IEEE speech recognition symposium (Erman, 1974). Of necessity, all practical MPCS systems function in restricted problem domains, dealing with utterances in specialized languages with limited vocabularies, for which it is possible to formalize the properties of the language. The divergence of problem domains, coupled with the novelty and complexity of the research, makes it difficult to assess the relative merits of different systems, except in a qualitative way. In the current state of the art, MPCS systems are best regarded as frameworks for research, rather than the prototypes of commercially viable systems, and the few performance figures given here should be interpreted as no more than benchmarks.

14.3.3. The Hierarchical Approach

The most obvious approach to the design of an MPCS system is to apply the KSs in series, in a fixed order. For instance, Fig. 14.15 shows a hypothetical system in which an acoustic–phonetic preprocessor attempts to reduce the utterance to a string of phonemes. An error set and dictionary are then used to compile a "word tree" summarizing all the possible sequences of words which could have produced the phoneme string, with probability measures for each pathway. Syntax then removes the ungrammatical paths from the word tree and updates the probabilities of those remaining, and semantics decides which sentence is most appropriate. One difficulty with this idea is immediately apparent: it involves passing large amounts of data, stored in complex data structures between every stage. Green and Ainsworth (1972) implemented the first two stages of such a system. They found that if the probability threshold for inclusion of a word in the tree was set low enough to ensure that the correct path was retained, the word trees (even for small vocabularies and short utterances) became unmanageably large.

Of recent MPCS projects, the *ARCS* (automatic recognition continuous speech) system, developed at IBM (Tappert *et al.*, 1973; Tappert, 1974; Paul and Rabinowitz, 1974) comes closest to the hierarchical model. The aim is recognition rather than understanding of speech; the system produces an orthographic transcription of a spoken sentence. A number of variations of ARCS have been reported: only the basic method is described here. An overview is shown in Fig. 14.6.

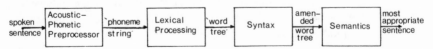

Fig. 14.15. A hypothetical continuous speech recognition system.

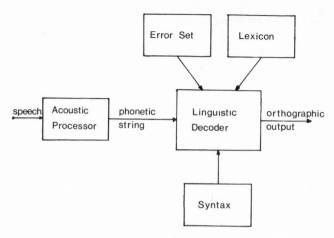

Fig. 14.16. Overview of the ARCS system.

The acoustic processor produces a broad phonetic transcription of the incoming utterance. This "machine-derived phonetic string" is input to the linguistic decoder, which uses the language syntax, an error set, and a phonetic lexicon to produce the orthographic output. Some phonological processing is incorporated in the acoustic processor, but semantics, pragmatics, and prosodics are not currently used.

Compared to an "ideal" phonetic form of the utterance, the machine-derived phonetic string will contain errors of substitution, omission, and insertion, introduced both by the speaker and by the machine. The possibility that a given machine error has occurred in a given place can be deduced from the error set. The probability that a given fragment of the phonetic string corresponds to a given word can be estimated by matching each of its variants and choosing the best match. The score for a match will depend on the product of the probabilities of the machine errors necessary to transform the word into the phonetic string fragment.

The problem is viewed as that of searching the syntax-defined tree of all possible utterances to find that path which requires the most likely set of error hypotheses. A likelihood measure for a given partial path can be obtained from the matching scores for its constituent words. Since the tree is too large to be fully explored, the linguistic decoder must employ a heuristic search technique which finds a high-scoring path in a reasonable time, with no guarantee that it is the best. Left-to-right sequential decoding algorithms using both depth-first (Fano or backtrack) and best-first (stack) techniques have been tried. A depth-first algorithm will follow the best choice down a given path until the path is complete or its likelihood falls below a threshold, when it returns to the last choice point and explores the next best alternative. Slightly better results have been obtained with best-first algorithms, which at every stage grow from the existing pathway with the highest likelihood.

In experiments reported by Tappert (1974), a best-first version of ARCS was evaluated on sentences taken from a 250-word vocabulary, with a finite-state grammar. A 54% accuracy in sentence recognition, corresponding to 91% accuracy in word recognition, was achieved for a single speaker after the machine had been trained.

Left-to-right parsing techniques originate from written-language process-ing and may not be the best approach to the MPCS situation, where the input is heavily corrupted by noise. For instance, severe contamination of the first part of the utterance could cause the temporary or permanent blockage of the correct path through the tree. Several of the systems to be described later attempt to build in both directions around "islands of reliability," where identi-fication can be made with more confidence. The ARCS system has been adapted to gain some of the virtues of a nondirectional approach by using "anchors" supplied by the acoustic processor to constrain the search (Paul and Rabinowitz, 1974).

Objections to the hierarchical approach stem from its inflexibility. In recognizing or understanding speech, we are essentially presenting the system with a problem to solve, rather than a noisy input to be decoded. The system's behavior should be problem-driven; the KSs should not be applied in a fixed, immutable sequence, but should be called into play as and when they are needed, in the context of the problem in hand. The purely hierarchical model seems inappropriate to a situation in which any KS can make mistakes. In the absence of feedback or feedforward, errors introduced at any stage will propagate through the system. This means that no KS can afford to be very selective in its output: unlikely possibilities must be retained in case they later become impor-tant. Doing this requires a good deal of expensive processing which will eventually be wasted. Thus in the hypothetical system of Fig. 14.15, lexical processing produces a complex word tree, almost all of which is subsequently discarded. In the ARCS system, searching for words is syntax-guided, but early application of semantic constraints, say, could further restrict the lexical matching. At the acoustic–phonetic level, is it necessary to search for every phoneme at every point in the utterance?

14.3.4. The Top-Down Approach

Top-down MPCS systems function according to the "hypothesize-and-test" paradigm. The acoustic data is used to evaluate or test predictions or hypotheses which have been actively generated by the "higher" levels of the system. A typical hypothesis might be the occurrence of a specified word or phrase within a specified time range. "Hypothesize-and-test" is similar to the old idea of "analysis-by-synthesis" (MacKay, 1952; Halle and Stevens, 1962), in which testing was to be done by matching the actual utterance against a synthesized version.

Top-down analysis may be thought of as growing a directed graph in which each node represents a partially decoded sentence. Each hypothesis defines a possible successor node to a given node (a way of filling a gap in its partially decoded sentence to create another). The merit of this construction depends on our *a priori* confidence in the hypothesis and on the results of the test. The process stops when a node representing a complete and acceptable sentence is found. Design problems include how to generate the hypotheses and assess their *a priori* chances of success, how to reconcile this with the results of the testing, how to choose a new starting node when a hypothesis fails, and how to avoid duplication of effort (different paths through the graph will not be independent and looping is a possibility). (Also see Chapter 5.)

The voice-controlled data management system (VDMS), developed at SDC, is a good example of the top-down philosophy (Barnett, 1973; Ritea, 1974). This is a speech understanding system in which a spoken data management language (currently 150-word vocabulary, described by 35 syntactic rewrite rules) is used to access a data base. In the work reported, the data base contained information about submarine fleets, a typical utterance being "Print type where missiles greater than seven." An overview of VDMS is shown in Fig. 14.17.

Speech is input to an "acoustic–phonetic processor" which forms an array of data called the "A-matrix." Each row of the A-matrix corresponds to a 10-msec speech segment and contains a rough segment label (e.g., vowel-like, strong fricative), one or more refined segment labels (roughly phonemic) with associated probability measures, and various acoustic parameter values.

The "lexical matching procedure" verifies or rejects predicted words by matching against the A-matrix. The lexicon supplies an idealized phonemic form of the word which is expanded into a list of possible variants by phonological rules. Each variant is mapped against the acoustic data. The matching is done on a syllable-by-syllable basis, using a further set of rules dealing with interactions over syllable boundaries. The parser uses "predictive linguistic constraints" – grammatical and semantic – to hypothesize phrases and hence words in the utterance. A nondirectional method is used: the parser can predict either to the left or right of already recognized phrases. The parser contains four major modules:

1. *Classifier*, which assigns a syntactic category to each word accepted by the lexical matching procedure
2. *Bottom driver*, which predicts how found phrases might fit into complete sentences
3. *Top driver*, which takes a predicted phrase and from it derives either a word to be matched or a shorter phrase to look for next

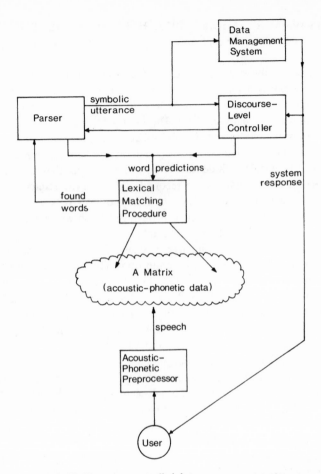

Fig. 14.17. The voice-controlled data management system.

4. *Side driver*, which determines which part of an incomplete phrase
(suggested by the bottom driver) the top driver should inspect next
(Complete phrases which do not cover the entire utterance are returned
as input to the bottom driver, so that parsing can be driven recursively.)

VDMS makes great use of pragmatic constraints. In the "discourse-level"
controller, a "user model" determines from the man—machine dialogue what
"mode" the user is in (e.g., "interactive query," "report-generation," "system
log-in") and from this predicts additional grammatical constraints on the form
of his next utterance. A "thematic memory" keeps track of content words
which have occurred in the dialogue and predicts words which might occur in
the next utterance, on the assumption that recently mentioned concepts are

likely to be used again. These predictions can be sent directly for lexical matching to initiate processing of an utterance.

In initial testing of VDMS, an average of 52% of sentences in realistic dialogues were correctly understood. Two male speakers were used: it is necessary to compile vowel-formant tables for each speaker.

In the Stanford Research Institute (SRI) speech understanding system (Walker, 1974; Paxton, 1974), top-down control is taken one stage further. The acoustic preprocessor does not attempt any phonetic labeling, other than a rough "phoneme-type" classification for each 10-msec interval. Predictions are tested by calling "word functions"; there is a separate word function for every word in the vocabulary. A word function is a set of subroutines, written after detailed examination of the acoustic data for the word in a variety of contexts, which return the probability of the word being present.

The SRI system simulates a spoken dialogue with a "robot" about the assembly and repair of small mechanical devices, e.g., repairing a leaky tap. The task domain is simulated by a "world model" and analysis of an utterance (e.g., "what little brass parts are in the box?") produces a program that operates on the word model. The execution of this program constitutes "understanding."

Analysis is controlled by a parser which operates a best-first strategy: every step on every path is assigned a "priority," path priorities are calculated from the products of the step priorities, and the system follows the highest priority path. Four kinds of step can be made: *syntactic steps* reflect the selection of a particular grammatical construction, and *lexical steps* correspond to the choice of a particular word from a class. Each lexical step is followed by a *word verification step* (a call to a word function), and a facility designed to avoid duplication of effort among competing paths produces "*interparse-cooperation*" *steps*. Step priorities for word verification steps depend on the goodness of the word-function match and how well it aligns with adjacent words, while for syntactic and lexical steps the priorities depend on KSs — syntactic and semantic cues and world model information.

In processing 71 utterances, with 42-word functions available, 62% were correctly understood. The system is rapidly evolving; for instance, it is planned to expand the KSs used in calculating priorities and to add dialogue, task models, and prosodics. The above results can be interpreted as a lower bound on the capability of the system.

The effectiveness of the top-down approach clearly depends on the predictive power of the higher levels of the system, on the capability of syntax, semantic, and pragmatics, or making intelligent guesses. Thus top-down systems function best in problem domains where there are strong linguistic constraints on what the speaker is allowed to say. Prediction is perhaps most difficult in the initial stages of processing an utterance, when more possibilities are open. Prosodic information is potentially of great importance in simplifying the hypothesis-making: Lea (1974) has proposed a "prosodically guided speech understanding strategy." In this scheme, a preliminary analysis, which relies heavily on prosodic features of the utterance, yields both syntactic and lexical hypotheses which initialize top-down processing, so that the parser needs to consider only a small subset of possible structures from the outset. The likely grammatical form of the sentence is derived from phrase boundaries, rhythm, and stress patterns, while probable words are predicted from acoustic—phonetic analysis and lexical matching of stressed portions of the utterance. As yet, only the preliminary analysis modules in this system have been implemented.

Another speech understanding system which is developing rapidly is BBN SPEECHLIS (Woods, 1974; Bates, 1974; Nash-Webber, 1974; Rovner et al., 1974). The problem framework is again that of verbal accessing of a data base. The data base concerns chemical analysis of the APOLLO 11 moon rocks, for which a natural language question and answer system called LUNAR already exists. LUNAR permits a wide range of sentence forms, e.g.,

> "What is the average concentration of rubidium in high-alkali rocks?"
> "List potassium/rubidium ratios for samples not containing silicon."
> "In what samples was titanium found?"

The full LUNAR system operates with a 3500-word lexicon: currently SPEECHLIS uses a 250-word subset.

A feature of the development of SPEECHLIS is the use of "incremental simulation" (Woods et al., 1974), whereby parts of the system not yet automated are replaced in simulations by humans in order to gain some feeling for how these parts might work. For instance, this technique enabled the authors to decide that there was no point in looking on a "bottom-up" scan for short function words ("a," "of," "the," etc.), which are usually unstressed, because of the high probability of accidental matches. However, syntax should be able to predict, later, where function words might occur and semantics can decide which are most likely.

SPEECHLIS does make use of an initial word scan, but only for content words of three or more phonemes and words which are pragmatically likely to begin an utterance. Before this, acoustic—phonetics assembles, with the aid of phonological rules, a "segment lattice" which represents probabilistically the alternative phonetic segmentations of the utterance, and the possible identities of each segment. Word matching is accomplished by a "lexical retrieval" program which finds words in the lexicon which could fit fragments of the segment lattice and a "word verification" component which, given a particular word and a particular location, returns a matching score. Phonological rules and a "similarity matrix" — a kind of error set — are used in the matching algorithms.

The robust word matches found in the initial lexical scan, together with their scores, are entered in a "word lattice." The word lattice data is used to initialize internal data objects called "theories": each theory represents a hypothesis that a certain nonoverlapping collection of words are in the utterance, and includes syntactic and semantic structures linking these words and associated scores. The theories grow and change as SPEECHLIS finds evidence for or against them. Eventually, a theory representing a complete understanding of the utterance is evolved.

Semantic knowledge in SPEECHLIS is embodied in a network which represents the associations between words and concepts. Semantics can use this network to propose refinements to a theory. For instance, a word match for "chemical" suggests the concepts CHEMICAL ANALYSES and CHEMICAL ELEMENTS, and semantics can formulate "proposals" to look for words like "analysis," "determination," "element." If "analysis," say, is found, a new theory will be formed containing the two word matches and their semantic link: In turn, this theory could lead to a search for an instantiation of the concept ROCK and so on.

As new theories are created by semantics, each is examined to determine whether syntax might be able to develop it further by postulating grammatical structures linking the words in the theory. Syntactic knowledge is embodied in a transition network grammar, which allows for prediction to the right or left of any given phrase. For a theory "... people done chemical analyses ...," syntax might suggest that an auxiliary verb appears somewhere in the utterance (probably at the beginning) to modify the past participle "done."

Theory-building activities in SPEECHLIS are governed by a "control module" a principal mechanism in this is the creation of "monitors." A monitor is a trap set for evidence which, if found, might effect a particular theory. When a monitor is noticed, an "event" is created, pointing to the monitor and the new data. The basic control strategy is to form queues of proposals, theories, and events, ordered by their likelihood scores, and to select elements from the top of these queues for processing.

14.3.5. The Heterarchical Approach

The notion of an artificial intelligence consisting of a number of independent, specialist modules which interact like a committee of experts, rather than a master and his slaves, has gained widespread acceptance in machine intelligence research (Winston, 1972). The idea has been applied to the MPCS problem in the HEARSAY system, developed at Carnegie-Mellon University, Pittsburgh, Pa. (Reddy *et al.*, 1973a Lesser *et al.* 1974; Reddy *et al.*, 1973b.)

In HEARSAY, each KS is modeled as a self-contained procedure with three functions:

1. Deciding when it has something useful to contribute

2. Making contributions by the mechanism of originating hypotheses

3. Testing hypotheses made by other KSs

This system structure is inherently more flexible than the top-down model. Which of the KSs originate hypotheses, and which of the KSs verify them, depends on the current problem context rather than on any predetermined order. The HEARSAY framework in addition makes it easy to evaluate the contributions made by individual KSs (by disconnecting them) and easy to add new KSs into the system.

In the HEARSAY I system (Reddy *et al.*, 1973a, Reddy *et al.*, 1973b), operational in 1972, three independent "recognizers" were implemented: a diagram of the recognition process is shown in Fig. 14.18. The recognizers cooperate by individually suggesting and collectively verifying word hypotheses to fill the gaps in a "currently accepted partially recognized utterance" (CAPRU). No hypothesis is adopted unless it is acceptable to all the recognizers. Recognition proceeds in alternate hypothesizing and verifying phases, controlled by a "recognition overlord" (ROVER). At the end of each hypothesizing phase, ROVER sends the hypothesis from the "most confident" recognizer to the others for verification. ROVER also stores other hypotheses on a stack to enable backtracking and updates CAPRU after verification.

Preprocessing produces a "partial symbolic utterance description" (PSUD), including an estimate of the phonetic identity of each 10-msec time segment,

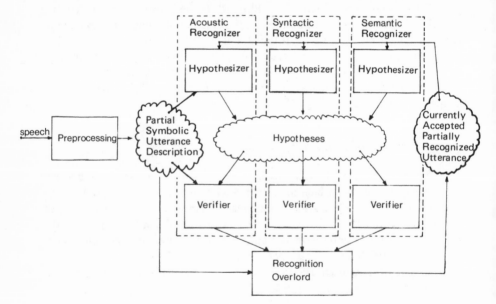

Fig. 14.18. Overview of the HEARSAY I system.

plus segmentation results and parameter values. This acoustic data, together with acoustic-phonetic, phonological, and lexical knowledge, is used by the "acoustic recognizer" to predict and verify syllables and words. Predictions are made by retrieving words from the lexicon, which contain features which are present in PSUD. Acoustic verification is similar to the "word matching" idea described above.

The "syntactic recognizer" can predict words to the left or right of phrases present in CAPRU. Only local context is used in hypothesizing, but syntactic verification requires complete parsing of the partially recognized sentence.

Both semantic and pragmatic knowledge is embodied in the "semantic recognizer." The problem domain chosen was "voice-chess" — recognizing spoken chess moves (e.g., "knight to king's bishop three") in the context of a given board position. A "chess program" generates a list of possible moves in the current board position, together with an estimate of the likelihood of each move. In conjunction with current context (CAPRU) and other semantic cues, this enables the prediction of likely words in the utterance. In voice-chess, the semantic recognizer thus has a powerful capacity and usually generates the hypotheses, though the acoustic recognizer will take over when robust acoustic cues are present.

HEARSAY I was quite successful in the voice-chess domain. In one run, 16 out of 19 utterances were understood correctly. The CMU team are currently working on HEARSAY II (Lesser *et al.*, 1974), an ambitious development of HEARSAY I with the same heterarchical organization. HEARSAY II is to be implemented on a 16-processor minicomputer system, enabling the KSs to function in a self-activating, asynchronous, parallel manner. Cooperation among the KSs is to be achieved through a global data base or "blackboard," which allows the hypothesis-and-test paradigm to function on a variety of levels. The blackboard is a three-dimensional structure, the dimensions being level of representation (parametric, phonetic, lexical, etc.), time into the utterance, and alternative hypotheses for a given level and time. Hypotheses are represented by nodes in this structure, and interdependencies between hypotheses can be specified by links between the nodes.

14.4. CONCLUSIONS

From this brief survey of research in automatic speech recognition, it will be seen that the focal point of interest has gradually shifted from the relationship between the acoustics and phonetics of speech to the incorporation of this relationship as one part of a problem-solving automaton. For the first time in 20 years there is a feeling that the right questions are being asked and rapid progress is being made.

A device for the recognition of isolated words has now reached the stage where it is being tested in practical applications. Despite its limitations of being able to deal with the voice of only a single speaker and a small vocabulary, there are situations where it may be extremely useful. For example, where an operator's hands and eyes are already occupied, as in the sorting of parcels, the voice provides a convenient medium for entering data into a machine. Even in situations where the hands are available, it has been suggested that operators prefer voice input to alternative methods, such as via keyboards.

Machine perception of continuous speech has not yet reached the stage where it can be tested in practical applications. It is already beginning to affect other fields. The group developing the HEARSAY system are computer scientists who have tackled the problem of speech recognition because they believe that this problem pushes the performance of current machines to their limit, and thus suggests the direction in which computer science should develop. Their 16-processor system bears witness to the truth of this belief. The field of linguistics is also being developed more rapidly because of the needs of machine analysis and recognition of speech.

It is worth remembering that the "hypothesis-and-test" paradigm on which many of the latest systems are based borrows much from psychological theories which were proposed to explain experiments and observation in perception, especially visual perception (MacKay, 1967; Gregory, 1970). The full circle of the wheel is now almost complete: the systems developed for machine recognition of continuous speech are being suggested as possible models for the perception of speech by humans.

It is perhaps too early to write a review on this subject. It may turn out that success in one problem domain may not be capable of generalization to others, just as the early recognizers failed when more speakers were tested. It is during the period when progress abounds, however, that interest is at its greatest.

REFERENCES

Ackroyd, M. H., 1974, Commercial applications of speech recognition, *IEE Colloquium on Speech Synthesis and Recognition*, Digest No. 1974/9, p. 7.

Ainsworth, W. A., 1972, Duration as a cue in the recognition of vowels, *J. Acoust. Soc. Am.* 51:648–651.

Ainsworth, W. A., 1973, Intrinsic and extrinsic factors in vowel judgements, *Auditory Analysis and Speech Perception*, Academic Press, London.

Barnett, J., 1973, A vocal data management system *IEEE Trans. Audio Electroacoust.* AU-21:185–188.

Bates, M., 1974, The use of syntax in a speech understanding system, *IEEE Symp. Speech Recognition*: 226–233.

Bezdel, W., and Chandler, H. J., 1965, Results of analysis and recognition of vowels by computer using zero-crossing data, *Proc. IEEE* 112:2060.

Broadbent, D. E., and Ladefoged, P., 1960, Vowel judgements and adaptation level, *Proc. Royal Soc. B* **151**:384–399.

Davis, K. H., Biddulph, R., and Balashek, H., 1952, Automatic recognition of spoken digits, *J. Acoust. Soc. Am.* **24**:637–642.

Dudley, H., and Balashek, S., 1958, Automatic recognition of phonetic patterns in speech, *J. Acoust. Soc. Am.* **30**:721–732.

Fant, C. G. M., 1960, *Acoustic Theory of Speech Production*, Mouton, s'Gravenhage.

Forgie, J. W., and Forgie, C. D., 1959, Results obtained from a vowel recognition computer program, *J. Acoust. Soc. Am.* **31**:1480–1489.

Fry, D. B., and Denes, P., 1958, The solution of some fundamental problems in mechanical speech recognition, *Language and Speech* **1**:35–58.

Fujisaki, H., and Kawashima, T., 1968, The roles of pitch and higher formants in the perception of vowels, *IEEE Trans. Audio Electroacoust.* **AU-16**: 73–77.

Gerstman, L. J., 1968, Classification of self-normalized vowels, *IEEE Trans. Audio Electroacoust.* **AU-16**:78–80.

Green, P. D., 1971, *Temporal characteristics of spoken consonants as discriminants in automatic speech recognition*, Ph.D. Thesis, University of Keele.

Green, P. D., and Ainsworth, W. A., 1973, Towards the automatic recognition of spoken Basic English, *Machine Perception of Patterns and Pictures*, Inst. of Physics Conf. Series No. 13, p. 161–168.

Gregory, R. L., 1970, *The Intelligent Eye*, Weidenfeld and Nicolson, London.

Halle, M., and Stevens, K. N., 1962, Speech recognition: a model and a program for research, *IRE Trans. Information Theory* **IT-8**:155–159.

Jakobson, R., Fant, C. G. M., and Halle, M., 1952, *Preliminaries to Speech Analysis*, MIT Tech. Report No. 13.

Klatt, D. H., and Stevens, K. N., 1973, On the automatic recognition of continuous speech, *IEEE Trans. Audio Electroacoust.* **AU-21**:210–217.

Lavington, S. H., 1968, *Measurement systems for automatic speech recognition*, Ph.D. Thesis, University of Manchester.

Lawrence, W., (1953), The synthesis of speech from signals which have a low information rate, *Communication Theory* (W. Jackson, ed.), Butterworths, London, 460–469.

Lea, W. A., Medress, M. F., and Skinner, T. E., 1974, A prosodically-guided speech understanding strategy, *IEEE Symp. Speech Recognition*, 38–44.

Lesser, V. R., Fennel, R. D., Erman, L. D. and Reddy, D. R., 1974, Organization of the HEARSAY II speech understanding system, *IEEE Symp. Speech Recognition*, 11–21.

Licklider, J. C. R., and Pollack, I., 1948, Effects of differentiation, integration, and infinite peak dipping on the intelligibility of speech, *J. Acoust. Soc. Am.* **20**:42–51.

Lindblom, B. E. F., and Studdert-Kennedy, M., 1967, On the role of formant transitions in vowel recognition, *J. Acoust. Soc. Am.* **42**:830–843.

MacKay, D. M., 1952, Mentality in machines, *Proc. Aristot. Soc. Suppt.*, **26**:61–86.

MacKay, D. M., 1967, Ways of looking at perception, *Models for the Perception of Speech and Visual Form* (W. Wathen-Dunn, ed.), MIT Press, Boston, 25–43.

Nash-Webber, B., 1974, Semantic support for a speech understanding system, *IEEE Symp. Speech Recognition*, 244–249.

Nelson, A. L., Werscher, M. B., Martin, T. B., Zadell, H. J., and Falter, J. W., 1967, Acoustic recognition by analog feature-abstraction techniques' *Models for Perception of Speech and Visual Form*, (W. Wathen-Dunn, ed.), MIT Press, Boston, 428–439.

Newell, A., Barnett, J., Forgie, J. W., Green, C., Klatt, D., Licklider, J. C. R., Munson, J., Reddy, D. R., and Woods, W. A., 1973, *Speech Understanding System* North-Holland Publishing Co.

Öhman, S. E. G., 1966, Perception of segments of VCCV utterances, *J. Acoust. Soc. Am.*, **40**:978–988.

Oshika, B. T., Zue, V. W., Weeks, R. V., Neu, H., and Aurbach, J., 1974, The role of phonological rules in speech understanding research, *IEEE Symp. Speech Recognition*, 204–207.

Paul, J. E., and Rabinowitz, A. S., 1974, An acoustically based continuous speech recognition system, *IEEE Symp. Speech Recognition*, 63–67.

Paxton, W. H., 1974, A best-first parser, *IEEE Symp. Speech Recognition*, 218–225.

Peterson, G. E., and Barney, H. L., 1952, Control methods used in a study of the vowels, J. Acoust. Soc. Am. **24**:175–184.

Pollack, I., and Pickett, J., 1964, The intelligibility of excerpts from conversation, *Language and Speech* **6**, 165–171.

Potter, R. K., Kopp, G. A., and Green, H. C., 1947, *Visible Speech*, van Nostrand, New York.

Purton, R. F., 1968, Speech recognition using autocorrelation analysis, *IEEE Trans. Audio Electroacoust.* **AU-16**:235–239.

Reddy, D. R., 1967, "Computer recognition of connected speech, *J. Acoust. Soc. Am.*, **44**:329–347.

Reddy, D. R., Erman, L. D., and Neely, R. B., 1973, A model and a system for machine recognition of speech, *IEEE Trans. Audio Electroacoust.* **AU-21**:229–238.

Ritea, H. B., 1974, A voice-controlled data management system, *IEEE Symp. Speech Recognition*, 28–31.

Rovner, P., Nash-Webber, R., and Words, W. A., 1974, Control concepts in a speech understanding system, *IEEE Symp. Speech Recognition*, 267–272.

Sakai, T., and Doshita, S., 1962, The phonetic typewriter, *Proc. IFIP Congress*, Munich.

Tappert, C. C., 1974, Experiments with a tree search method for converting noisy phonetic representation into standard orthography, *IEEE Symp. Speech Recognition*, pp. 261–266.

Tappert, C. C., Dixon, N. R., and Rabinowitz, A. S., 1973, Application of sequential decoding for converting phonetic to graphic representation in automatic recognition of continuous speech (ARCS), *IEEE Trans. Audio Electroacoust.* **AU-21**:225–229.

Teacher, C. F., Kellett, H., and Focht, L., 1967, Experimental, limited vocabulary, speech recognizer, *IEEE Intern. Conv. Record* (Part III), 169–173.

Walker, D. E., 1974, The SRI speech understanding system, *IEEE Symp. Speech Recognition*, pp. 32–37.

Winograd, T., 1972, *Understanding Natural Language*, Edinburgh University Press, Edinburgh.

Winston, P. H., 1972, The MIT robot, *Machine Intelligence* **7**:431–463.

Wiren, J., and Stubbs, H. L., 1956, Electronic binary selection system for phoneme classification, *J. Acoust. Soc. Am.* **28**:1082–1091.

Woods, W. A., 1974, Motivation and overview of BBN SPEECHLIS, an experimental prototype for speech understanding research, *IEEE Symp. Speech Recognition*, pp. 1–10.

Woods, W. A., and Makhoul, J., 1974, Mechanical inference problems in continuous speech understanding, *Artific. Intell.* **5**:73.

15

Editorial Introduction

C. D. Binnie, G. F. Smith, and B. G. Batchelor introduce the electroencephalo-gram (EEG) as a noninvasive technique for monitoring cerebral function. The EEG has found many uses in medicine, including:

1. Location of tumors
2. Detecting different "levels" of sleep/anesthesia
3. Detecting brain death
4. Sensing whether autistic children can see/hear

Despite its widespread use, the EEG is far from being well understood; its origins are obscure and its waveform structure resembles random noise (to the casual observer). The EEG recording suite produces almost as much paper as a computer room! To circumvent the tedium of viewing this by eye, many workers have sought to apply pattern recognition methods. The application of pattern recognition techniques to the EEG has been in progress for several years and seems to be bearing fruit. One of the most successful studies was concerned with predicting the outcome following cardiac arrest. A human operator viewed the EEG tracing, on paper, and answered a number of simple questions about it. (These were specified on a proforma.) Pattern recognition methods were used to compute the more probably outcome, either death due to brain damage or recovery of normal cerebral activity. Despite the ghoulish prospect of having a machine making life/death decisions, this project was very successful; the case-management of cardiac arrest patients at the London Hospital has gained very considerably from this work. In this chapter, we also see other, almost immediate prospects for the application of pattern recognition to medicine.

This chapter has obvious parallels with those by Thomas (Chapter 13), and Ainsworth and Green (Chapter 14), since all three have the analysis of time-varying signals as their central theme. [Parks (Chapter 10) also touches on this topic.] The work reported here grew out of an attempt to apply the techniques described in Chapter 4 to the EEG. We soon realized that this would provide only some of the procedures needed. Although not developed here, it is clear that the work by Bell (Chapter 5) would find application in analyzing the replies on a proforma, such as that mentioned earlier, in connection with cardiac arrest. There are also (tenuous) connections between this chapter and those by Amos (Chapter 11), Ullmann (Chapter 2), Halé and Saraga (Chapters 7 and 9), who all discuss the use of Fourier methods in pattern recognition. (The last three mentioned chapters use methods on spatial patterns, while Binnie *et al.*, Ainsworth and Green, and Thomas apply them to temporal patterns. The differences are trivial.)

15

Pattern Recognition in Electroencephalography

C. D. Binnie, G. F. Smith, and B. G. Batchelor

15.1. INTRODUCTION AND GENERAL DESCRIPTION OF THE HUMAN ELECTROENCEPHALOGRAM

When large numbers of nerve cells in the surface layers of the brain are involved in synchronous electrical activity the signals produced can be picked up by electrodes applied to the surface of the scalp. The activity recorded, which is termed the *electroencephalogram* (EEG), has an amplitude generally of some $10-100 \mu V$ and is usually recorded in a frequency band from 0.5 to 50 Hz. The EEG is studied for a variety of purposes and can be used as an aid to the diagnosis of structural diseases or functional disorders of the brain or as a research tool in human psychology and psychopharmacology. For a standard introductory textbook of electroencephalography and as a source of key references, the reader should consult Kiloh *et al.* (1972).

The electrical activity differs from one area of the scalp to another and, for many purposes, it is necessary to employ a multichannel display to monitor the activities of different regions. The most conspicuous feature of the EEG of a normal waking adult, resting quietly with closed eyes, is the alpha rhythm,

C. D. Binnie · Instituut voor Epilepsiebestrijding, Heemstede, Netherlands
G. F. Smith · Neuropsychology Unit, Department of Applied Psychology, University of Aston, Birmingham, England *B. G. Batchelor* · Department of Electronics, University of Southampton, Southampton, England

regular activity of about 10 Hz recorded chiefly from the back of the head (Fig. 15.1). This activity is reduced or abolished during drowsiness, arousal, and visual attention. Other faster and slower activities are commonly present, and for descriptive purposes these are divided into arbitrary frequency bands, designated by Greek letters:

> *delta:* below 4 Hz
> *theta:* not less than 4 but less than 8 Hz
> *alpha:* 8–13 Hz inclusive
> *beta:* more than 13 Hz

The EEG changes with age, state of awareness, and a variety of other physiological factors. The EEG of the infant is dominated by delta and theta activity; the alpha rhythm appears during early childhood (Fig. 15.2). By adult life, delta activity is virtually absent and theta considerably less prominent than the alpha rhythm. During drowsiness, the alpha activity gives way either to theta or beta components, and as the subject falls more deeply asleep, delta activity comes to dominate the picture.

As might be expected, recording a biological signal in the microvolt range presents considerable technological difficulties; details of EEG recording techniques will be found in Cooper *et al.* (1974) and Binnie *et al.* (1974). Suffice it for the present account that, despite the use of complex and costly equipment and careful attention to technique, particularly in the preparation and application of electrodes, the EEG tracing will normally be contaminated with a variety of other signals, chiefly of biological origin. These arise from eye

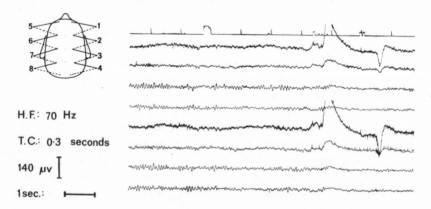

H.F.: 70 Hz

T.C.: 0·3 seconds

140 μv

1 sec.:

Fig. 15.1. The EEG of a normal adult. On the channels recording from the posterior regions of the head (3, 4, 7, and 8) alpha activity is seen. 3 s before the end of the excerpt, the eyes are opened; the alpha rhythm is attenuated and an eye-opening artifact is seen on the anterior channels, followed 1½ s later by a blink.

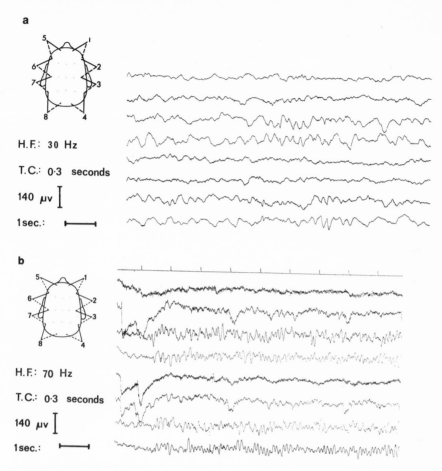

a

H.F.: 30 Hz

T.C.: 0·3 seconds

140 μv

1sec.:

b

H.F.: 70 Hz

T.C.: 0·3 seconds

140 μv

1sec.:

Fig. 15.2(a) EEG of a child age 4 months. The record is dominated by theta and delta activity; no alpha rhythm is seen. (b) EEG of a normal 7-year-old child. A well-developed alpha rhythm is now present over the posterior regions (channels 3, 4, 7, and 8) but a large amount of underlying theta and delta activity is still present (cf. Fig. 15.1).

movement, scalp muscles, the heart, etc. (Fig. 15.1). Although these artifacts may be of some interest as indices of the physiological state of the subject, they can obscure, or indeed be mistaken for, EEG activity and their identification represents a banal but important application of pattern recognition in electro-encephalography.

The signals displayed in a routine EEG recording arise from spontaneous cerebral electrical activity, but other phenomena can be elicited by stimulation of the subject. For instance, a transient stimulus such as a click or a flash of light will produce a characteristic pattern of waves lasting for some 300 ms (Fig. 15.3).

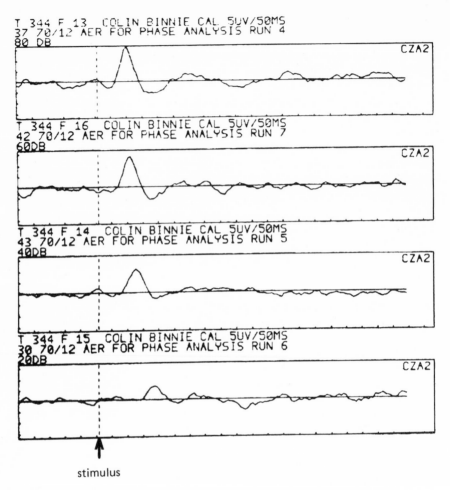

Fig. 15.3. Auditory evoked response. Cerebral responses of a subject with normal hearing to auditory stimuli (1 kHz) tone pulses of intensities, from upper to lowest tracing, of 80, 60, 40, and 20 dB. Note diminishing amplitude and increasing latency as intensity of stimulation is reduced. At 10 dB (the subjective auditory threshold of this particular individual) no response was obtained. (Time calibration in 50 ms units, amplitude 5 μV, stimulus delivered at vertical dotted line.)

These evoked potentials can be used both for investigating cerebral physiology and, for clinical purposes, as a means of testing the integrity of any sensory pathway. For instance, the suspicion that a dumb child is deaf may be tested by determining whether or not sound stimuli elicit evoked responses.

15.2. CLINICAL AND RESEARCH APPLICATIONS

The EEG is used as an aid to clinical diagnosis in those nervous disorders associated with abnormal cerebral function, either primary as in epilepsy or secondary to structural disease of the brain. EEG abnormalities may be generalized or confined to a small area of the scalp; they may consist of a continuing disorder of cerebral electrical activity or of episodic disturbances, sometimes lasting only a few tens of milliseconds. There is considerable individual variation in the EEG and also, as indicated above, marked changes are seen as a function of age and state of awareness. Nevertheless, the normal findings for subjects at a particular age and in a particular state of arousal have been known, admittedly in highly subjective terms, to clinical electroencephalographers for many years. More recently quantitative statistical descriptions of the normal EEG have become available (see Matousek and Petersen, 1973a).

Disturbances of ongoing activity* are often manifest by a change in the frequency composition of the EEG: typically the proportions of theta and delta activity are increased. Since the proportions of the slower components decrease in amount with age and with arousal, it may be necessary to know the age and state of awareness of the subject before the record can be judged to be normal or abnormal. In addition to changes in the frequency composition of the EEG, there may be localized alterations of amplitude. The normal tracing is, in general, symmetrical (although some activities are often of slightly greater amplitude over the nondominant cerebral hemisphere). Thus disease predominantly affecting one side of the brain (as, for instance, a stroke) may produce

---

*EEG reflecting continuous cerebral activity.

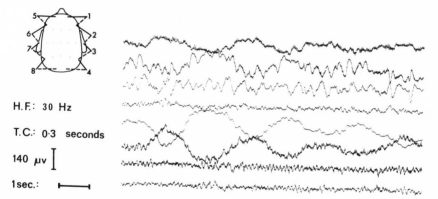

H.F.: 30 Hz

T.C.: 0·3 seconds

140 μv

1 sec.:

Fig. 15.4. Gross right-sided EEG abnormality following cerebral hemorrhage. Note abolition of normal rhythms over the right hemisphere (channels 1 to 4) and their replacement by slower activities in the delta and theta ranges.

asymmetries, which place the record outside normal limits. Figure 15.4 shows a grossly abnormal record: theta and delta activity have much greater power on the right than the left, whereas normal faster activities are greater on the left than on the right. The patient had suffered a stroke affecting the right side of the brain a few days before the record was taken.

Transient phenomena are seen particularly in the EEGs of people with epilepsy. During a seizure various abnormal waveforms occur, typically spikes or complexes of spikes and slow waves (Fig. 15.5). However, between the attacks, people with epilepsy may show sporadic discharges of similar type. Like the disturbances of background activity, paroxysmal abnormalities may be generalized or confined to one region of the brain. Generalized paroxysmal discharges occur typically in patients whose epilepsy is of "constitutional" origin and not accompanied by demonstrable structural abnormality (Fig. 15.5): localized discharges (Fig. 15.6) occur where the epilepsy is the result of disease or damage of the underlying brain, for instance a scar.

In many EEG departments the largest single group of patients referred for investigation are those in whom some structural abnormality of the brain is suspected, for instance, tumor, abscess, or hemorrhage. These may produce both localized abnormalities and more diffuse disturbances, due to raised pressure, or damage to structures deep in the brain and depending on the rate of progress of the condition and the stage of any healing process. The information obtained from the EEG in this group of patients is of limited clinical value, but electroencephalography has the important advantage of being a harmless noninvasive

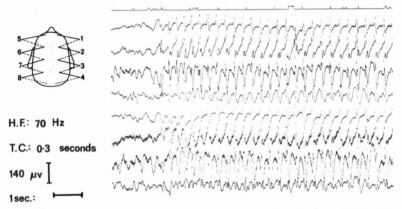

H.F.: 70 Hz

T.C.: 0·3 seconds

140 μv

1sec.:

Fig. 15.5. 3-Hz generalized spike wave discharge during a petit mal seizure. Petit mal is one of the "primary generalized epilepsies" in which abnormal activity is recorded simultaneously over all regions of the scalp and in which there is no obvious localized structural abnormality of the brain (cf. Fig. 15.6).

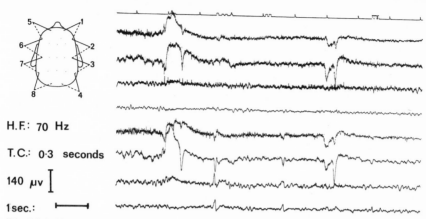

H.F.: 70 Hz

T.C.: 0·3 seconds

140 μv

1 sec.:

Fig. 15.6. Temporal lobe epilepsy. Left temporal spikes in a 40-year-old man with scarring of the left temporal lobe, which gave rise to seizures characterized by difficulty in speech, visual hallucinations, and confusion.

technique. Thus its chief role in the investigation of structural disease of the brain is for the early detection of abnormalities at a stage of the illness where there are insufficient clinical grounds to warrant more elaborate, costly, and possibly dangerous methods of investigation. Repeated EEG records may be taken in order to detect changes over a period of time. These may make it possible to differentiate between a progressive condition, such as a tumor, and one which is followed by gradual recovery, such as a brain hemorrhage.

Epilepsy represents another major clinical application of electroencephalography and is one in which the EEG makes a unique contribution. It is not common for a patient to have a seizure at the time of the recording, but the presence of paroxysmal activity, while not conclusive, lends support to any clinical evidence of epilepsy. Unfortunately, the converse does not hold, and some patients known to have fits may at times have apparently normal EEGs. From the standpoint of treatment, it is of particular importance to establish whether the seizures are due to a disordered function of the entire brain, or whether they arise from a local structural abnormality. The distribution of paroxysmal activity in the EEG often helps to resolve this problem.

There are a number of degenerative or infective diseases of the brain which may be difficult to identify clinically but which are accompanied by fairly characteristic EEG patterns. Thus the EEG may assist the diagnosis and also help to distinguish a deterioration in mental function due to generalized cerebral degeneration, from one due to a discrete, treatable lesion such as a tumor.

The activity of the brain is affected by a variety of general medical disorders which interfere with the supply of oxygenated blood or produce biochemical disturbances. The EEG bears a close quantitative relationship to certain biochemical abnormalities and can be used for monitoring the progress of the underlying disease, for example, in liver failure. There is also an increasing number of patients who suffer reversible acute disturbances (such as cardiac arrest or a profound drop in blood pressure) which may produce severe brain damage. Here the EEG is of value for predicting the eventual outcome and is used as an aid to the determination of brain death (Binnie *et al.* 1970; Prior, 1973).

Mention has been made above of the clinical application of evoked potentials for testing the integrity of sensory pathways. The main application of such techniques at present is in the assessment of hearing in children with disorders of communication. There is also increasing evidence that structural diseases of the brain may produce characteristic abnormalities of evoked responses, notably in multiple sclerosis. Other special clinical EEG investigations include recording from the exposed surface of the brain at operation (electrocorticography) and recording during sleep. The latter may help to highlight localized abnormalities, particularly in patients with epilepsy.

The EEG has not, in the main, proved a very useful tool for basic research on cerebral function, except possibly in relation to sleep (see Section 15.6.2). There have nevertheless been several studies relating the morphology of the EEG to personality, and cerebral dominance; changes in the EEG are demonstrable during the performance of various different psychological tasks. Such studies of normal waking subjects generally rely upon precise quantitative analysis of the record. For this purpose visual assessment is usually inadequate and electronic data processing must be used.

The EEG is often a more useful tool in those research applications which depend upon the detection of serial change, so that the subject can act as his own control. Examples include studies of cyclical alterations of the EEG and of biochemical variables related to the menstrual cycle and to some episodic psychiatric illnesses. The area to which the EEG has been most successfully applied in this manner is psychopharmacology. A change in the EEG following administration of a drug amounts to prima facie evidence that the substance has an action on the nervous system. Specifically, it is claimed that the various different groups of drugs used in psychiatry can be distinguished by characteristic effects on the EEG (Itil, 1974).

Night-long recordings during spontaneous sleep have provided a great deal of valuable information about the physiology of sleep. They are also used in studies of agents, again particularly drugs, which may interfere with sleep patterns in various ways.

15.3. THE MOTIVE FOR USING PATTERN RECOGNITION IN ELECTROENCEPHALOGRAPHY

It can be seen from the preceding sections of this chapter that over the years a large body of knowledge has been amassed concerning the visual interpretation of the EEG. The lack of anything other than the most superficial of theories relating EEG phenomena to brain function and structure has meant that most of this knowledge has been acquired in an *ad hoc* manner in the main by simple visual analysis of the waveforms and empirical observation of their clinical or psychological correlates. The absence of any viable alternative to visual analysis has meant that it remains the most widely used interpretative tool in electroencephalography. Electroencephalographers have not been slow to recognize its limitations. The EEG signal is very complicated and frequently a large amount of data has to be scanned; under these conditions an interpreter's performance may vary, not only from day to day and from subject to subject, but throughout the same recording! There is also the problem of ensuring standardization, or indeed communication, between different observers. Further, the human visual processing system almost certainly cannot detect some characteristics of the EEG which may be of clinical significance.

The direct application of pattern recognition techniques to emulate (and, hopefully, to improve on) the visual processing capability of the human observer in the detection of specific waveforms is obviously of great importance. (In fact, surprisingly little work has been carried out in this area, despite the fact that it should be amenable to the techniques developed for handwritten character identification.) However, there are more fundamental reasons why pattern recognition techniques are of importance in electroencephalography and which make them potentially more useful than many other automatic analysis methods.

Electroencephalographers have long known that successful EEG interpretation depends on taking into account, and attaching relative importance to, many different factors: a multivariate approach is required. The analysis of complicated multivariate data is the forte of pattern recognition and has led to hopes that it might profitably be employed in automatic methods of EEG interpretation.

15.4. FEATURE EXTRACTION

One of the greatest problems encountered when applying pattern recognition techniques to the analysis of the EEG is deciding on the method of feature extraction to be used. The difficulty lies in the fact that the exact relationship between the EEG and the operation and structure of the brain, is at present, only very superficially understood. Consequently, there are seldom any reliable *a priori* grounds for deciding which features of the signal (if any) are important to the investigation of any new clinical or psychological problem. Despite, or possibly because of this a large number of feature extraction processes have been suggested for EEG analysis (although few have been used in a direct pattern recognition context).

One of the best-established techniques is spectral analysis and this has, up to the present, been the most important as far as pattern recognition in the EEG is concerned. (See also Chapters, 11, 13, and 14.) This method of describing the EEG assumes that each epoch of the signal can be represented by a sum of harmonically related sine waves. It has been customary in EEG analysis to consider the distribution of power in the sinusoids.

Manual methods of spectral analysis were used first (Dietsch, 1932). Analog hardware, consisting of narrow bandpass filters and integrators, was used later (Baldock and Walter, 1946). (It is perhaps inappropriate to equate spectral analysis and hardware filtration methods, although there is a close relationship between them.) Since the rediscovery of the fast Fourier transform (FFT) by Cooley and Tukey (1965), digital computation has come to the fore. The application of the FFT to the EEG is reviewed by Dumermuth and Keller (1973). Spectral analysis, in real time, of up to eight channels of EEG signals is just within the capabilities of present day minicomputers, providing that integer arithmetic is used. Clinical EEG records often employ more than eight channels, and fixed-point arithmetic on a 16-bit machine gives barely adequate precision. Hence, at the moment, there is still a place for special-purpose hardware.

Numerous variations have been suggested on the spectral analysis theme, many of an intuitive nature. Various workers have used descriptors such as Q-factor, bandwidth of spectral peaks, mean frequency, variation of mean frequency. A number of authors have suggested ways of combining banded power-spectrum values; Matousek and Petersen (1973b) describe an ingenious method along these lines. From a large number of normal subjects, they developed criteria which allowed them to estimate the age of a subject from various measured EEG spectral parameters. It has long been noticed by electro-encaphalographers that certain types of cerebral abnormality are characterized by EEGs which, although abnormal for the subject under scrutiny, would be normal in a younger person. Faced with quantifying the EEG from such a suspected abnormal subject, they calculate his estimated age and divide this by his real age. Then they use this age-dependent EEG quotient as a descriptor of the EEG. Wennberg and Zetterberg (1971) used descriptors based on a parametric model of the power spectrum.

In order to take into account the relationship between different EEG channels, Walter (1963) suggested the use of coherence. This is a measure which expresses the degree of phase dependence between two signals at a given frequency. The time-domain analog of coherence is the correlation coefficient.

Many people have questioned the relevance of spectral analysis of the EEG. As Hjörth (1973) points out, ". . . the concept of frequency is questionable as to its relevance to the description of phenomena, which are known not to be periodic nor to have a sine shape." Hjörth (1970) has proposed three descriptors of the EEG which he calls activity, mobility, and complexity. Although they can be related to moments of the power spectrum, they are calculated in the time domain and do not depend on the concept of frequency. They are relatively quickly and simply implemented in software, and hardware is available for the computation of the parameters for 16 channels simultaneously.

An approach which is akin to that of the human observer is period analysis. The simplest method, whether implemented manually, by digital computer, or with analog hardware, involves detecting those points where the signal passes through a zero-voltage level (or through a near-zero threshold). The intervals between successive base-crossings may be plotted on a histogram, which provides a representation of the frequency distribution of waves of different frequencies. A weakness of simple methods of period analysis is that activities of different frequencies mask one another. Large fast waves cause frequent base-crossings and mask slower activity, whereas large slow waves carry any superimposed faster components away from the baseline, so that they are not detected. To some extent, these difficulties can be overcome by separately analyzing the primary signal and its first derivative. The latter ensures recognition of fast activity, since every peak or trough in the signal has a zero gradient and therefore produces a base-crossing of the first derivative (Saltzburg *et al.*, 1957). Various extensions of period analysis employ maxima and minima to define wave boundaries and measure peak-to-peak amplitude and other features of each wave. From this information, it is possible simultaneously to derive an approximation to the power spectrum and to identify various transient phenomena such as spike-wave complexes or even artifacts (Binnie, 1975).

Finally, mention should be made of the use of a human observer for feature extraction. In this method, various factors in the EEG are rated by an interpreter on previously defined scales, thus yielding numerical descriptors which provide the input for a decision-making process. As we shall see later (Section 15.6.4) this has been used to good effect in at least one particular study. Although the technique suffers from the disadvantages of visual analysis, it does mean that, providing a suitable classification procedure is chosen, an electroencephalographer can at least be confident of the objectivity of any decision obtained.

At present there is no concensus as to which is the best feature extraction method for a given investigation. As further knowledge becomes available concerning the physiological processes underlying the EEG, so, hopefully, the suitability of the various methods will become more apparent.

15.5. STEPWISE DISCRIMINANT ANALYSIS

In subsequent sections of this chapter, there are several references to stepwise discriminant analysis, a technique which has, up to now, been one of the most widely used pattern recognition methods in EEG analysis. In order to avoid interrupting the discussion in the following sections, a brief explanation is included at this point.

The popularity of stepwise discriminant analysis stems from the fact that it combines a method for positioning linear decision surfaces between classes, with a process for selecting from a large number of features a small set of "important" ones. Many workers have used the program of Dixon (1970) to which the reader is referred for more information than it is possible to include here. As its name suggests, the method proceeds in steps. At each stage a number is calculated for each feature to express the amount of "discriminatory power" that the feature provides. The feature with the largest "discriminatory power" is then included, along with the features selected in previous steps, in a conventional linear discriminant analysis (Wilks, 1962). This method is different from the procedure described in Chapter 4; it assumes that the classes have normal distributions. The stepping process is terminated when various criteria are satisfied: for instance, when a satisfactory level of discrimination between the classes has been achieved or when the remaining features not included in the discriminant analysis can be expected to contribute little.

At the first step, the number used to express the discriminatory power of a feature is the classical F-ratio for testing the statistical significance of the difference between the means of two sets of observations. At subsequent steps, F is adjusted for each feature, based on its correlation with the previously selected feature. This avoids the selection of correlated features, which together provide little more discrimination between classes than each does singly.

A disadvantage of stepwise discriminant analysis is that it is based on the assumption that the descriptors are normally distributed for each class. It is difficult to justify such assumptions in advance, although the results achieved by the procedure may provide *a posteriori* vindication.

15.6. PATTERN RECOGNITION APPLICATIONS

15.6.1. Transients and Evoked Responses

In EEG terminology a transient is any isolated wave, or sequence of waves, which is distinguished from the background activity. These phenomena are often of clinical importance particularly in epilepsy. The automatic detection of transients may therefore be of value in its own right as a means of signaling their occurrence to the clinician or recording their incidence over an extended period of recording. However, the output of a transient detector can also be used as a descriptor of the signal for a further level of analysis.

A simple method for detecting transients in the EEG is by means of a matched filter. Although this technique is usually covered by communication theory, it is also employed in a pattern recognition context (Fukunaga, 1972). An example of this method is reported by Lloyd *et al.* (1972), who describe the application of a matched filter to the detection of spike-wave complexes. (Figure 15.7 illustrates the results obtainable by this method.) In general, the matched filter is not a very powerful or reliable technique for detecting EEG transients. The chief reason is that matched filters are usually designed under the assumption, often unjustified, that the spectrum of the activity masking the transient is uniform and markedly different from that of the transient

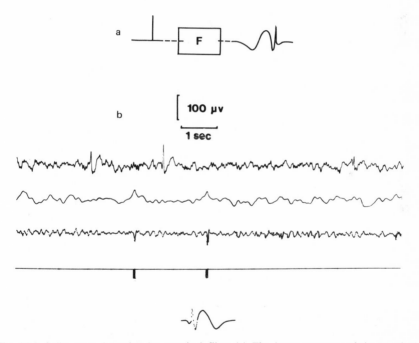

Fig. 15.7. Spike-wave detection by matched filter (a) The input response of the matched filter is the time inverse of a spike wave complex. (b) A matched filter (implemented on a digital computer) was set up to identify spike-wave complexes with the outline shown at the bottom of the figure. The raw EEG (top tracing) is passed through the filter giving the output shown in the second tracing. Whenever a spike-wave complex occurs this is followed after $1\frac{1}{2}$-s time delay by a peak on a filter output. These peaks are highlighted by calculating the second derivative of the filter output (third trace). A spike-wave complex is identified whenever the second derivative passes through a predetermined threshold (bottom).

phenomena. However, Lloyd *et al.* found that much improved results could be obtained by thresholding, not simply the output of the filter, but rather its second derivative.

Other approaches to recognition of EEG transients commonly employ the more usual approach to pattern recognition of preliminary feature extraction followed by some form of decision-making process. Any abrupt or short-lived change in a feature may be regarded as evidence that a transient has occurred irrespective of the naked-eye appearance of the tracing. More usually, a human observer is employed to define particular discontinuous phenomena which are to be recognized. The features may be selected on the basis of the known properties of the waveforms in question or the criteria may be developed empirically. Thus in the left-hand half of Fig. 15.8 the amplitude and the duration of individual waves has been plotted; a spike-wave complex is identified by the occurrence of two waves, one of long and one of short duration, of unusually great amplitude in relation to their period. To the right of Fig. 15.8 is shown the result of a more empirical approach, plotting Hjörth's normalized slope descriptors in a three-dimensional space; values relating to paroxysmal

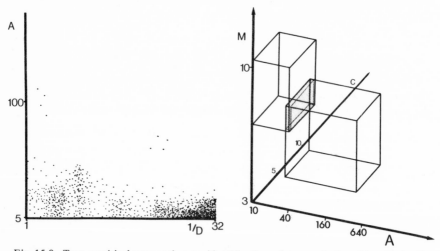

Fig. 15.8. Two empirical approaches to identification of paroxysmal activity. *Left*: The amplitude and duration of individual waves is measured and amplitude (in microvolts) plotted against reciprocal duration (i.e., frequency). Four spike-wave complexes have occurred and these have produced 4 short-duration and 4 long-duration waves of conspicuously greater amplitude than others of the same period. *Right*: Slope descriptors (Hjörth, 1970) calculated from 1-s epoch of a single EEG channel containing normal and paroxysmal activity and displayed in a 3-dimensional feature space. All the sections of EEG designated normal gave values within the upper left-hand cuboid; all values containing paroxysmal activity fell in the lower right-hand cuboid and there was little overlap between the two groups.

discharge (identified by human observer) were found to occupy a distinct region of the space.

Transient phenomena generally occupy a very small part of any EEG tracing and this presents particular problems when hyperspace methods are used. If, for instance, only 1% of the recording contained epileptic activity then a method which classified 1-s epochs with 95% reliability, with errors randomly divided between false positives and false negatives would be useless. The false-positive classifications of paroxysmal activity would exceed the number of actual paroxysmal discharges in the ratio 5:1. In such a situation it may be necessary to employ biasing methods in order to reduce the number of false-positive classifications albeit at the price of reducing the overall reliability of recognition (Batchelor, 1974).

In a direct pattern recognition context Serafini (1973) has described a nonsequential technique for discriminating between spike-wave complexes and some other EEG waveforms. However, as mentioned in Section 15.3, the detection of transients in the EEG is not an area in which established pattern recognition techniques have been widely used.

Evoked responses present a special case of transient phenomena but, as the timing of their occurrence is predictable, their identification is not generally regarded as a pattern recognition problem. Nevertheless, stepwise discriminant analysis has been applied successfully to the study of visual evoked responses (Donchin, 1969). Usually, the detection of evoked responses which are of substantially smaller amplitude than the ongoing EEG activity, depends upon averaging the signal following repeated stimulus presentations. Donchin was able to discriminate between two different types of responses without averaging, and there is hope that pattern recognition may eventually provide an improved means of detecting evoked responses without the need for large numbers of stimulus presentations.

15.6.2. Sleep

The EEG recorded from a sleeping subject shows considerable changes from that in waking and in different levels of sleep. In order to facilitate quantification, sleep has been divided into five stages on the basis of the EEG. To ensure consistency between different laboratories and investigations, most workers "stage" sleep by the standardized criteria of Rechtschaffen and Kales (1968); these may be summarized as follows (see Fig. 15.9):

Stage 1:   A relatively low-voltage, mixed-frequency EEG without rapid eye movements (REMs)

Stage 2:   12–14 Hz sleep spindles and K complexes on a background of relatively low-voltage, mixed-frequency EEG activity

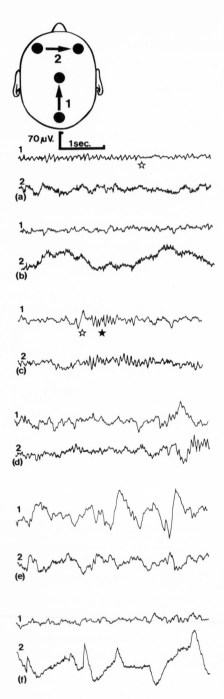

Fig. 15.9. Channel 1 is an ordinary EEG channel (CZOZ). Channel 2 is a composite eye movement monitor channel. (a) The EEG of a subject resting with his eyes closed is usually characterized by the sinusoidal-like waves of the alpha rhythm. As the subject becomes drowsy the waves disappear (✰). (b) Stage 1 sleep is characterized by a relatively low-voltage, mixed-frequency EEG without rapid eye movements (REMs). (c) Stage 2 sleep is characterized by *k* complexes (✰) and 12–14 Hz sleep spindles (★) on a background of relatively low voltage, mixed frequency EEG activity. (d) Stage 3 sleep is characterized by moderate amounts of high amplitude, slow-wave activity. (e) Stage 4 sleep is characterized by large amounts of high amplitude, slow-wave activity. (f) Stage REM sleep is characterized by a relatively low-voltage, mixed-frequency EEG in conjunction with episodic REMs.

Stage 3:   Moderate amounts of high-amplitude, slow-wave activity
Stage 4:   Large amounts of high-amplitude, slow-wave activity
Stage REM: A relatively low-voltage, mixed-frequency EEG in conjunc-
             tion with episodic REMs and low-amplitude electromyogram

To free the human observer from the tedious job of inspecting and classifying long sleep records, many workers have suggested ways of automating the sleep staging process. Many of the methods proposed concede no direct alliance to the pattern recognition school as such, but most can be interpreted in this light. The most common types of method are variations on the system shown in Fig. 15.10. In effect, these methods are really simple decision-tree classification or (hierarchical) procedures (see Chapter 5). Smith and Karacan (1971) described a system using this approach in which the outputs from analog filters were used as descriptors. Frost (1970) has described a similar method using period analysis. In a direct pattern recognition context, Martin *et al.* (1972) report a procedure they developed for sleep staging in which the percentage agreement between the automatic method and a consensus of human observers was 82%. (It should be pointed out that the agreement between different observers is typically of this order; Martin *et al.*, for instance, report a figure of 89% for the average percentage pairwise agreement among three interpreters.) Martin and coworkers used a period analysis approach to measure the percentage of delta activity in an epoch and by setting thresholds, thus detecting stages 3 and 4 sleep. Epochs not classified as stages 3 or 4 were then subjected to a decision-tree classification procedure, utilizing the discriminant analysis, iterative design technique of Viglione (1970). The result of this was a three-way classification: "stage 2," "awake," or "stage 1 *or* REM." Stages 1 and REM were then distinguished by scanning eye-movement monitoring channels.

Although the use of decision trees is the most popular reported method of

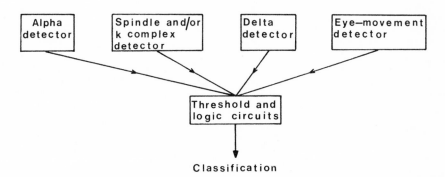

Classification

Fig. 15.10. The typical decision-tree (hierarchical) approach in automatic sleep classification.

automatic sleep staging, the feature-space approach has not been neglected. Larsen and Walter (1970) describe a method of automatic sleep classification, using stepwise discriminant analysis. As descriptors, they used banded spectral values and report the overall accuracy as 79%.

15.6.3. Psychological State

Walter et al. (1967) report an experiment in which they used step-wise discriminant analysis in an attempt to distinguish between five different psychological states, on the basis of the EEG in four subjects. The EEG phenomena associated with some states of consciousness, such as sleep, are easily recognized visually and hence are well documented. However, it is quite possible that there are other states of consciousness with their own attendant EEG phenomena which are not easily recognized. Although the reliability achieved by Walter et al. was not impressive, these results do seem to indicate that, by using sophisticated analysis techniques such as pattern recognition, hidden order might be detected.

The five psychological states considered by Walter et al. were:

1. *Eyes closed*, resting
2. *Eyes open*, resting
3. *Eyes closed*, performing a tone-recognition task
4. *Eyes open*, performing a visual discrimination task of 3 s duration
5. *Eyes open*, performing a visual discrimination tasks of 1 s duration

Segments of EEG from each state were subjected to spectral analysis. The variables calculated in each of four channels were: delta, theta, alpha, and beta power; the mean frequency within each band, and the bandwidth within each band. Coherences between the channels in each band were also calculated.

Using four features, Walter et al. found that they could discriminate between the five states, for all subjects combined, with an accuracy of 49%. Taking each subject alone, and using his personal four best descriptors, they found that an accuracy between 60 and 70% could be achieved.

15.6.4. Cerebral Anoxia

There are substantial numbers of patients admitted to intensive care units every year after circulatory arrest. Some of these patients, although still alive, fail to regain consciousness rapidly after resuscitation and present the problem to the clinician of deciding if there has been any brain damage and the degree of recovery possible. Binnie et al. (1970) have reported the results of an investigation in which they used visually extracted features of the EEG and linear discriminant analysis to predict fatal anoxic brain damage in those patients who failed to regain consciousness rapidly after cardiac arrest. They obtained

93 EEGs from 41 subjects in which the outcome after cardiac arrest was reliably known; either as recovery of consciousness without apparent intellectual or neurological deficit, or as death from pathologically proven anoxic brain damage. All of the subjects exhibited an absence of complicating factors, such as head injury or drug intoxication. For each EEG, 49 descriptors were obtained by visually rating the presence, prominence, and distribution of a wide range of features. Two further features were also included in the study — the state-of-awareness of the subject during the recording, as indicated by neurological evidence, and the time between resuscitation and recording. Linear correlation between the features was utilized to reduce the number to 36, and conventional linear discriminant analysis (Wilks, 1962) was then used to position a hyperplane decision boundary between the two classes. Figure 15.11 illustrates the results which were obtained. Only one of the 93 EEGs would have been misclassified using the decision surface calculated.

Prior (1973) has described the application of nonlinear decision surfaces to the above data. The adaptive technique of Batchelor (1969) was used to train a hypersphere classifier. All of the 93 EEGs were correctly classified using just 13 variables.

Maynard and Prior (1973) have given details of a predictive instrument which was designed with the above data (also see Batchelor, 1974). By means of decade switches the values of certain variables relating to patient are indicated to the machine. The distance of the point representing the patient in a feature space from a decision surface can then be read on a meter. Provision is made to signal points which are not typical of previous experience. (See Section 4.5.4).

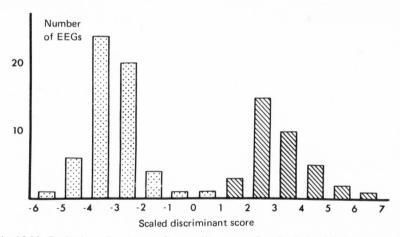

Fig. 15.11. Prediction of outcome after cardiac arrest. Result of visual feature extraction and linear discriminant analysis of EEGs recorded as the subject was resuscitated after cardiac arrest. The EEGs of those who died (stippled bars) were linearly separable from the EEGs of those who recovered (cross-hatched).

15.6.5. Dyslexia

Dyslexia manifests itself as an inability to acquire reading and spelling skills which match a subject's intelligence. Sklar *et al.* (1973) used stepwise discriminant analysis and found that they could distinguish between 12 dyslexic and 13 normal children on the basis of various EEG spectral parameters. Their work is important because it may provide a means of detecting dyslexic children at any early age when special care can most successfully be directed to them.

15.7. CURRENT RESEARCH

In the previous sections, we have seen how pattern recognition is finding applications in electroencephalography. The collaboration between the two fields is still developing, but there are signs that it will prove a fruitful cooperation. It is impossible to predict far ahead, but a glimpse of the immediate future may be obtained from a brief description of our own current research program.

15.7.1. Feature Space Mapping

The fundamental problem of analyzing the EEG is that there are too many data, representing rather a small amount of significant information. How can this large volume of data be condensed into a few succinct statements, which retain the essential information?

We believe the old adage that "seeing is worth ten thousand tellings." With this in mind, we have been investigating methods of displaying hyperspace data structures, so that they can be comprehended after only a small amount of training. The human being is a superb pattern recognizer of low-dimensional data, and may be able to detect patterns of structure and/or movement on a display which would not be apparent to a pattern recognition machine. Thus, the introduction of a human operator may well provide something novel in pattern recognition systems: *very high intelligence.* Apart from this, we believe that it is sensible to concentrate on developing visual display methods as a means of involving doctors and EEG technicians in the process of making decisions. This is important for three reasons:

1. It builds their confidence in pattern recognition.
2. It makes use of their experience.
3. It encourages EEG workers to think using pattern recognition concepts.

We shall only briefly describe the techniques available, since a detailed discourse could occupy the whole book. The most popular mapping techniques are based on projecting the pattern points from hyperspace onto a pair of carefully chosen reference vectors. Thus, if the pattern points, in hyperspace, are represented by X_i ($i = 1, \ldots, N$) and the reference vectors by P and Q, then the resulting 2-space map is found by plotting $(P \cdot X_i)$ vs $(Q \cdot X_i)$, for $i = 1, \ldots, N$. The reference vectors P and Q may be found in a variety of ways, the most popular being to choose the two eigenvectors of the descriptor variance— covariance matrix which possess the two largest eigenvalues. (P and Q may not be orthogonal if they are selected in some other way.) This was the method used to draw Fig. 15.12. Another group of techniques calculates some measure of "distance" from P and Q to each of the X_i. One obvious way of choosing the reference vectors P and Q here is to let P be the mean vector of one class and Q that of the other. (This assumes, of course, that our problem is essentially a two-class discrimination.) An extension of this idea and a much more powerful one is developed in Section 4.3.3.3. The advantage of this is that it incorporates some of the most powerful hyperspace-analysis methods with the ergonomic advantages of a visual display (Batchelor and Hand, 1975). Other methods of note are those due to Sammon (1969) and the *minimal-spanning-tree*, which draws a 2-space map with points linked together by lines, so that the (hyperspace) clustering is evident to a human observer (Zahn, 1971; Lee, 1974).

Whatever display technique is finally selected, it is important that the two-dimensional map should contain most of the desired structure and little of the nonessential detail found in the hyperspace. Some information loss is inevitable and it is wise to monitor this wherever possible. For example, for a discrimina-

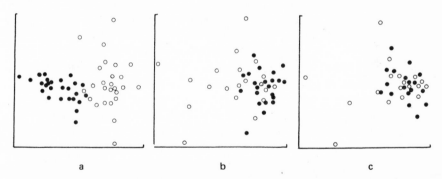

a   b   c

Fig. 15.12. Feature space mapping; projections from 24-space onto two eigenvectors (corresponding to the two largest eigenvalues) of the descriptor variance—convariance matrix. Three maps illustrating the variety of response to eye closure in the EEG of the resting subject. Points marked with an open circle describe the EEG with the eyes open, points marked with a filled circle with the eyes shut. Map (a) is typical of those subjects whose EEG after eye closure is markedly different from that before; in the map the eyes-shut and eyes-open regions are quite separate. Map (b) comes from a subject whose EEG was moderately changed by eye closure. Map (c) is from one who exhibited almost no EEG differences between the eyes-open and the eyes-shut states.

tion problem, the estimated Bayes' error probability may be calculated in the hyperspace and from the map. (See Section 4.3.3.1.)

The objective of feature space mapping is to demonstrate the important aspects of the data structure found in the hyperspace, in a readily understood form. We believe that feature-space mapping is an ideal way of presenting the results of data analysis to a doctor. He can annotate the map, for example, by drawing boundaries which define the limits of variation found in a particular disease (Fig. 4.13). Moreover, he can label regions of the map as "confusing," "rare," "clearly disease X." Most pattern recognition workers have a deep desire to find an entirely automated solution to the complete problem. Accepting more restricted, short-term objectives may, in fact, eventually achieve more. This is especially true in medicine where the professionals have a high "inertia," which avoids undue haste in adopting unproved techniques. Feature-space mapping is an attempt to combine some of the modern data-analysis methods with human intelligence.

15.7.2. Automation of Clinical EEG Analysis

Undoubtedly the most challenging application of pattern recognition in electroencephalography is attempting automatic clinical EEG interpretation. This is a task of a completely different order of complexity from the examples considered so far, most of which only require a few types of pattern to be identified, usually on the basis of one or two channels of EEG information. Routine clinical EEG assessment by contrast involves examining activity recorded at over 20 electrodes and the range of possible phenomenology is very large. In view, therefore, of the probable high cost of such an exercise its possible value must be seriously examined. Kellaway (1973) suggests (and few professional electroencephalographers would disagree with him) that the great majority of EEGs recorded both in the United States and the United Kingdom are interpreted at a very low level of competence by inadequately trained personnel. Consequently, the clinical service provided is of a much lower standard than can reasonably be expected and much of the expenditure on this quite costly examination is wasted. Particularly in those countries where medical salaries are high, an automatic interpretive system could be justified if it either provided a higher level of reliability than that which generally prevails, or could be used to screen EEGs and identify those requiring inspection by a trained electroencephalographer. Some workers in this field are devoting very considerable resources to research and development in automatic EEG interpretation but, once a particular system was validated, the cost of special purpose hardware would be relatively small.

Traditional visual EEG interpretation involves two stages, closely following the usual pattern recognition procedure of feature extraction followed by decision-making. The first of these tasks is performed by the human visual system less consistently than by a computer, yet with a flexibility which is difficult to match by any automatic means; the ability of the trained observer to distinguish genuine cerebral activity from artifact (itself a pattern recognition task) is particularly difficult to imitate. On the other hand, the human observer is much less skillful at making complex decisions, on the basis of observations on large number of variables, than he may care to admit. The study of Binnie *et al.* (1970) (which used linear discriminant analysis of visually encoded data) exemplifies an ideal situation in which man and machine perform the tasks to which they are most suited.

Unfortunately this approach is not suitable for routine use as it depends upon a level of consistency in the human observer which cannot be guaranteed and indeed which is most unlikely to be maintained over a period of years. Thus, it seems logical to us that computer-assisted EEG interpretation must in the event use automatic methods for both feature extraction and decision-making. This is clearly not a general view; many workers have used computers to perform a preliminary numerical analysis of the EEG and then present the results graphically to the user for interpretation. Compressed spectral arrays (Bickford *et al.*, 1971) are particularly popular (Fig. 15.13). Other workers have interposed a stage of interpretation between feature extraction and presentation of results. The age-dependent EEG quotient used by Matousek and Petersen (1973b) exemplifies this approach (Section 15.4). Similarly Gotman *et al.* (1973 and 1975) used the quotient: (delta + theta)/(alpha + beta) to characterize EEG abnormality. The spectral values in the different frequency bands are multiplied by weighting factors determined by trial and error, to achieve optimal discrimination between the EEGs of patients and those of volunteers.

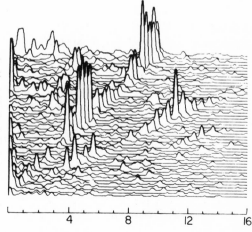

Fig. 15.13. Compressed spectral array of EEG during flicker stimulation at a steadily increasing frequency. Power spectra of successive epochs have been plotted so as to give the impression of a third dimension, time, receding from the observer. The fundamental and harmonic components of the response are clearly seen standing out like mountain ranges and skewed obliquely toward the higher frequencies as the flicker rate increases. (From Bickford, 1973.)

The use of a direct pattern recognition approach to this task is surprisingly difficult. Apart from some logistic and methodological problems, it is a simple matter to obtain a set of normative data consisting of EEGs from healthy people. The statistical distribution of a particular set of features based on these records can be established, and by appropriate choice of features and transformations, the assumptions involved in some decision-making techniques can be satisfied. The same is unlikely to be true of the class of abnormal EEGs which has to be distinguished from the controls. The EEGs of patients do not form a homogeneous group and differ from normality in a variety of ways, so that the distribution of the features in this class is likely to be complex and to present problems for many decision-making techniques. Here the hyperspace approach can be modified to handle "the one class case," that is to determine the probability of observations belonging to the single category of EEGs from healthy people (cf. Section 4.5.4).

We are currently collecting a data-base of EEGs from "normal" volunteers. To avoid self-selection (leading possibly to an excess of neurotics or indeed of people with possible cerebral disease) and to ensure a wide range of ages and socioeconomic groups, the subjects have been obtained by EEG surveys of the staff in two places of work, a somewhat isolated mental hospital which is the main employer in its locality, and the staff of a tobacco factory and its associated research and development establishment. After feature extraction (see below) the EEGs are being screened for homogeneity to determine whether the population falls into easily distinguished groups with different EEG characteristics.

Preliminary results show, as expected, marked differences with age and, less predictably, a striking dependence of the EEG pattern upon smoking habits and sex. As the data-base grows, it may need to be stratified so that any patient's record can be compared to a control group, matched for age, sex, smoking habits, and possibly personality and cerebral dominance.

A number of different strategies are being used for feature extraction and compression. Power spectra obtained from 24 channels of EEG, recorded under eyes-opened and eyes-closed conditions, have been compressed into 4 or 7 frequency bands or used to calculate quotients, as employed by Matousek and Petersen (1973b) and by Gotman et al. (1973). Various different feature compression methods are being tested to combine the results of the eyes-opened and eyes-closed conditions and to extract information about the asymmetries or regional differences. Feature extraction has also been performed by *slope descriptor analysis* (Hjörth, 1970), and wavelength amplitude methods are being developed to provide both spectral descriptions of the EEG and to identify transients. Simple pattttern recognition techniques are also being developed to identify artifacts and to exclude contaminated sections of the record before feature extraction is performed.

A slightly unconventional one-class decision-making process is currently being used. This constructs an estimator of probability density function, describing the distribution of points in a feature space, describing the activity of each channel or each region of the head. The estimated values of these probability density functions can be presented visually to good effect. Twenty-four circles are located at the electrode sites on a plane view of the subject's head and each is given a radius proportional to the value of the estimated probability density function. Figure 15.14 shows the results obtained from a patient with a tumor in the right central region. The estimated probability density function can be thresholded to discriminate between "normal" and "abnormal."

In due course, the technique can be developed by using the estimated probability density functions as descriptors. This will allow us to provide a global assessment of the EEG and eventually to discriminate between different diagnostic categories.

15.7.3. Notes for the Guidance of Future Work

Developments in this field are often slow, due to the difficulty and cost of collecting data. It is well to remember that an EEG recording requires a person to visit the hospital, or be moved from a ward in the hospital to the EEG recording suite. (If the patient is fortunate, there is the possibility of moving the recording equipment to him, rather than moving him to the equipment.) In round figures, an "average" EEG recording occupies two people for an hour and a half! A large teaching hospital with a worldwide reputation may only encounter one case of a certain rare disease during each year. It is important, therefore, to avoid algorithms that require 10^6 cases to guarantee convergence. Of course, we are overstating the point, but it is important to remember that cases are people,

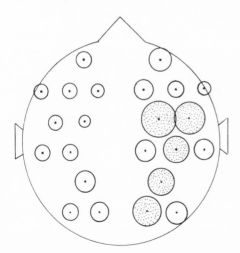

Fig. 15.14. Topographic display of EEG abnormality. The size of the circles relates inversely to the probability of the EEG activity, at the corresponding site, being found in a normal subject. Those circles representing a probability less than 0.02 are stippled. More simply, they represent EEG abnormality and the greater their size the more severe is the anomaly. The patient had a right-central glioma. Spectral analysis was used for feature extraction. The recognition of abnormality was based on measuring the mean distance in the feature space of the values of the patient's EEG from those of the 10 nearest neighbors from a normal population.

that disease is misery, and that some things cannot be done by computer (yet!). We feel that there is no justification in ever forgetting that there is a multitude of problems that are waiting for the application of low-level technology. If all theoretical work in pattern recognition stopped now, there would still be a vast amount of research to be done with existing techniques, by pragmatism, and by experimentation. We hope that, by writing this chapter, we have made some of the possibilities evident to a wider audience and that this will encourage greater effort in the application and development of pattern recognition techniques and machines specifically designed for EEG problems.

REFERENCES

Baldock, G. R., and Walter, W. G., 1946, A new electronic analyser, *Electron. Eng.* **18**:339–342.

Batchelor, B. G., 1969, Learning Machines for Pattern Recognition, Ph.D. thesis, University of Southampton.

Batchelor, B. G., 1974, *Practical Approach to Pattern Classification*, Plenum Press, London and New York.

Batchelor, B. G., and Hand, D. J., 1975, On the graphical analysis of pdf estimators for pattern recognition, *Kybernetes* **4**:239–246.

Bickford, R. G., 1973, Application of compressed spectral array in clinical EEG. In: *Automation of Clinical Electroencephalography* (P. Kellaway and I. Petersen, eds.), p. 318, Raven Press, New York.

Bickford, R. G., Fleming, N. I., and Billinger, T. W., 1971, Compression of EEG data by isometric power spectral plots, *Electroenceph. Clin. Neurophysiol.* **31**:631–636.

Binnie, C. D., 1975, A comparison of different methods of period analysis, *Electroenceph. Clin. Neurophysiol.* **38**:662.

Binnie, C. D., MacGillivray, B. B., and Osselton, J. W., 1974, Traditional methods of examination in clinical neurophysiology. In: *Handbook of Electroencephalography and Clinical Neurophysiology*, 3C (A. Remond, ed.), Elsevier, Amsterdam, p. 126.

Binnie, C. D., Prior, P. F., Lloyd). S. L., Scott, D. F., and Margerison, J. H., 1970, Electroencephalographic prediction of fatal anoxic brain damage after resuscitation from cardiac arrest, *Brit. Med. J.* **IV**:265–268.

Cooley, J. W., and Tukey, J. W., 1965, An algorithm for the machine calculation of complex fourier series, *Math. Computation* **19**:297–301.

Cooper, R., Osselton, J. W., and Shaw, J. C., 1974, *EEG Technology*, Butterworths, London.

Dietsch, G., 1932, Fourier analyse von Electroenkephalogrammen des Menschen, *Pflüger's Arch.* **230**:106–112.

Dixon, W. J., 1970, BMDØ7M — stepwise discriminant analysis. In: *Bio-medical Computer Programs*, University of California Press, San Francisco, Calif.

Donchin, E., 1969, Discriminant analysis in average evoked response studies: The study of single trial data, *Electroenceph. Clin. Neurophysiol.* **27**:311–314.

Dumermuth, G., and Keller, E., 1973, EEG spectral analysis by means of fast Fourier transform. In: *Automation of Clinical Electroencephalography* (P. Kellaway and I. Paterson, eds.), Raven Press, New York.

Frost, J. D., 1970, An automatic sleep analyser, *Electroenceph. Clin. Neurophysiol.* **29**:88–92.

Fukunaga, K., 1972, *An Introduction to Statistical Pattern Recognition*, Academic Press, New York.

Gotman, J., Skuce, D. R., Thompson, C. J., Gloor, P., Ives, J. R., and Ray, W. F., 1973, Clinical applications of spectral analysis and extraction of features from electro-encephalograms with slow waves in adult patients, *Electroenceph. Clin. Neurophysiol.* **35**:225–235.

Botman, J. R., Gloor, P., and Ray, W. F., 1975, A quantitative comparison of traditional reading of the EEG and interpretation of computer-extracted features in patients with superatentorial brain lesions, *Electroenceph. Clin. Neurophysiol.* **38**: 623–639.

Hjörth, B., 1970, EEG analysis based on time-domain properties, *Electroenceph. Clin. Neurophysiol.* **29**:306–310.

Hjörth, B., 1973, The physical significance of time-domain descriptors in EEG analysis, *Electroenceph. Clin. Neurophysiol.* **34**:321–325.

Itil, T. M., 1974, Modern Problems of Pharmacopsychiatry: Eight Psychotropic Drugs and the Human EEG, Karger, Basel, München, p. 337.

Lee, R. C. T., 1974, Sub-minimal spanning tree approach for large data clustering, *Proc. 2nd Int. Jt. Conf. Pattern Recognition*, Copenhagen, p. 22.

Kellaway, P., 1973, Automation of clinical electroencephalography: the nature and scope of the problem. In: *Automation of Clinical Electroencephalography*, (P. Kellaway and I. Petersen, eds.), Raven Press, New York, p. 318.

Kiloh, L. G., McComas, A. J. and Osselton, J. W., 1972, *Clinical Electroencephalography*, Butterworths, London, p. 239.

Larsen, L. E., and Walter, D. O., 1970, On automatic methods of sleep staging by EEG spectra, *Electroenceph. Clin. Neurophysiol.* **28**:459–467.

Lloyd, D. S. L., Binnie, C. D., and Batchelor, B. G., 1972. Pattern recognition in EEG. In: *Interdisciplinary Investigation of the Brain* (J. P. Nicholson, ed.), Plenum Press, London.

Martin, W. B., Johnson, L. C., Viglione, S. S., Naitoh, P., Joseph, R. D., and Moses, J. D., 1972, Pattern recognition of EEG-EOG as a technique for all-night sleep stage scoring *Electroenceph. Clin. Neurophysiol.* **32**:417–427.

Matousek, M., and Petersen, I., 1973a, Frequency analysis of the EEG in normal children and adolescents. In: *Automation of Clinical Electroencephalography* (P. Kellaway and I. Petersen, eds.), Raven Press, New York, p. 318.

Matousek, M., and Petersen, I., 1973b, Automatic evaluations of EEG background activity by means of age-dependent EEG quotients, *Electroenceph. Clin. Neurophysiol.* **35**:603–612.

Maynard, D. E., and Prior, P. F., 1973, A predictive instrument to assist with interprediction of EEGs of patients resuscitated after cardiac arrest, *Electroenceph. Clin. Neuro-physiol.* **34**:744.

Prior, P. F., 1973, *The EEG in Acute Cerebral Anoxia*, Exerpta Medica, Amsterdam.

Rechtschaffen, A., and Kales, A., 1968, *A Manual of Standardized Terminology Techniques and Scoring System for Sleep Stages of Human Subjects*, U.S. Government Printing Office, Washington, D. C.

Saltzberg, B., Burch, N. R., McLennan, M. A., and Correll, E. G., 1957, A new approach to signal analysis in electroencephalography, *IRE Trans. Med. Electron.*, 8:24.

Sammon, J. W., 1969, Non-linear mapping for data structure analysis, *Trans. IEEE* C-18:401.

Serafini, M., 1973, A pattern recognition method applied to EEG analysis, *Computers Bio-med. Res.* **16**:187–195.

Sklar, B., Hanley, J., and Simmons, W. W., 1973, A computer analysis of EEG spectral signatures for normal and dyslexic children, *IEEE Trans. Biomed. Eng.* **20**:20–26.

Smith, J. R., and Karacan, I., 1971, EEG sleep-stage scoring by an automatic hybrid system, *Electroenceph. Clin. Neurophysiol.* **31**:231–237.

Viglione, S. S., 1970, Applications of pattern recognition technology. In: *Adaptive Learning and Pattern Recognition Systems: Theory and Applications* (J. M. Mendal and K. S. Fu, eds.), Academic Press, New York, pp. 115–162.

Walter, D. O., 1963, Spectral analysis for electroencephalograms: Mathematical determination of neurophysiological relationships from records of limited duration, *Exp. Neurol.* **8**:155–181.

Walter, D. O., Rhodes, J. M., and Adey, W. R., 1967, Discriminating among states of consciousness by EEG measurements, a study of four subjects, *Electroenceph. Clin. Neurophysiol.* **22**: 22–29.

Wennberg, A., and Zetterberg, L. H., 1971, Application of a computer-based model for EEG analysis, *Electroenceph. Clin. Neurophysiol.* **31**:457–468.

Wilks, S. S., 1962, *Mathematical Statistics*, Wiley, New York.

Zahn, C. T., 1971, Graph-theoretical methods for detecting and describing Gestalt clusters, *Trans. IEEE* **C-20**:68.

16

Editorial Introduction

Our approach in this book has been practical; we have given scant attention to the philosophical aspects of our subject which Newman discusses here. The seemingly naive questions:

> "What is a Pattern?"

and

> "What does Recognition require?"

are far from simple to answer in general terms. As Humpty Dumpty said, these words can mean whatever we want them to mean; a "pattern" can be anything we care to call a pattern. Pattern recognition cannot be defined in a formal sense, but the tasks we have discussed in this book all have one thing in common: they are all ill-defined. Pattern recognition is concerned with fuzzy concepts, such as "alike," "similar," "often," "reasonable," which we have used repeatedly in the foregoing pages. Statistics is one branch of mathematics capable of handling vague quantities. Primitive set theory is another, since the sets in question may be defined in whatever way we may choose. One of the fundamental notions of set theory is *mapping*, which Newman equates with recognition. Mapping is a very generalized concept, including most of the specific examples of recognition discussed earlier in this book; each of the following is a special type of mapping:

1. Partitioning a hyperspace (Chapter 4), equivalent to evaluating a discriminant (mapping) function
2. Evaluating a Boolean function (Chapters 3 and 5)
3. Emulating a human teacher: providing the same output as the teacher, for a given input (e.g., speech recognition, OCR)
4. Diagnosing disease (mapping patients on to a set of disease labels)
5. Graph isomorphism (mapping graphs on to the set {TRUE, FALSE})

The mapping, whatever form it takes, must be capable of being applied to unseen members of the input set and behaving in a way which we regard as being sensible. The mapping operator must be designed using a training set and be capable of generalizing to classify the test set. Thus, mapping is already a familiar concept and Newman sets these ideas into a formal framework.

Newman regards "patterns" as being structured relations, which is a common form of pattern description, as we have seen. Although it is ostensibly concerned with scene analysis, this chapter has numerous lessons for a wider audience. Indeed, the essential unity of pattern recognition, which is so difficult to establish, is clearly evident here. One of its essential features is the need to employ vague, fuzzy concepts. This distinguishes pattern recognition (and its kindred subjects, for example, picture processing, artificial intelligence, and cybernetics) from all other branches of technology, where precision is paramount. Engineers, scientists and mathematicians are venturing on to soft, pliable, new ground. This chapter helps us to see that the way forward is not quite so strange as we might imagine.

Scene Analysis: Some Basics

E. A. Newman

Broadly speaking this chapter covers the techniques used in scene analysis — broadly speaking because "scene analysis" is not at all well defined. The rough divisions within this chapter are as follows.

1. The three different concepts which are sometimes meant by the expression "scene analysis": (a) The *robotics* problem, which can be thought of in terms of perceiving words consisting of bricks, wedges, and other similar shapes assembled in meaningful groups, (b) The problem of, for example, detecting and recognizing an animal hiding in the undergrowth, (c) The recognition of objects such as human faces

2. Two essentially different ways of looking at pattern recognition in terms of what it is used for and in terms of what it is (Section 16.2)

3. Why the design of automatic means of carrying out "scene analysis" requires a very good understanding of the basic nature of pattern recognition (Section 16.3)

4. Some examples of work on scene analysis (Section 16.4)

5. Summing up (Section 16.5)

E. A. Newman • Computer Science Division, National Physical Laboratory, Teddington, Middlesex, England

16.1. WHAT IS SCENE ANALYSIS? — ROBOTICS PROBLEM

Work in this field is usually classified as being on *artificial intelligence*. It is concerned with the sort of capability needed in a robot that looks for bricks and stacks them or that assembles blocks together to make a building. Essentially it sees its "world" as being built from parts, and even the parts from parts. It is important that recognition techniques used in this "world" should use this fact. In one case failure to do this would have increased the number of classifications that needed consideration from 30 to 3×10^8. This case is mentioned in one of the examples in Section 16.4.

16.2. VIEWS OF PATTERN RECOGNITION — SUMMARY

As with many other things "pattern recognition" can be described (1) in terms of the purpose for which it is used; (2) in terms of what it is; and (3) in terms of how it is done [this is dependent on (1) and (2)].

16.2.1. How Pattern Recognition Is Used

Good examples of a description in terms of what it is used for come from the use of *optical character recognition* (OCR). Often when OCR is used, something or someone knows what the "character to be recognized" is meant to be. "Recognition" of the character is carried out in order to allow some process to be performed. Otherwise there is perhaps no point in the "character recognition" anyway. Whether or not the "character" is recognized correctly can then be tested by overall success. The only other way in this context of testing recognition accuracy is to refer back to the creator of the character.

16.2.2. What Pattern Recognition Is

Here one is looking for some intrinsic quality of the thing being recognized. In doing so one has to make many basic postulates, often without being aware of it. To discuss the subject sensibly one must know what is meant by pattern and by recognition. These terms are used by different people to refer to a variety of concepts. For this reason authors using these terms should define them. Rarely do they do so. The following aspects can be considered: (1) recognition, what does it mean?; (2) a pattern as a *relation*, thus containing parts; (3) a pattern as a *structured relation* (usually structure is multileveled); and (4) *classification* as a way of looking at sets of different patterns from a point of view that makes all those in a given set the same. In scene analysis the patterns under consideration are complex. One is often trying to recognize some particular part of the overall scene, or even a part of a part. Classification of the sort needed in scene analysis must either explicitly or implicitly assume that the pattern is highly structured.

16.2.2.1. Recognition

"Recognition" implies more than one "cognition." The shorter Oxford English Dictionary defines "to recognize" as "to know again: to perceive to be identical with something previously known." It defines cognition as "the action or facility of knowing; knowledge; consciousness," and "to know" as "to be aware; as appraised of."

Clearly recognition is a subjective thing. It involves the state of being aware. There would seem to be only one way for the (would-be) "objective" thinker to get out of this dilemma: to assume that he is aware of the situation described below and then to deny that he had assumed he is so aware.

When a man studies a "recognition system" he "sees" the system and also a set of stimuli that could affect the system. He is aware of all these things; he can distinguish one from the other. He is also aware that the system can be in a number of different states. Given all this — not otherwise — he could say that the system recognized each of the different stimuli if he observed that it was always put into the same state when presented by the same stimulus, but into different states for different ones. If, however, a subset of stimuli all produced the same state, then he could say that the system recognized this subset as a class.

On this basis recognition is a duality: it involves both what is "recognized" and what "recognizes." If one defines a pattern P as anything that is recognized by a recognizer R, then P is recognized "under" R. In particular, interest is usually on a set of patterns recognized under R. More might be asked than this of "recognition," in that the system should have a set of recognizable outputs each of which must of course correspond to one or more internal states of the system. "Recognition" could be taken to imply that the system produces suitable output states. This "definition" of "recognition" implies a number of postulates.

POSTULATE 1. Even for the would-be "objective" thinker, the ultimate sanction to "reality" is his own awareness. This is the postulate he has now to deny, in order to be "objective."

POSTULATE 2. The recognizer makes "observations" in each of which it "recognizes." There are a number of observations, in general, associated with any recognizer, each being an individual "isolated" event. In each observation the recognizer observes a "view" V.

POSTULATE 3. If the recognizer is in some state S when making a given observation, then some substate S' of this state is associated with the "view."

If some substate P of S' is common to more than one observation P is due to a "pattern" within the view, and the pattern is recognized under the recognizer.

Of course postulates 2 and 3 do not have to be part of any particular person's concept of a pattern; there are no "basic truths" about such things. In practical cases, their validity can be tested. But most people's concept of a pattern contains these postulates.

16.2.2.2. A "Pattern" Is a "Relation"

Given the idea of "recognition" discussed above, a pattern is, by implication, anything that is recognized. For most people the idea of a pattern is less general than this. Typically it is a distribution of some set of measures over some continuous one-, two-, or three-dimensional space. Generalized to a greater number of dimensions, this is precisely the mathematical concept of a functional relation such as:

$$(Y_1, Y_2, Y_3, \ldots, Y_m) = f(X_1, X_2, \ldots, X_n)$$

$Y_1, \ldots, Y_m; X_1, \ldots, X_m$ being members of infinite, well-ordered sets. [*Note:* The relation is functional since the value of $(X_1, X_2, \ldots, X_n)$ determines that of $(Y_1, Y_2, Y_3, \ldots, Y_n)$, while that of the latter does not necessarily determine that of the former.] From a physical point of view, the pattern consists of some measurable property V some value of which is associated with each of a set of positions A. That is, a pattern can be expressed by giving V as a function of A. A recognizer then has to detect this function. From this point of view a pattern is a set of pairs, one item in each pair being unique to it — and thus "identifying" it. This item serves the same kind of purpose as an "address," be this in human terms or computer ones.

In the general case, therefore, a pattern P is:

$$V = P(A)$$

Where V and A can both be vectors. Intuitively, patterns which occur in the context of scene analysis are *continuous*. That is to say, P is a continuous function.

In practice, patterns are nearly always treated as *finite*, as though a pattern P was defined as the relation:

$$V = R(A)$$

where $V = R(A)$ is a finite binary functional relation. That is to say, the relation R is a finite set of different pairs. One element of each pair is chosen from the set A and one from the set V. The element chosen from A defines that from V, but that from V does not always imply that from A. In practice, the elements within set A will not be independent of one another. Relations will hold between them. Nor will they necessarily be indivisible. The same is true of set V. To give a practical example, suppose one assembles a "pattern" out of a set of cubes of various colors. Each cube is defined uniquely by its x,y,z coordinates, hence the color is defined uniquely for each (x,y,z). The "position" or "address" element of A is a three-dimensional vector. The "value" v_i of V is one of a set of "colors." The pair $(v_i{:}a_i)$ is a pattern element. The position determines the color; the color does not uniquely determine the position. There are relations existing within set A. These are the relations that cause the "positions," "points," or "addresses" to make up the kind of 3-space with which we are familiar. Figure 16.1 illustrates this. The complete "pattern" consists of nine cubes. The dotted one has the "address" $(3, 3, 2)$. If it is soft and red, then associated with it is

$$(\text{cube, dotted, soft, red})$$

and the pattern element or "point" is

$$[(\text{cube, dotted, soft, red}) (3, 3, 2)]$$

Under these circumstances then a pattern becomes a set of distinguishable pairs, and it is then possible to resolve the system into a set of distinguishable *properties*. Often such properties are called *features*, but this usage is not very meaningful. The Cartesian product of set V and set A gives – under these conditions – the set of all possible "features." A pattern can be represented by "naming" each possible "feature," and then giving the value 1 to those "features" in the pattern, and value 0 to the rest. This gives the so called *binary pattern*. It is obvious that such a "binary" pattern ignores nearly all the important aspects of any practical pattern from which it could have been derived.

16.2.2.3. Patterns Are Structures

Many patterns of interest consist, from a recognition point of view of largely independent subpatterns. These, although possibly complex internally, can be treated almost as independent elementary units (primitives). Thus a pattern can be seen as a hierarchy of subpatterns. In ordinary life this is obvious.

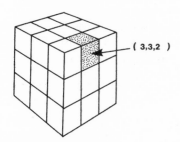

(3,3,2)

Fig. 16.1. The cube whose position is $(3, 3, 2)$ is shown dotted. It is also soft and red. Then the pattern element, or point, is:

$$[(\text{cube, dotted, soft, red}) (3, 3, 2)]$$

A sentence, for example, evidently consists of words containing letters. In group theory, the idea that "factor groups" can be group elements in their own right is an important one. But even today, the significance of this "pattern" factorization is not universally realized. A recognizable subpattern is often also called a "feature." This use of the term is a far better one than the virtually meaningless but more common use of the term discussed earlier.

The practical importance of the existence of subpatterns depends on the theoretical observation that every new level that is legitimately introduced reduces the required number of recognition steps to the logarithm of what they were before. Exploiting hierarchical structure, where it exists, enormously reduces the number of steps required. The "recognition" process at each level in the hierarchy is essentially of the same kind. But experience has shown that the further down a multilevel hierarchy one goes, the "rougher" the subpattern can be. Thus, in a handwritten note, individual letters need scarcely be separable or recognizable. In most practical pattern recognition tasks, discrimination at lower levels can be quite crude.

16.2.2.4. Classification

In practice, we do not wish to recognize every different pattern as different. Our real wish is to treat certain sets of patterns as though they were identical. This is the process of classification. Sometimes such patterns are said to be "identical" if they all contain the same set of properties, and "alike" if they have most of their "properties" in common.

Other postulates are now involved in recognition. These are postulates about "classes." Not all people agree about what constitutes a pattern class. The following postulates fit some people's ideas.

POSTULATE 4. Assume two recognizers R_1 and R_2, and R_1 recognizing "patterns" and R_2 states of R_1. A class of patterns under (R_1, R_2) are patterns which differ under R_1 but are the same under R_2.

POSTULATE 5. Assume that a pattern of interest is associated with three concepts:

1. With a set of properties $\{C_i\}$ which are of interest but which are not accessible
2. With a set of properties $\{R_i\}$ which are not of direct interest, but which can be used to predict which if any of the properties C_i exists. Sometimes it is the *absence* of some possible property which is of predictive value.
3. With a "name" used for communication. It is then the properties C_i which determine the class, the "mix" of properties R_i (i.e., the pattern) which is used to predict the class, and the name that is used to communicate the existence of the class.

In practice, there are two ways by which patterns can be classified. Either one can list all possible patterns and associate each in the list with its class, in which case one "recognizes" by looking up the unknown in the list or else one can use some concept of *likeness*. In the latter case one lists a relatively small set of typical class members, *class paradigms*, and classifies patterns according to how like they are to one or another paradigm. Figure 16.2 illustrates this. If a pattern *P* is

$$V = P(A)$$

where V is a continuous function of A, then clearly it is very unlikely that the first way will be practical simply because A is then an infinite set, and the possible set of patterns is likely to be also infinite. The exception would be if each pattern class consisted of instants of very different patterns, each one precisely some function $P_i(A)$. So, in practice, given continuous patterns, likeness concepts must be used.

16.3. WHY SCENE ANALYSIS IS NOT TRIVIAL

It was suggested above that practical recognition techniques always predict some "property" in a situation in which this property is not directly accessible. This can only be done by finding "patterns" of accessible or measurable properties which correlate with the required inaccessible state. In one sense the purpose is to predict, and this is as true of scene analysis in particular as of pattern recognition in general. But the simple schemes of recognition are not often efficient for the purpose of scene analysis. In an attempt to show why this should be so, "recognition" will now be dealt with in more detail. This discussion is in three parts: (1) to find correlating properties and to measure them; (2) how to find which patterns of these correlating properties correspond to the various inaccessible states which have to be detected — the classification process; and (3) where simple schemes do not work.

16.3.1. Finding and Measuring Suitable Properties

Finding the right "measures" to use in simple scenes is often intuitively easy, which is fortunate, for the number of possible candidates even in a simple situation is very great. Many scenes are very complex and for such situations, although intuition helps and certainly greatly limits the field of investigation, it is far from obvious which are the right measures to use.

Workable "recognition" techniques operate on the postulate that the class of future "unknowns" will be discoverable from past experience. At its most primitive this means that future patterns will be identical to those of the

"All" patterns (32 in all)	Classes A, B, C (6 members each)		Paradigms
00000 11111			
00001 11110			
00010 11101	A	⌈ 00000 10000 01000 00100 00010 ⌊ 00001	A₁ 00000
00100 11011			
01000 10111			
10000 01111			
00011 11100			
00110 11001	B	⌈ 11100 01100 10100 11000 11110 ⌊ 11101	B₁ 11100
01100 10011			
11000 00111			
10001 01110			
00101 11010	C	⌈ 00111 00011 00101 00110 01111 ⌊ 10111	C₁ 00111
01010 10101			
10100 01011			
01001 10110			
10010 01101			

Fig. 16.2. A member of a class differs from its paradigm by not more than one digit. (The Hamming distance between a paradigm and a class member is 0 or 1.)

past both in detail and class association. Recognition then simply involves listing all the patterns that have been associated with each class and finding in which list the unknown occurs. The straightforward way of dealing with this is to "store" all the patterns for the class, and check, one by one, whether the unknown is one of these. This is the method illustrated by Fig. 16.2 and known as "table look-up."

In practice, this cannot be done, for there is never enough past experience to enable even a reasonable proportion of possible future unknowns to have been listed. The future, even when predictable, need not follow the past in a simple way. Even if it did follow simply the task of searching through every list, it would be much too expensive. In practice, use is made of the concept of two patterns being *alike*. A future pattern is associated with a given class if it is "like" those that have been so associated in the past. Where this is not convenient, it can be sufficient just to compare unknowns with sets of patterns deemed to be *typical* of their class; that is, with class paradigms. A pattern class is now a set of "like" patterns or rather a set of sets, one for each paradigm. The selection or design of these typical patterns is a major task in itself but if a small set can be suitably chosen the total amount of work is very much reduced. The question is now one of determining likeness. There are many ways in which this could be attempted.

16.3.2. Recognition Techniques

Once a decision has been made about suitable measures, the recognition process consists of actually measuring them, and then of manipulating them to obtain the classification. Many pattern recognition workers look upon automatic recognition as a three-stage process:

1. "Taking measures." This is often called "scanning."
2. "Manipulating the measures" to obtain a distinguishable set of "properties." The result is a binary pattern. The process is often called preprocessing.
3. Final classification, working on the binary pattern.

In principle, such a breakdown fits most classification schemes, provided one allows feedback between the steps.

Final Classification and the Detection of Likeness. Once the "measures" made when attempting the automatic classification of an "unknown" have been reduced to a set of properties by preprocessing, the classification problem reduces to the detection of binary patterns and the use of these to predict class.

A common measure of likeness between two binary patterns is the *Hamming* distance, which is the number of properties possessed by one or the other but not both of the patterns. Zero distance is equivalent to identity. Simple

Hamming distance likeness fills some of our instinctive ideas of likeness. However, given properties which are most readily derived from primary measures, members of a given pattern class are not always shown to be alike when Hamming distance is used. (See Fig. 16.3.) For one reason, many of the measures are in no way relevant, and for another, the basic measures that one can take can be associated with class in a very complex way. Sophistication may be added to these ideas in several ways. Some properties can be looked upon as being more important in judging likeness; they have greater "weight" than others. Properties can be looked upon as being present in varying amount, and weights can be looked upon as depending on which set of paradigms an unknown is

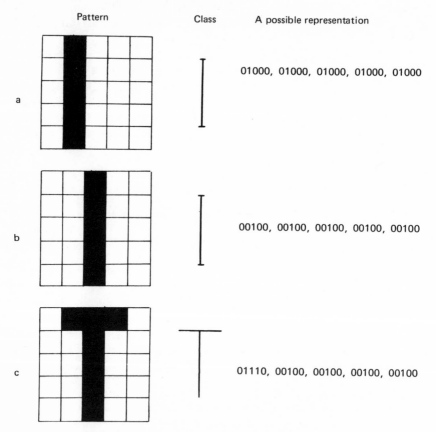

Fig. 16.3. A case where Hamming distance is inappropriate, giving a misleading indication of similarity. Hamming distances between the patterns shown:

    a, b: 10
    b, c: 2
    a, c: 10

being judged as most alike. These extensions lead to Bayes' rule discriminations and to "potential" surface separation. "Weights" are then determined statistically. (See Chapter 4.)

16.3.3. Where Simple Schemes Are Not Workable

Often, from the point of assessing the class of a pattern, Hamming distance proves to be a very poor measure of likeness. This is especially true where the properties on which the Hamming Distance Criterion is used at a very low level, where, for example, every "black" point in a chessboard picture is judged to be a different property. Then, even a very high degree of sophistication is often of no avail. Worldwide, a great deal of academic pattern recognition work has studied the use of extremely sophisticated Hamming distance derivatives operating on sets of "black" and "white" parts of patterns. Such work often has little to offer the practical pattern recognition machine-maker. This is really to be expected, for the view is clearly oversimplified.

In fact, it is very important to know the nature of the relationships between properties. To see that this is so, consider the following.

Suppose that a set of m detectable "features" is being used to indicate the occurrence of r "hidden" properties. That is to say, the set of "binary patterns" producible from features is to be classed into r classes. The calculation of the number of ways in which this can be done is complex. The possible number of different "patterns" is 2^m. If we know nothing about relationships between features, we must assume that the 2^m patterns are distributed into the r classes in an arbitrary way. If $n = 2^m$ there are many more than r^n possible distributions. (It can relatively easily be shown that the number is greater than r^n.) For example, if the "properties" consist of 10 items of different colors, there are 1024 different patterns. These could be used to represent letters of the alphabet in many more than 26^{1024} different ways!

The task of discovering the real distribution from so many possible ones is much too great to be practical. So we tend to assume that very simple relations exist between properties and classes, such as the Hamming one. But as mentioned above, in real life this is often much too simple. Practical pattern recognition that actually works is therefore done by intuition resting on very extensive practical experience.

There are other methods of judging "likeness," different from using Hamming distance on "features" found by low-level preprocessing. For example, one means of judging "likeness" is to find "distortions" of various "power." If one pattern can be converted to another by a "weak" distortion, the two patterns are alike. Some tasks which could reasonably be called scene analysis are possible using "one-level" likeness: here the goal is of the kind involved in the detection of an animal hiding in the undergrowth. But effective classification often requires use of a hierarchy, in which a difference at high level is more significant than one at low level.

16.3.3.1. "Likeness" Judgment by the Use of "Distortion"

Many people find it conceptually helpful to think in terms of "operators," something that "alters" or "distorts" one pattern into another. Basically this introduces nothing new, it is merely an aid to intuitive understanding, but it does allow one to think of likeness measures more complex than Hamming distance. The idea of a distortion operator is most readily applied to "patterns" defined as relations. Thus, if

$$V = P(A)$$

is a pattern, then

$$V = D[P(A)]$$

is a distorted pattern.

This is not quite what we mean by distortion. Nearer would be

$$V = P[D(A)]$$

The difference between $V = D[P(A)]$ and $V = P[D(A)]$ is illustrated by the case of a black ink drawing on a white rubber sheet. If we make the ink wet, and it runs, distortion is as the former. If we stretch the rubber it is as the latter.

In these cases we once again consider whether unknown patterns are like one or other of a set of class paradigms. However, likeness is judged by the existence of a suitably defined operator, which will convert the unknown into one of the paradigms, or alternately, of an operator which will convert a paradigm into the unknown.

16.3.3.2. "One-Level" Likeness

Where the goal is of the kind involved in the detection of an animal hiding in the undergrowth, successful results have been obtained without recourse to multilevel techniques. The techniques used will not be described in any detail, for they are fully covered in Chapters 7 and 8. Here we shall only consider those aspects arising from some of the concepts touched on earlier in this chapter.

The necessary "preprocessing" is often quite complex and the pre-processor is usually, although not always, designed using a great deal of intuition. The "scene" is usually thought of as a monochrome picture. Thus, the basic "pattern" is a relation;

$$I = R(A)$$

where I is a set of intensities and A a set of "positions" or "addresses." Often, the "addresses" make up a two-dimensional array. The "scene" I could be, for example, a monochrome photograph.

Scanning and preprocessing result in a new relation:

$$F = D\{P[E(A)]\}$$

where the "feature" values are usually constrained to be members of the set $(0, 1)$, although sometimes they cover a wide range. The new "address" set is $E(A)$, and forms the feature space. The "mapping" functions E and D are usually much too complex to be estimated statistically, but are obtained by experiments based on hypotheses derived by experience and intuition.

If the address set contains n items, it is a frequent practice to represent a pattern related to it as a point in an n-dimensional space. This space is often called a *feature space*. The n dimensions represent the n "addresses," and the position of the point gives all the "values."

The above statement could readily mislead. When people talk about "dimensions," there is usually an implication that each "dimension" forms an ordered set, and probably a well-ordered set. In particular, they are usually thinking of a set in which it is meaningful to say, for example, that six items of the set lie between item S and item T. In such a "dimension" one can intuitively think of the "difference" between S and T as being something "like" the number of items between S and T in the set, and likeness as being the reciprocal of difference. (See Fig. 16.4.)

Any set can clearly be used to represent any other set having the same number of elements. But, in general, relations between the members of one set cannot be expected to give any clue to relations between the members of the represented set. Since even the members of a one-dimensional set are distinct,

(a)

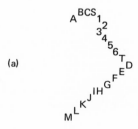

(b)

```
ABCS123
FEDT654
GHIJKLM
```

Fig. 16.4. Demonstrating adjacency in different data structures. S and T are adjacent in (b) but are separated by a distance of 7 units in (a). (a) A string (one-dimensional) of characters. (b) An array (two-dimensional) containing the same string as in (a) but folded back at 34 and FG.

such sets are often used in pattern recognition work to represent sets containing far more complex relations. (See Fig. 16.5.)

Clearly a "pattern" consisting of n pairs, the "value" element in the pairs being chosen from a set of m items, could be represented by n elements chosen from an $n \times m$ one dimensional set, or by one item from an m^n-dimensional set. For example, consider a pattern made from 140 cubes, some red, some yellow, some green, stacked in a $4 \times 5 \times 7$ block. The "pattern" is defined by 140 pairs, one for each cube. One item in a pair is an "address." This is different for each pair. It might give the position in the block, or refer to a distinct marking on each cube. The other item is a "value," one of three red, yellow, or green. A set of 420 unordered pairs can be made by pairing every "value" with every address. The pattern (the $4 \times 5 \times 7$ block) can be represented by a suitable subset of these 420 pairs containing 140 of them. Only a minute proportion of all possible 140 item subsets of the 420 are possible patterns. There are 140^3 possible patterns. The given pattern could therefore be represented as one item chosen from a set of 140^3. This gives another representation. (See Fig. 16.6.)

What is usually done is to represent the pattern as a "point" of an n-dimensioned set, each dimension having m values. The hope is that "position" of this point will prove suggestive of class. Where the pattern is a set of features, then each dimension has only two values 0 or 1, and the representation is then sometimes called a *feature space*. Otherwise it is more often called a *decision space*. This latter term is also used with other meanings, which will not be discussed here.

Since pattern classification can only be based on past experience, a frequent practice is to note the past frequency of occurrence of patterns of a given class in various small regions within the decision space. (See Chapter 4.) One has to take finite regions, for no two patterns could really be expected to occur at exactly the same position in the space.

Under these circumstances, it is expected that it will be possible to label each of the small regions according to the class which had most past occurrences in that region. In practice, one must sometimes expect that several classes will have the same occurrence number. More sophisticated labelings are also possible, for example, indicating the relative proportions of occurrences. The "labeled" decision space can be represented by a space of one extra dimension, having as many items in its set as there are classes. The hope is that this new space can

Part No.	1	2	3	4	5
Part name	2 BA bolt	4 BA bolt	0 BA bolt	5-in. nail	1-in. nail

Fig. 16.5. Parts 1, 2, 3 are bolts, but their sizes are not in part number order. Parts 4 and 5 are quite different from parts 1, 2, and 3.

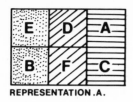

REPRESENTATION.A.

[1,1,RED] [1,2,RED] [2,1,GREEN]
[2,2,GREEN] [3,1,YELLOW]
[3,2,YELLOW]

REPRESENTATION.B.

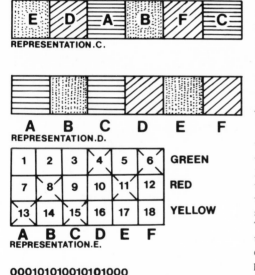

REPRESENTATION.C.

A   B   C   D   E   F
REPRESENTATION.D.

1	2	3	4	5	6	GREEN
7	8	9	10	11	12	RED
13	14	15	16	17	18	YELLOW

A   B   C   D   E   F
REPRESENTATION.E.

000101010010101000
REPRESENTATION.F.

Fig. 16.6. Consider a 2 × 3 frame, in each location of which can be put a red, green, or yellow square. Given this, 729 (3^6) patterns are possible One possible representation is to indicate red by stipple, green by diagonal lines, and yellow by horizontal lines, and the locations by a set of squares arranged in the same way as in the frame. It is assumed that the frames' locations are also marked by the letters A, B, C, D, E, F. Given this, one of the 729 patterns − pattern number 25, say − can be represented as in (a). Representation (a) shows the colors in each square of the frame for pattern 25 and the letters are marked on the appropriate locations. Pattern 25 could be represented by coordinates, giving representation (b). Representation (c) uses one dimension to represent the location. For the first three positions, $y = 2$, and for the last three $y = 1$. Within each of these triples, the value of x is given by position. In representation (d), the letters have been removed. This associates color with label (A to F) and indicates the labels by their (alphabetic) order in the sequence. In representation (e), the color-indicators and number sequence have both been removed. This indicates color by y coordinate, the label of the location by x coordinate, and marks (by x) the six pairs defining the pattern. (f) is a representation of (e) in which position represents the place in that representation, and 1 is a "crossed" place. All are valid representations of the pattern − as is the number 25, which is the number (from the range 1 to 729) *arbitrarily* associated with the pattern. Some of the representations throw away a great deal of relational information.

be used to indicate the class of a new pattern. This will only be possible if past experience is "similar to" the future. Even where this is true the indication will be sound only if occurrences "cluster." (That is, if the same results would have been obtained if occurrence of classes had been determined for subregions of the small regions.) Under these circumstances, the decision space can hopefully be partitioned so that the partition into which any new pattern occurs is a good indicant of its class, while the number of partitions indicating any one class is fairly small. Whether this turns out to be the case depends critically on the "preprocessing" that is used. (See Fig. 16.7.)

It is, of course, important to find rules both for partitioning the decision space, and for calculating the partition in which any new pattern occurs. Good partitioning rules allow accuracy to be obtained given a practical number of samples. This is particularly important when preprocessing is poor, so that the partitions per class are large and their boundaries inconveniently shaped.

16.3.3.3. The Use of Hierarchies of "Features"

In principle, multilevel likeness rules could be implemented by the use of a very complex preprocessor. But the designer of such a preprocessor would have to understand a great deal about such rules, and this is a subject that is not yet well understood. Multilevel likeness schemes can make very good use of feedback between levels and by doing so can greatly reduce the work required. The subject is perhaps best thought of in terms of some actual examples. The examples use pattern recognition techniques to make decisions about real-life situations. Usually these situations are represented by suitable patterns, and in what follows, the patterns are often represented by relations. It is not always understood that their purpose is solely to aid communication and recording. Many people find the representation of patterns by relations very helpful in this respect.

16.4. EXAMPLES OF WORK IN THE AREA OF SCENE ANALYSIS

The two cases considered are "robotics" and faces.

16.4.1. "Robotics"

Scene analysis, as used in the field of "robotics," is in essence a task carried out as part of a practical job, although up to now this task has always been artificial and experimental. The purpose is to manipulate real objects present in real places. In the usual experimental situations the bodies present are usually cubes, "bricks," wedges, etc. and are distributed about in a room, perhaps in such a way as to make arches. (See Chapter 8.)

In the case to be considered in more detail later, the bodies are quite varied. Typical examples are given in Fig. 16.8. It is clear that the set of bodies could, in one sense, be accurately described as a distribution of density of matter in a three-dimensional space. In the terminology used above, it would then be a pattern:

$$V = P(A)$$

where the "value" set is a set of densities, and the "address" set the various positions in the space. This is, of course, a silly way of looking at the situation. Real "meaning" lies at much higher levels. In general, at the highest level of structure in the system one would expect to have recognizable conglomerations of bodies such as arches, then bodies, and then structures such as edges, surfaces, and corners. It is not always quite obvious at which levels different "features" are best assigned. Bearing in mind the principles of duality in 3-space, one might expect surfaces and corners to be at a level below edges. A 3-space is defined by a plane and a point not on that plane; or by a point and a plane not on that point, or by two skew lines. The first two definitions are "dual" in that "point" and "plane" are interchangeable. The lines are "self-dual." Point and plane are interchangeable in the axioms. However, one would obtain the best clues about arches from relations involving bodies: about bodies from surfaces, corners, and edges.

It is important to know what are the relatively independent levels in the system, since these give clues to what is possible. For instance, it is well known that for "reasonable" solids the sum of the numbers of corners and faces is always two more than the number of edges. This is not true for solids such as one consisting of two cubes joined at just one corner, or along just one edge. For many of the solids one might expect to find in "robotics" applications it is true. It is an example of a relation that holds at high level. If the levels are correctly chosen each constrains those above and below. Hence a point might only be of recognition value if it is part of a line, and a line if it is part of a surface.

If a point cannot be part of any line (because of line constraints) then doubt is put on the lower level. By carrying out a multilevel recursion, one can "home" into a good high-level decision. From the point of view of many tasks that might be needed, the initial relation

$$V = P(A)$$

is, although too detailed in many ways, grossly insufficient in others. It ignores such matters as force patterns between parts of color or chemical nature, and a host of other things. All that could be found about the system could be put into the form of relations, but this is clearly not necessarily a good thing to do.

In the robotics problem, one important property has certainly been ignored. In principle, in this case a body could be defined as a part of the whole that could be separated from the rest by the use of "reasonable" force, but which could not itself be separated into parts in this way. In the experiments that are of interest here, the primary measurements are made by taking a "view"

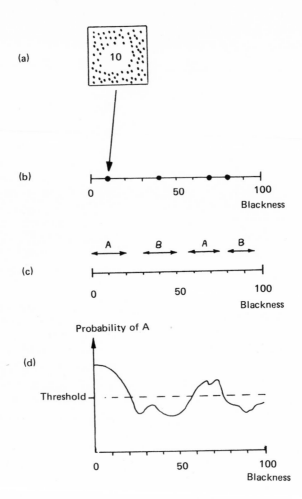

Fig. 16.7. (a) 1-"point" pattern. Value (square, darkness = 10). Patterns are distinguished by their darkness. (b) Representation of a set of four such patterns (darkness: 10, 40, 70, 80). (c) Classes A, B as zones. (Patterns having darkness between 0 and 20, or between 57 and 75 are in class A. Those with darkness between 30 and 50 or 80 and 95 are in class B.) (d) Showing "probability" of A. Here patterns in class A can have any darkness, but some darknesses are more likely than others. (e) 2-"point" patterns. (f) Set of four such patterns. (g) Classes A, B as zones.

of the "room," using, for example, a television camera. This provides a two-dimensional "scene." What is being measured is the light intensity pattern due to reflection from objects in the room. The goal is to use these measures to locate and indentify real objects in the real room. This "photograph" will have different intensities according to the point of reflection. Some parts will correspond to body surfaces, some to shadows. In such a scene, given the kinds of

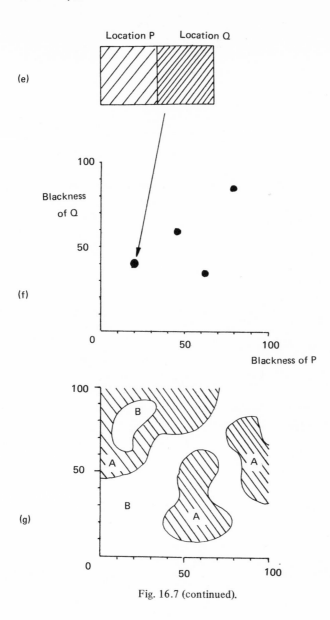

Fig. 16.7 (continued).

bodies being viewed, the boundaries between regions (they are all lines) are fairly readily detected. But not all regions correspond to surfaces of bodies; some, for example, are shadows; and not all lines are boundaries between two surfaces of a body. Some are boundaries between shadows and body surfaces, and others are "cracks," the "edge" produced when one body sits on another.

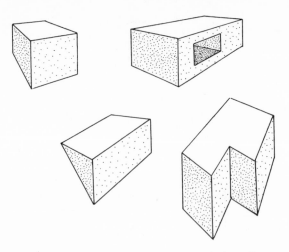

Fig. 16.8. Objects often used in scene analysis studies.

The lines will meet at points. Often these points correspond to corners of bodies, but again not necessarily so. For example, the points might be part of a shadow or surface dirt.

However, there are many high-level relations that can be taken into account. Shadow regions, in general, do relate to body surfaces, "shadow" lines to edges of bodies, and shadow points to corners. Usually, the first stage is to scan the "photograph," thus providing the prime pattern. In principle, this could perhaps give color and brightness as a continuous function of position in a 2-space. In practice, the output of the scanner is usually in the form of a finite relation. The 2-space is divided into a finite set of "regions" — these are the "address" part of the relation, and a mean brightness or mean color associated with each region. This "value" vector is also usually quantized. (See Fig. 16.9.)

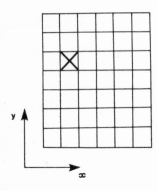

Fig. 16.9. This "scene" contains 6 × 7 (42) locations or addresses. The point X has the "address" (2,5). The "value" at X is a vector; for it can have one of three colors, (blue, green, or yellow) and 1 of 10 brightnesses (1 to 10). The location X is, say, green and has brightness 7. The "point" X is therefore (green, 7; 2,5) and is an element in a 42-point binary relation.

Each element of this primary relation is often called a "point." This is bad practice in many ways because such a "point" has little to do with a point in a geometric sense. One interesting difference between the "photograph" and the real scene is that in 2-space lines and points are dual, and at the same "level," whereas in the 3-space points and planes are dual and at the same level, whereas a line is at a higher level. Lines and points can be found in the "pattern" by suitably processing the primary relation. Since the lines will meet at points, it is possible to derive a new pattern, a new relation:

$$L = A(P)$$

where the "value" L is information about lines, and the "address" set P is a set of points. The relationships between L and P and the "values" and "addresses" from which they are derived are quite complex. To find what they should be statistically is an impossible task.

Often points in the "photograph" correspond to corners of bodies, but again not necessarily so. For example, the points might be part of a shadow. However, there are many high-level relations that can be taken into account. Shadow regions, in general, do relate to body surfaces, shadow lines to edges of bodies, and shadow points to corners. In sorting out evidence of correspondence, it pays to make extensive use of knowledge of high-level relations, for these can sort out errors and ambiguities that always occur at low-level interpretations.

An Example:

A great many organizations throughout the world have carried out valuable work in the robotics area. However, since workers at the Massachusetts Institute of Technology (MIT) have done as much as any, what follows, very roughly, goes along their line of development.

Let us start with a scene that has been scanned so as to give a binary pattern, and in which further processing has discovered points and lines. In general, the first features to be extracted will be lines, and points will be detected at the meeting of lines. In a 2-space, one would expect lines and points to be at the same level. In practice, it is usually easier to derive lines as relations between the elements in the primary relation, and then to find points from these lines. Remember that here a point is defined as the meeting of lines, and lines as a "meeting" of points.

At a still higher level, one has "features" consisting of a set of lines meeting at a point. The features are classified not only by the number of lines, but at a more detailed level by the vertex angle of the sector in which all lines could fit. Each point in the scene can be related to a number of lines (typically two to four). Three lines can be so placed that there is a line-free angle of greater than $180°$ about the point – this gives an arrow – or so that there is an angle of $180°$ giving a T, or as a fork. Four lines can also be arranged in a similar way, giving a "peak," or a "cross," or as other figures, such as a $\curlyvee$ or a K. Examples are given in Fig. 16.10. (Also see Chapter 8.)

Hence, for example, each "corner" can be related to a number of lines, each set of lines being related within the set. Furthermore all "true" points will have at least two lines meeting at them (no loose ends), and any line will meet at least one other line at each end.

Every "genuine" line, in contrast to an artifact, will be the interface between two regions. These might correspond to two surfaces of the same body or they might not. Where they do, Adolfo Guzman of MIT would call them "links." When they do not, they correspond to surfaces of two different bodies or to the surface of a body and one of "the room," to a surface and shadow, or even to two shadows. In some of his work, Guzman argued that a pair of lines at a point implied two hidden surfaces (it could mean a shadow), the "shaft" of an "arrow" was at the edge between two surfaces of a body, and a "fork" implied three edges of a body meeting at a corner. Figure 16.11 illustrates this.

Such implications cannot be relied upon, since shafts of "arrows" do not always link two surfaces of one body, and a "fork" is not always round three edges of a body. Therefore many other rules are introduced. If one looks at the problem from the highest level, (that is, where all the features are bodies) it can be seen that there are severe constraints on the relations that can exist between lines and their causes, given any set of lines meeting at a point. In a similar manner, there are constraints on the ways that "shadow" surfaces and real surfaces can meet.

It must be remembered that the "pattern" that is examined in the situa-

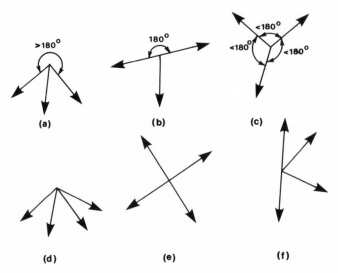

Fig. 16.10. (a) Arrow; (b) "T"; (c) Fork; (d) Peak; (e) Cross; (f) "K".

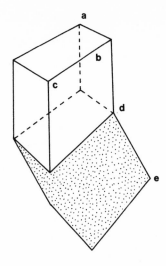

Fig. 16.11. Where simple rules work (a–c) and where they do not (d–e). a, two hidden surfaces; b, edge between two visible surfaces; c, junction point of three visible surfaces; d, meeting of one visible surface and shadow; e, corner of shadow.

tion under discussion is a projection of a scene in 3-space taken from one point of view.

No work of any importance is carried out in isolation, and the workers at MIT have learned a great deal from other organizations. In later work at MIT, David Waltz, making use of work by Max Clowes, David Huffman, and others, has used many of the properties that lines in the projection can have. Among these are the following:

1. A line can be an artifact of the projection.

2. A line can be the boundary between a shadowed part of a surface and an unshadowed part.

3. A line can correspond to the edge between two surfaces, in which case, from a given point of view only one of the surfaces might be visible. If both are, the edge could be convex or concave. The surfaces might be physically separable or not, since these could belong to different bodies.

4. Where one body partially obscures the view of another, a line can separate a region corresponding to an edge-bounded surface of one body from a region corresponding to an unobscured part of a surface of another. (See Fig. 16.12.) Waltz also labeled lines according to the illumination on the regions separated by them. They could, for example, be directly illuminated, or shadowed by facing away from the light, or shadowed by some other body. The labeling could have indicated the color of the separated regions, but originally it did not. In Waltz's scheme, the number of combination of properties a line might have is quite large, but in fact only about 50 are allowed by the constraints. For example, a line cannot correspond both to a convex and to a concave edge. If n lines meet at a point, then roughly 50^n labelings can be associated with this

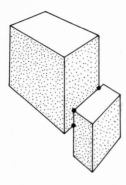

Fig. 16.12. Partial occlusion of one body by another. Difficulties in analysis occur at the points indicated by the heavy black dots.

point. However, a higher level of examination shows that only very few combinations of line types can be associated with any given point. Of the 2500 (50^2) 2-line points only about 80 are physically possible, of the 125,000 (50^3) "arrows" only about 70 are possible, and where four lines meet within a sector of less than π radians only 10 of the 6×10^6 (50^4) are plausible. (Also see Chapter 8.)

The constraints available become very strong, and the interlevel iterative game is very powerful.

16.4.2. Recognizing Faces

The next example is taken from the work of Takeo Kanade of Kyoto University. Although at first sight it could appear to be quite different from the work considered above, a little thought shows it has much in common. The aim is to recognize people, given pictures of their faces. The system uses a number of levels of processing. The face to be recognized is, of course, an object in space. The following processes take place.

1. The scene is projected onto measuring apparatus, and the measured picture is looked upon as being made of a matrix of squares, each square having a different reflectivity. (See Fig. 16.13.)

2. Using the terms already introduced in this chapter, the scene is basically looked on as the relation:

$$V = P(A)$$

where V is degree of reflectivity, and A the position in a matrix. An element of the relation is a reflectivity-matrix pair. Any element can be examined by reference to its "address."

3. Simple distortions are performed on the relation, and thus new ones derived. For example, resolution can be reduced by averaging the reflectivities of suitable submatrices, as seen in Fig. 16.14.

0	1	1	0
0	1	2	0
0	1	2	1
0	0	2	1
0	0	0	0
0	1	0	0

Fig. 16.13. Example of a 4 × 6 matrix. In practice, the matrix is much larger.

a

1/2	3/4
1/4	5/4
1/4	0

b

2/9	5/9	5/9	3/9
3/9	8/9	1	6/9
2/9	8/9	10/9	8/9
1/9	5/9	7/9	6/9
1/9	3/9	4/9	3/9
1/9	1/9	1/9	0

Fig. 16.14. (a) Six squares obtained by averaging over four elements in the matrix of Fig. 16.13. (b) Each square is given a value equal to the average of its value and that of the eight squares surrounding it. ("Squares" outside the matrix in Fig. 16.13 are given a value of 0.)

4. Elements of the derived relations are used to find features, such as the top of the face, and to relate these to position.

5. A top-level relation is obtained, and this is used for recognition.

There is extensive feedback between all levels.

16.4.2.1. Derivation of Primary Relation

Two means of "scanning" have been used, a CRT scanner, and a TV camera. The CRT scanner is very flexible and powerful. When it is used the view is normally first photographed, and this photograph examined. Thus the primary scene is recorded on film. It is treated as a 1024 × 1024 matrix of small square regions. The primary pattern is of the form:

$$I = R(A)$$

where the address set has 2^{20} items, since it is a $2^{10} \times 2^{10}$ array, and is structured in an ordered way, being a subregion of a two-dimensional region. I is the intensity, and has one dimensional order. The effective number of brightness or intensity levels that can be discriminated on the film is very considerable, but the scanner, for any given setting, can only output 32 levels. However, these levels can be made to relate to actual film intensity in four different ways, accordingly to scanner setting. One of these is linear; one enhances the white (that is, output levels are closer together in the black region than the white); one enhances the black; and one the middle regions. The scanner settings are under computer control. The scanner allows any pair in the pattern relation to be recorded for further processing. The pair gives the intensity value at a given place. The TV camera does not use a film intermediary. It treats the scene as consisting of 240 × 300 elements.

16.4.2.2. Lowest Level Processing

Some of the terms used in this section are directly due to Takeo Kanade. Such terms are in quotation marks.

This level is done in a computer. The simplest operation considers a "rectangular" subset of the total region. (See Fig. 16.15.) Since the address set

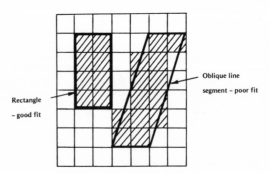

Fig. 16.15. Extraction of rectangles from an image. Notice the "quantization noise," due to the coding of partially covered squares in the line segment.

in the primary relation is looked on as a 1024 X 1024 matrix, any address can be given an (x,y) location. All those elements in the pattern associated with the submatrix with "corners" (x,y), $(x + W,y)$, $(x,y + H)$, and $(x + W, y + H)$ can be extracted. This is called a rectangle. In a similar manner, an oblique "rectangle" of elements H elements long, W elements wide, can be extracted from any part of the array. The obliqueness required can be specified. This extraction is called a "line segment." It is also possible to obtain sequences of elements, either as a raster scan or as a circular scan. Operations can be carried out at a level above this, say, at a sublevel up. Two examples are:

1. To extract the $n \times n$ rectangle centered on each basic address and calculate the average intensity associated with these $n \times n$ elements (not possible at the edges). This gives a new set of values which can now be paired with each of the original addresses to give a new relation with 1024 X 1024 elements. This is shown in Fig. 16.16, for a 3 X 3 rectangle operating on 7 X 7 elements. (Similar operations on pictures are described in Chapters 6, 7, and 12.) Subjectively this process produces a defocused effect.

2. Another process which can be carried out is the *Laplacian*. For this, one first extracts the 3 X 3 matrix centered on each address (x,y). For each address, a value L is calculated, thus giving a new relation. To calculate the value of L for an address (x,y) one finds the values a,b,c,d,z associated with addresses $(x,y - 1), (x,y + 1), (x + 1,y), (x - 1,y), (x,y)$ and calculates L where

$$L = a + b + c + d - 4z$$

The derived relation shows where edges (subjectively acceptable edges) occur in the preprocessed pattern.

Processes can also be performed on patterns at a higher sublevel still. An example of this is the process which provides the "integral projection of a slit." The process first extracts a "line segment," as shown in Fig. 16.17. This line segment is called a "slit" in this context and is regarded as consisting of W

○	○	1	2	1	2	○
○	1	2	2	2	1	1
○	3	3	2	2	1	1
1	2	3	4	1	2	8
1	8	2	1	4	3	1
1	6	2	4	3	1	8
1	2	3	4	5	1	2

Fig. 16.16. Local averaging. The square with coordinates (3,3) will be given the value 22/9, using 3 X 3 matrix centered on it. Every other square is given a new value in a similar way.

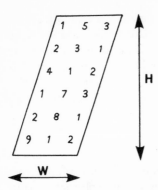

Fig. 16.17. Extracting line segments. In practice, the line segment has ragged edges (see Fig. 16.15). Column 1: (1, 2, 4, 1, 2, 9); column 2: (5, 3, 1, 7, 8, 1); column 3: (3, 1, 2, 3, 1, 2). W = 3; H = 6.

columns each of H elements. Each column of the slit is extracted as a "line segment," and each is associated with an address c. The set c of addresses is considered to be ordered in the same way as the column (See Fig. 16.18). For each column, the values of the elements in it are added together and the result associated with the appropriate address. This gives a new relation having H elements, with a one-dimensional set of addresses. Usually "integral projection of a slit" is carried out on a pattern having binary values. This is illustrated in Fig. 16.19.

16.4.2.3. After "Preprocessing" Step 1

The preliminary to the next level of processing is to extract a suitably chosen part of the whole scene, so as to center the face (approximately). This usually is based on a 422 × 626 array and is processed to give a 140 × 208 element relation to be described. These sizes are convenient ones to use in the system. Every ninth address of the original is taken, in such a way that, if the

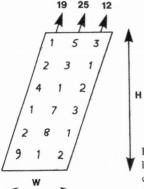

Fig. 16.18. Linear projection of a slit along its columns. The linear projection is (19, 25, 12) and is obtained as follows: column 1: 1 + 2 + 4 + 1 + 2 + 9 = 19; column 2: 5 + 3 + 1 + 7 + 8 + 1 = 25; column 3: 3 + 1 + 2 + 3 + 1 + 2 = 12.

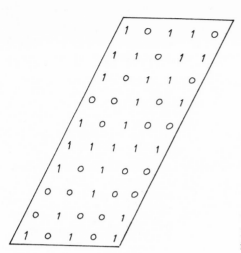

Fig. 16.19. Linear projection of a slit having only binary values. The projection is (7, 3, 8, 4, 5).

starting address is (0,0), then the selected addresses are $(2 + 3n, 2 + 3m)$; where n is $0, 1, 2, \ldots$, and m is $0, 1, 2, \ldots$. For each new address, the 5×5 matrix centered on it is taken, and used in the "defocus" operation. This is shown in Fig. 16.20. When the TV input is used, it is processed in a similar way, and a 140×208 element relation is again produced. Next the procedure converts the picture into an "outline" by using a suitable differentiating operator, normally the Laplacian. Any picture can be converted from multitone to binary by saying that any "gray" lighter than some previously selected threshold level is white, and darker than that level is black. This process is at this stage carried out on the outline picture obtained by the previous operation. In other words, the processing starts with primary measures making a primary pattern in the form of the binary relation

$$V = R(A)$$

In this relation the address set has 422×626 members, and is a set of squares arranged as a two-dimensional array, and the value set is derived from the set of quantities of light being reflected from the squares. This primary relation is processed to give a 140×208 addressed relation of the form:

$$E = D(A)$$

which roughly shows where edge elements occur.

16.4.2.4. Finding Features

Analysis at a next higher level makes use of "slits." The "slit" used is a "rectangle," typically covering a substantial part of the face. By the use of

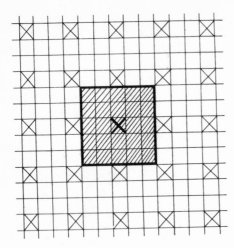

Fig. 16.20. Centers selected for defocussing. X's mark the selected addresses. The 5 × 5 matrix (crosshatched) is the defocusing "window" for one of the selected addresses.

"integral projection of slits," and operating on the higher-level relations obtained, the "outline" picture can be processed to produce yet "higher-level" information, such as the positions of the top of the head and the middle of each cheek line. This provides reference points which indicate where other high-level features, such as the nose, mouth, and eyes are expected to be located. A search for these gives further reference points, which can be used to look for eyes or chin. A typical "integral projection" is shown in Fig. 16.21. The first high-level move is to find the top of the head. This is recognized by finding the highest position on the picture of a horizontal "slit" so that its "integral projection" has a nonzero value for some address. Then the sides of the face are found. This is done by trying horizontal "slits" at about the right place below the top of the head. The cheeks are recognized by the production of a suitable pattern in the "integral projection." Vertical "slits," suitably placed, locate the nose, mouth, and chin. There is now available sufficient information to predict the probable line of the chin. This is checked using a suitable set of angled "slits." After this, detailed aspects of nose, cheeks, and eyes can be located. Various techniques can be used for this. For example, a one-dimensional pattern can be produced by scanning with "slits" of various dimensions and directions, and these can be used to detect directions of lines in the outline picture. Short line elements can be assembled to make longer lines. As a result of all this, a relation is established between position, and certain high-level features, such as eyes, cheeks, chin, nose, and mouth. This is used to distinguish one face from another.

16.4.2.5. Recognition Using Relations

The most important aspect of the whole procedure is the way in which high-level information is used to add certainty to the low-level decisions, by the use of features from different levels. At first, the lower levels are examined only to a sufficient degree to make plausible postulates about higher-level features. The constraints in relations involving these are then used to decide where to direct attention to the lower levels. The scheme moves up and down among the various levels, to make maximum use of knowledge about the constraints existing at each level. In fact, the analysis can refer right back to the primary film several times during the analysis. Special areas (indicated as being in need of extra study, as the result of higher-level analysis) can be rescanned using different scanning parameters.

In practice, this technique is greatly superior to the simple-minded approach. In the latter, it is first assumed that no relations exist between elements of the address set in the primary pattern. In other words the "address space" has as many dimensions as there are address elements. The various "face" classes are then discovered by finding the probabilities of various values being associated with the various addresses for each class. The superiority of the multi-level approach arises for reasons perhaps more readily seen with means of error correction of transmitted digital information. If nothing is known about the cause of error, the best results are obtained by assuming them to be random. But if the pattern of errors is known, any system making use of this is generally better than one based on the assumption of randomness.

16.5. SOME CONCLUSIONS

Patterns Are Structured. Those "patterns" which have meaning for us have structure. They are also "functional" relations. That is to say, they consist of a set of pairs, one item in the pair being chosen for a set of "values" or "attributes" and the other from a set of "addresses" or positions.

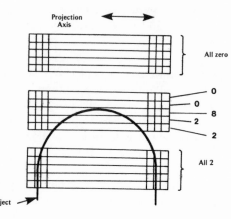

Fig. 16.21. Three integral projections along a moving 40 × 5 slit.

Many of the patterns we deal with we inspect by means of our eyes; and we interpret what we see according to our understanding of the real world. This real world, as we think of it, is very basically connected with a four-dimensional set of entities which we call space and time. We then identify events by associating them with one or other space–time "instance." Thus in this world, these space–time instants act as a locator or address, and the event is an attribute of some sort. This gives us a pair. If we consider the entire world from this point of view, it consists of a set of pairs, in which any given address is unique to one pair. Hence the functional relation arises. The space–time set itself is very highly structured, and so has internal relations that constrain it heavily. For instance, it is difficult to conceive of a being whose sequence of awareness jumps about ours in a random way. From his point of view anything that seemed "normal" from our point of view would seem to him to be random noise. We see a cork bobbing up and down in the water. Its behavior makes a pattern against time. If the height is h and the time is t,

$$h = f(t)$$

If we only inspect the situation once every second, we have a relation

$$H = R(T)$$

If we put the pairs in the order of time, which we can if and only if we assume the time set is well ordered, then we see the height set reflects this order, for the set of heights repeat. Thus

$$h = f(t) = f(t + p)$$

where p is some constant period.

In practice, the real world is very highly structured. We see houses, cars, roads, walls, windows, doors, roofs, bricks, and mortar. All the various levels interact a little but, in principle, any item at any level exists and is recognizable in its own right. If we take a projected view of the world, say we take a photograph of it, we find that levels in the projection are more mixed up, there is more interaction both between items within and between levels. However, the parts of the real world can to some degree be seen to correspond to parts of the photograph — and to some extent these parts are sometimes separable. Any complex scene is virtually unclassifiable unless we take these hierarchic relations into account. In principle, perhaps, since all of the interrelations can be expressed in terms of relations and operators on relations, all can be discovered by the aid of cunning statistics and clever recognition theory. In practice (even given the fastest computing equipment ever likely to exist) such is the immensity of the combinatorial problem that the life of the universe is too short for a good job to be done!

We Make Assumptions. Many of the operations we carry out in pattern recognition are meaningless out of context. This context is based on postulates about structure. For example, neither the "defocus" operation used in the face recognition task, nor the "laplacian," nor yet the "linear projection" would make any sense if the address set in the basic relation were not "two-dimensional." We are very liable to make basic assumptions about underlying relations in familiar kinds of pictures without being aware of this. If we then use our intuition, we are likely to be consistent. If we use "rigorous" mathematical analysis, we can much more readily make nonsense by ignoring our underlying assumptions.

Some Final Remarks. A lot of work has been carried out on self-adaptive or learning systems, and this kind of concept has been applied to scene analysis. A self-adaptive system is closely related to a feedback one, since it improves its performance esssentially by measuring how far it is from perfect, and then making changes to improve its performance. To do this, it must measure performance (which is not necessarily the same as measuring error rate). Many self-learning systems start with a given fixed preprocessor (hence a given decision space) and work simply either by modifying the rules for associating zones of the decision space to classes, or where the regions are implicit in a decision procedure rather than explicit, by altering the procedure. However, if this first stage of analysis is wrong the regions resulting are too complex for any reasonable adaptation procedure to operate. It is often best to use a self-adaptive system of a hierarchical kind. The use of high-level constraints to modify lower-level decisions to remake them is an example of rapid self-adaptation. It is possible to carry out the adaptation process at several levels.

In the terms previously used, every level of analysis is a decision process, working on "features" detected at a lower level. Its purpose is to "recognize" the features which are the raw material of a higher-level recognition system. The set of features found at a given level in a scene are sometimes found not to be appropriate. Higher-level constraint rules can then usually be found which show this to be the case.

Thus, analysis at a higher level can indicate that feature recognitions carried out at a lower level were erroneous. As a response, the overall recognition process could set in motion a "rapid" adaptation, taking place within an overall recognition. But the evidence also can suggest long-term modification to the decision rules, or that the actual features being looked for at the lower level are wrong and hence cause another level of adaptation. Even longer-term evidence can be accumulated to indicate that the rules for determining constraints are also wrong and are in need of modification.

All those scene analytic systems that work well for really difficult scenes use a hierarchy of decisions and use several levels of "feature." Most systems that purport to use just one level of feature are in fact multileveled, since the more complicated preprocessors, in effect, carry out several levels of feature analysis. However, the most powerful systems tend to use multiple levels overtly, rather than to try to push them away as something rather nasty. Where these systems are used in an adaptive system such as P. H. Winston's work at MIT, which is a good example of learning at high level, they tend to make use of multilevel adaptation also. Winston's "learning" system sits on top of the recognition systems developed by Waltz and others as mentioned above. The system aims at identifying structures built from toy blocks, such as those bodies used in the other MIT work. In essence the structure is to be defined by a relation in which the "values" are bodies and in which the "addresses" relate to positions and orientations.

REFERENCES

Cayley, A., 1889–1897, *Collected Mathematical Papers,* Cambridge University Press.
Clowes, M. B., 1968, Transformation grammars and the organisation of pictures seminar, Paper No. 11, CSIRD Division of Computing Research, Canberra.
Clowes, M. B., 1971, On seeing things, *Art. Int.* **2**: 79–116.
Griffith, A., 1970, Computer recognition of prismatic solids, *MAC Tech. Rept.* 73, MIT, Cambridge, Mass.
Guzman, A., 1967, Some aspects of pattern recognition by computer, *MAC Tech. Rept.* 37, MIT, Cambridge, Mass.
Guzman, A., 1968*a*, Decomposition of a visual scene into three dimensional bodies, *Proc. Fall Joint Computer Conf.,* pp. 291–304, IFIPS.
Guzman, A., 1968*b*, Computer recognition of three dimensional objects in visual scenes, *MAC Tech. Rept.* 59, MIT, Cambridge, Mass.
Hibbert, P., and Colin Vossen, H., 1952, *Geometry and the Imagination,* Chelsea Pub. Co., New York.
Hodes, L., 1961, Machine processing of line drawings, MIT Lincoln Laboratory.
Huffman, D. A., 1968, Decision criteria for a class of impossible objects, *Proc. 1st Hawaii Int. Conf. Systems Sciences.*
Huffman, D. A., 1976, A duality concept for the analysis of polyhedral scenes, In: *Machine Intelligence,* Vol. 8 (E. W. Elesch and D. Michie, eds.), pp. 475–492, Halsted Press, New York.
Penrose, L. S., and Penrose, R., 1958, Impossible objects. A special type of illusion, *Brit. J. Psych.* **49**: 31.
Ullman, J. R., 1973, *Pattern Recognition Techniques,* Butterworths, London.
Waltz, D., 1972, Shedding light on shadows, *Vision Flash 29,* Artificial Intelligence Laboratory, MIT, Cambridge, Mass.
Winston, P. H., 1968, Holes, *A I Memo 163,* Artificial Intelligence Laboratory, MIT, Cambridge, Mass.

17

Social Aspects of Pattern Recognition

B. G. Batchelor

Here is a last piece of advice. If you believe in goodness and if you value the approval of God, fix your minds on whatever is true and honourable and just and pure and lovely and praiseworthy . . . You will find that the God of peace will be with you. – Letter of Paul to the Church at Philippi, Ch. 4, v. 8–9. Translation by J. B. Phillips.

17.1. INTRODUCTION

Like all other research, pattern recognition is inexorably guided by the pressures of human society; no research worker can, for long, ignore the social implications of his work. Neither should he expect to abdicate his responsibility in speaking plainly of the dangers of his discoveries getting into the hands of evil people. The first industrial revolution brought many benefits and many miseries to the human race. The second industrial revolution is upon us and pattern recognition is a major part of it. The lessons of the first revolution should enable us to avoid the same mistakes during the second. [J. Rose has developed this theme at length in a recent book (Rose, 1974).] The purpose of the present chapter is to pose some of the questions which must concern all of us if we are to avoid horrors similar to those foreseen in Orwell's *1984.*

B. G. Batchelor • Department of Electronics, University of Southampton, Southampton, England

In Chapter 1, the objectives of pattern recognition research were discussed and we pointed to the possibility of performing the job of a copytypist by machine. Unemployment leads to social disquiet, and we have hitherto implicitly assumed that making copytypists redundant is a worthwhile objective! Is it? Can copytypists be effectively retrained? Do they really want other, "less boring" jobs or do they prefer to accept employment which is emotionally undemanding? There is no clear answer to this type of question, yet the development of optical character recognition makes this particular problem ever more urgent. (Copytypists are not the only potential victims of our work, as we shall see.) Pattern recognition research needs the guidance of social scientists, since it is not self-evident where our attention should be directed. This chapter is a plea for help and a statement of some of the dangers which the editor can foresee. We shall not presume to give answers here, since the questions are as yet ill-defined.

17.2. THE CASE AGAINST PATTERN RECOGNITION

The single feature of PR that makes it potentially so "dangerous" is that it is concerned with processing large volumes of data at high speed.

Political/Social Deviation. A psychiatrist* once suggested to the editor that many suicides might be anticipated and hence prevented by monitoring potentially suicidal patients in a mass-screening program. Suppose that this laudable objective were realized and that to cope with the large numbers of people involved, suitable PR techniques were developed to analyze the data. It is only a small step from this to detecting political deviants, which a totalitarian government might find to be very convenient to employ for its own self-preservation. This is perhaps a little ambitious for present technology, but many years ago it was proposed that PR might be employed for the more acceptable objective of suggesting the names of criminals whose *modus operandi* fitted that of a particular crime. The differences among these three projects are moral rather than technical.

Mass Screening for Genetic Disease. A few weeks before writing this, the editor witnessed a public debate on the morality of the enforced sterilization of children whose intelligence is very low and whose moral sense is severely underdeveloped. The ability to detect genetically "incomplete" individuals would be a powerful weapon in controlling the spread of this type of disease to future generations. However, many people, including members of certain very populous religious groups, find any form of control abhorrent. Whether eugenic policies are adopted at conception or termination of life is of little more than academic

---

*Dr. B. Barraclough, M. R. C. Clinical Psychiatry Unit, Graylingwell Hospital, Chichester, England.

interest to some people; both are repulsive. Suppose karyotyping (see Chapter 12) of fetal chromosomes were practiced on a routine basis. How many pregnant women would want to abort their child solely on the grounds that it was of the "wrong" sex? Even if we were to accept the morality of abortion on these grounds, there would probably be a gross distortion of the secondary sex ratio (boys:girls at birth), resulting in the birth of an excessively large number of boys. While this may not seem particularly offensive, the frustration at adolescence of those boys who will never find a partner may lead to social unrest and/or crime.

Case Management in Medicine. In a recent study (Prior, 1973), it was proposed that PR techniques be used to predict whether or not a patient would ever fully recover normal cerebral function after cardiac arrest. The objective was to guide decisions about case management and, in particular, to suggest which patients could make best/least use of resuscitation equipment. If a person has suffered a temporary cessation of cardiac activity, his brain may have been permanently damaged. In this case, he will be unable to recover however long he is attached to the resuscitator. On the other hand, a patient with an undamaged brain may die if he is disconnected from the resuscitator too early. The important point is that in this work a PR machine made a decision about a life/death problem. Properly used, as an aid to a medical decision by a doctor, the PR machine can help very considerably in making the most effective use of the resuscitation equipment. The problem occurs if there is overreliance on the machine. (There is no intention of suggesting here that Dr. Prior did this; her aim was to improve on a difficult medical decision and to make it more effective.) Such a machine may be employed "blindly" and applied to inappropriate cases. For example, an overenthusiastic use of the machine may result in its being applied to (say) cases of barbiturate poisoning, for which the machine was not designed and would therefore be quite inappropriate. [It must also be made clear that in the case being discussed, the machine's designer, Dr. D. E. Maynard, did consider the problem of *misapplication* of the machine and designed it to check on this (Batchelor, 1974).]

The lesson should be clearly stated: do not rely overmuch on PR machines. Be careful to prove them thoroughly.

Uses in Warfare. There is no doubt that PR is being used in the development of many new types of weapon. It is certainly being used in analyzing radar/sonar signals, for detecting ships, submarines, rockets, aircraft, tanks, etc. Many novel weapons, such as booby-trap bombs which explode when they detect a certain person's voice or a given spoken language, are feasible. Perhaps the use of mass spectrometers is being contemplated to "smell" the dye in enemy uniforms and use this as a trigger to detonate a bomb. Pattern recognition is, of course, ideally suited to electronic warfare, although this is perhaps an excusable activity, since it is essentially concerned with collecting information about enemy actions. One of the prime motivations for the development of syntactic PR has been the "need" to detect missile silos, troop movements, etc. from satellite photographs. Many other forms of sensor are possible, apart from those already cited, including:

1. Magnetic anomaly detectors
2. Acoustic signals
3. Thermal detectors
4. Seismic detectors
5. Television pictures

Our subject, we have repeatedly stated, is closely related to artificial intelligence AI). Army commanders are looking to AI for guidance about analyzing intelligence reports. (Please forgive the unavoidable ambiguity of usage of "intelligence.") War-gaming is a closely related activity to AI and, as we commented earlier, PR is an essential ingredient of the latter.

The most fearful prospect is that of totally mechanized warfare, with robot-like fighting devices which have little regard for their own "survival." Regrettably, PR makes this a little closer.

Crime. How could a criminal employ PR? Perhaps the most obvious area is fraud, based upon computer-file manipulation. Suppose that such a file is searched by a PR program, which attempts to find the "weak points," where a fraudulent modification to the file can go undetected for a long period.

Another possibility is the use of PR to detect "structure" in the movements of security staff, or in bank deliveries. While this can be partially overcome by randomizing the route taken by the watchmen (or money), the existence of PR-based methods for predicting the path does pose a serious new threat. The voice-operated assassin's bomb referred to above is another unpleasant weapon in the armory of the criminal. This weapon might even be refined to the point of sensing the "meaning" of what is being said and using this to select the most appropriate victim. What "better" way of eliminating one's political enemies? (It may be recalled that during the Vietnam war the comment was made by a war correspondent that the trouble with conventional bombing is that it is "not selective.")

Redundancy. We have twice mentioned copytypists in the foregoing pages. If optical character recognition ever becomes really cost-effective for manuscript, then we can also expect to see card-punch operators, type-setters, supermarket check-out staff, large numbers of bank and other accounts clerks being made redundant. Automatic speech recognition might eventually make audiotyping another unneccessary skill. Factory quality-inspectors, many machine operatives, and fruit and vegetable graders and sorters may also suffer from possessing unwanted skills.

It is not axiomatic that all unskilled people necessarily want to acquire a skilled trade. Neither is it self-apparent that skilled people want to be retrained, if they find their employment prospects eroded by technological progress. We may, perhaps, find ourselves in a situation where PR will not be used effectively, simply to avoid a world like that described by Huxley, where games and recreations were only approved if they were demanding of industrial production and human labor. A parallel here with the original Luddites is obvious; we may be forced to restrain the acceptance of PR techniques in industry and elsewhere to avoid the social chaos which might otherwise ensue.

Cost of Research. The bright hopes of the early 1960s have not been realized and PR has not yet proved itself to be generally useful. It is true that PR machines are now in use, in fixed-font (or limited font) optical character recognition machines, which are available commercially. It is also true that magnetic ink character recognition is in widespread use in reading checks. However, neither of these applications has made much use of PR theory; problems have been solved in an *ad hoc* manner, using a pragmatic philosophy. Pragmatism does not usually lend itself to the development of a general theory which, as we commented in Chapter 1, is still missing.

The cynic might ask whether the cost of PR research is being met by results that are of lasting value. To give some indication of the scale of PR research, the circulation list prepared by Dr. E. Backer of the Technical University of Delft, Holland, to publicize the Second International Joint Conference on Pattern Recognition (1974), contained 2500 names. (The editor has the feeling that the true worldwide PR research effort is several times as large as this.) In round figures, PR research is probably costing between \$10M–\$200M per annum, in salaries alone! Is this expenditure worthwhile? Shouldn't we be seeing much more, in terms of usable results, in return for this money? Shouldn't we be concentrating our efforts into "easy" application areas, as J. R. Parks advocates in Chapter 12?

Another point worth raising is that of organization: to be most effective, wouldn't it be sensible to establish a few specialist research centers which are properly equipped? In Britain it is common to find many (most?) of the research being conducted by small groups. For example, the research group of which the editor and D. W. Thomas are part, contains under 12 workers; of these, only two are established members of staff; the remainder are either (temporary) post-doctoral fellows or research-degree candidates. By the standards current in the U.K., this is a fairly large research group. (These remarks apply to PR research only.) PR laboratories are often poorly equipped; few laboratories can boast of special-purpose computers. Much of the progress is achieved through military contracts, which some people find distasteful. It seems to many workers a strange anomaly that medical research in PR "sits on the shoulders" of military projects which provide the bulk of the financial support.

17.3. IN DEFENSE OF PATTERN RECOGNITION

Most technological advances possess the potential for evil as well as good and, in this respect, PR is far from being unique. In the previous section, the editor deliberately played the part of "devil's advocate," so that the negative aspects of PR research can better be understood and avoided, if possible. To present a balanced view, we must now consider the prospects of beneficial developments in PR. It is well to repeat the plea made earlier for research to be conducted into the humanitarian and moral uses of PR. This is vitally important, in view of the power which PR research is trying to grasp; the ability to make complex decisions, at high speed, is the fulcrum of this research. This power, if abused, may unleash horrors on mankind of enormous magnitude, but if properly harnessed the world may reap great benefits. To counter the "attack" on pattern recognition in Section 17.2, we shall concentrate now on its more positive aspects.

In Chapter 1, we foresaw three possible advantages to be obtained through PR:

1. Economic
2. Social
3. Providing new abilities

The magnitudes of these potential benefits of PR are difficult to convey, but we must attempt to do this if this chapter is to achieve the impact which the editor feels is necessary. To do this, we shall cite a few of the more spectacular projects and by (a little) extrapolation attempt to convince the reader that the potential rewards for PR research are very great indeed. The word "potential" keeps appearing because PR is not yet able to boast of the spectacular success of which research workers dream. One reason for this is the relative paucity of research finding, compared to the real needs of PR research projects. Another reason is that it requires both courage and faith to embark on a full-scale implementation of some of the more ambitious proposals. [This is the reason for the editor's (sincere) cries of delight at the less ambitious *but successful* projects reported in Chapters 9–12.]

Medical Decision-Making. The work on prognosticating the outcome of cardiac arrest is a prime example of how PR may assist in making medical decisions. The project was begun because the existing methods, relying on human interpretation of a patient's history, were both inaccurate and often inapplicable, due to certain medical factors. The use of PR has resulted here in an improvement and a prognostic PR machine is in regular use at the London Hospital (Prior, 1973).

PR has perhaps its greatest (medical) contribution to make in mass screening. For example, it is now possible to distinguish between "headache" and "migraine" (Batchelor *et al.*, 1976) using PR techniques applied to data collected by questionnaire. Thus, it would be possible to incorporate a diagnostic classifier in a mass-screening program for migraine should this ever be felt to be medically desirable. Questionnaires can easily be read automatically by machine, since they can be designed to resemble checklists. As an alternative to questionnaires on paper, the data could be collected by a question-and-answer dialogue with a computer. In these days of microprocessors, a portable data-collection and diagnostic machine is perfectly feasible. Yet another alternative is to establish the dialogue using a static computer, a portable terminal, and the public telephone network. The terminal (or other data collector) could be moved to schools, colleges factories, etc. or be static in a clinic or a doctor's waiting room. The advantages of computer-based history-taking are most keenly felt in interrogating patients about their psychosexual problems; the terminal is inscrutable, calm, and unembarrassed, whatever the subject matter. Those areas of medicine which have been the subject of the greatest research by PR workers are thyroid disease, chemical pathology, ECG, EEG, and radiography, but relatively little work has been concentrated on analyzing *symptoms*.

One very important attribute of PR machines is that they can be designed to emulate (some of) the diagnostic skills of a doctor (or a group of doctors) and can be mass—produced to be used all over the world. The skills of many people may be "captured" in a single machine, which can be taken to represent a consensus among them. Thus, PR offers the prospect of amplifying the available medical manpower by emulating certain limited skills which are required on a routine basis.

Aerial Photography. Aerial photography has many potential uses, but we shall mention only two here, namely, crop identification and mineral prospecting. Satellites and rockets are both being used to obtain photographs of the earth's surface. They both produce very large quantities of data, which makes automatic analysis imperative. For example, a satellite scanner may encode 10^7 pixels, represented by 3—7 false-color components, in every 25-s epoch (50 m X 50 m ground-grid in a 160 X 160 km "window"). To cope with such high data rates, we must develop new types of computer systems, such as those described in Chapters 2, 3, 4, and 6. However, it is the magnitude of the potential benefits from using PR methods that we shall attempt to convey in this section.

Why is it important to be able to identify crops from a satellite/rocket? Suppose that it were possible to perform a worldwide survey to determine the amount of wheat growing. This would provide a very valuable facility, since it would enable governments to plan their cereal-planting program with knowledge of market trends. While this is far from being realized at the moment, governments are eager to have some method of monitoring their own nation's crop production so that they can achieve better management of their agricultural program. As just one example of how this can be achieved, the question of transport can be considered. If a nation can anticipate its export surplus well in advance, then it can order shipping at favorable rates. The government can also offer its surplus for sale at a time and price that are to its own advantage.

Minerals and water can often be detected from high-altitude photographs, since the presence of these valuable commodities can be sensed by the changes in vegetation which they produce. Countries with large land areas may possess vast mineral resources, whose existence is not yet even suspected. The use of aerial photography may help to locate these reserves, at least to the point of establishing a "suspicion," which ground-survey teams can confirm or dismiss. If this dream is realized, the large-area nations may reap enormous benefits from exploiting their resources more fully.

Improving Business Efficiency. If we accept that PR machines can be introduced into offices without large-scale redundancies, then there are enormous potential savings to be obtained. It is commonly observed that office automation (to date this can be equated with conventional business computers) has achieved very little reduction in manpower; the "slack" is usually accommodated by expanding the business (Rose, 1974). If this trend were continued with PR machines taking over routine tasks, then we could expect to see many people being employed more "intelligently" in expanding, lively companies. Thus, the optimist can see great benefits arising from optical character recognition, automatic speech recognition, automatic inspection and assembly, etc.; the pessimist sees only the dangers of unemployment. It is important to find how the actual progress may be made to approach the optimist's view, but this is a subject for government planners, company managers, and social scientists (and possibly lawyers).

Stimulating Research. Pattern recognition is having a significant impact on developments in computer architecture and hardware. New systems are being proposed to provide PR with better implementation techniques, as we have seen in Chapters 2, 3, 4, and 6. A recent paper described how a computer incorporating a *content addressable memory* may be used in PR (Navarro, 1976). In fact, PR provides one of the major applications areas for such a computer. Perhaps we shall soon be able to claim that PR has provided convenient ground for developing multimicroprocessor configurations.

The editor has often commented in the foregoing pages that PR demands advanced technology. The areas which are outstanding in this respect are:

1. Graphics terminals (see Chapters 11, 12, and 15)
2. Storage technology (see Chapters 2, 3, 4, and 6)
3. Transducers (see Chapters 2, 6, 7, 10, 11, 12, 13, and 14, especially solid-state cameras)
4. Arithmetic processing (see Chapters 2 and 6)

Pattern recognition is also influencing software and operating system developments:

1. Man—machine interaction (see Chapters 9, 11, 12, and 15)
2. Parallel (i.e., vector arithmetic) languages (see Chapter 4, Section 4.6)
3. Set theoretic languages

In Conclusion. We began in Chapter 1 by explaining why it is difficult to define pattern recognition. Perhaps we can now do so: *Pattern Recognition is the science/engineering art of finding data structure.* When expressed like this, we see that PR has a place amidst the other disciplines concerned with information processing. The boundary surrounding PR is fuzzy and progress in all of these fields is interrelated. It is clear that PR and all of these related subjects will have a great and increasing influence on our lives and those of our children. Let us make sure that the effect is a wholly beneficial one.

REFERENCES

Batchelor, B. G., 1974, *Practical Approach to Pattern Classification*, Plenum Press, London and New York.

Batchelor, B. G., Beck, D., Illis, L. S., Robinson, P. K., and Waters, W. E., 1976, Pattern recognition study of migraine and headache, *Proc. Conf. on the Applications of Electronics in Medicine*, April, 1976, pp. 203—213, Institute of Electronic and Radio Engineers, Southampton.

Navarro, A., 1976, The role of the associative processor in pattern recognition, *Proc. NATO Advanced Studies Institute,* Bandol, France, September, 1975.

Prior, P. F., 1973, *The EEG in Acute Cerebral Anoxia*, Excerpta Medica, Amsterdam.

Rose, J., 1974, *The Cybernetic Revolution*, Elek Science, London.

Index

A-matrix, 387
A posteriori vindication, 410
'A' weighting network, 336
Abnormal class, 4, 79
Accelerometer, 281
'Accept' class, 22
Acoustic, 3, 333, 365
  perception, 8
  phonetics, 382
  'world', 3, 9
Acoustical quanta, 350
Action group, 128
Activity, 346, 409
Adapt, Adaptation, *also see* Learning, 14, 461
Adaptive
  alignment, 257
  sample set construction, 83
  window, 343
ADC (Analogue to digital converter), 336
Addition, 152
Address, 433, 452, 460
  terminals, *see* (ME)
Adjacency matrix, 7
Admissible distortions, 32
Advantages, 468
Aerial photographs, 23, 29, 34, 190
A.I., *see* Artificial intelligence
Aircraft, 275, 341, 347, 465
Air traffic control, 274
Algorithm, 383, 390
Analogue
  descriptors, *see* Descriptors, 6
  hardware distance calculations, 93

Analogue (*cont'd*)
  CC, 95
  NNC, 92
  threshold logic units (ATL), *also see* Linear separator, Linear classifier, 380
  to digital converter, *see* ADC
Analysis
  cepstrum, 339, 350
  Fourier, *also see* Transform, 270, 277
  frequency, 276
  multivariate, *also see* Hyperspace, 319, 345, 407, 408
  scenes, 178, 191, 205, 429
  spectral, 408
  2-dimensional images, 276
Angles, 190
Angular frequency, mean, 345
Anti-aliasing filter, 335
APL (A Programming Language), 104
Archetype, *also see* Reference vectors, Paradigms, Representatives, 1, 69
Architecture, computer, 10, 13, 145
ARCS, *see* Automatic recognition of continuous speech
Area, 271
Array, *also see* Cell, 146
  hexagonal, 153
  photodetectors, *also see* Camera, 26, 261, 265
  processor, *see* Processor
  square, 153
  triangular, 153
Arrow vertex, 218

Artificial
   data, 108
   intelligence (A.I.), 9, 46, 119, 206, 231,
       383, 430
Asymmetric particles, 290
ATL, *see* Analogue threshold logic units
Attributes, *also see* Descriptors, 460
Audiotyping, 466
Autocorrelation, *also see* Correlation, 29,
       30, 276
Automatic
   digitization, *also see* Digitization,
       Transducing, 249
   drilling, printed circuit board, 231
   inspection, 253
   karyotyping, *also see* Chromosomes, 307,
       322, 465
   recognition of continuous speech (ARCS),
       384
   speech recognition, 365, 466
Automation, 178, 231
   of clinical EEG analysis, 420
Average period of oscillation, 351

Background envelope, 314
Bacteria, 31, 290, 296
Ball, 225
Band-pass filters (bank of), *also see* Filter,
       374, 380
Barren wastes, *also see* Voids, 212
Base-crossings, *also see* Zero-crossings, 409
Bayes optimal classifier, 12, 85
Best-first, 385
Biasing a classifier, *also see* Error rate, 413
Bimodal distribution, *also see* Multimodal,
       322
Binarization, 26
Binary
   inputs, *also see* Descriptors, 6, 9, 43
   pattern, 433
Bit-plane arithmetic, 166
Blivit, 221
Blob, 235, 239, 246
Bodywork, noise, 333
Booby-trap bombs, 465
Boolean
   expressions, 129, 135
   processor, *see* Processor
Boundary, *also see* Limits of variation, 73,
       211, 318
Brain models, 9

Breadth first, 130, 136
Breast cancer, 6
Brightness contrast, 257, 271
Bubble-chamber, 190, 273
Built in processors, *see* Processor
Bundling, components, 255
Bursts, energy, *see* Energy

Camera, *also see* Array, Television, 257
Card punching, 466
Cardiac arrest, *also see* EEG, 465, 468
Cars, 207, 226
Category, 16
Cathode ray tube (CRT), *also see* Visual
       display unit, Flying spot scanner, 19,
       256
CC, *see* Compound classifier
Cell
   accumulator, array processor, 152
   division, 304
   storage, logic arrays, 158
Cellular array processor, 152, 163, 197
Central moments, 345
Cepstrum, 339, 356
Character recognition, 20
Chemical 'world', 3
Chess, 393
Choosing, rules, 138
Chromosomes, 303, 465
   abnormal, 322
   acrocentric, 307
   centromere, 304
   chromatids, 304
   karyotyping, 307, 322, 465
   matchstick figure, 307
   metacentric, 307
   mitotic index, 308
   mosaic, 307
City block distance, *see* Distance
Class, *also see* Labelled, Partition, 74, 434
   abnormal, 422
   'accept', 22
   Gaussian distribution, 27
   'normal', 4, 79, 422
   'reject', 22
Classification, Classifier, *also see* Dis-
       criminate, Decision making, 2, 8, 10,
       73, 189, 205, 309, 319, 430, 434, 437
Classifier design methods, *also see* Learn-
       ing, Adaptation, 103
Clinical EEG, automation, 422

CLIP, 152, 159, 160, 166
Closed loops, 152
Cluster, 2, 44, 69, 322, 419, 443
Cluster analysis, *also see* Adaptive sample
    set construction, 9, 87
Co-articulation, 379
Coding, 183
Cognition, *see* Recognition
Coherence, 408
Color contrast, 257, 271
Column
    count, decision tables, 126
    weight, decision table, 128
Compactness, 68, 345
Compiler, 9
Complexity, 346, 409
Compound Classifier (CC), 78
Compressed spectral arrays, 421
Computer, architecture, *see* Architecture,
Computer, 70, 152, 170, 197
    also see Processor, Array, 72
    interactive, 10, 323, 418
    special purpose, 146, 197
    stochastic, 27
Concavities, 318
Condition, 123
Cones, 223
Confidence, 321
Confusing, 420
Connectedness, 52, 57, 311
Connectivity paradox, 153
Content addressable memory, 470
Context, 5
Continuous descriptors, 9
Continuous speech, machine perception,
    see Machine
Contour, 211
Contract, 148
Contrast, 26, 257, 273
Control, 10, 254
Convergence, 11
Conversion of decision tables, 131
Convex hull, 316
Copy-typing, 4, 464
Correct classification, *also see* Class, 72
Correlation, 18, 252, 270, 375, 408
    auto, *see* Autocorrelation
    cross, *see* Crosscorrelation
    linear, 18, 417
    optical, 18
    Van der Lugt, 19

Cost
    function, 11
    of PR research, 467
Counting, 254, 324, 307
Covariance matrices, 27
Cow, 1
Crankshaft period, 339, 343, 360
Crime, 466
Crop identification, 469
Cross correlation, 18, 23, 257, 270, 375
CRT, *see* Cathode ray tube
Cups, 207, 225
Cusps, 190
Cybernetics, 8
Cylinders, 223
Cylindrical harmonic function, 293
Cytogenetics, 303

Dash count, 126
Data
    -in terminal, 44
    management, voice controlled, 387
    -out terminals, 45
    processing, 10
    selection, 104
    structure, 89, 313
    de Morgan's rules, 135
Decision
    making, *also see* Classification, Discrimi-
        nate, 8
    lattice, 122
    reliable, 13
    rule, 119
    space, 442
    subtables, 124, 130
    surface, 73, 410
    table, 123
    theory, 226
    tree, 33, 119, 415
Defocus operation, 457
Dendrograms, 89
Density
    distribution, 296
    profiles, 319
Depth-first, 130, 385
Destruction of information, 5
Description of patterns, *also see* Descrip-
        tors, Vectors, Feature Lists, Graph, 1
Descriptors, 3, 6, 67, 101
Design methods for classifiers, 2, 11, 103,
        461

Design of decision trees, 123
Detection
  features, 32
  vehicles, 333, 340
Detector
  blob, 239, 246
  Maltese cross, 243
Development of decision rules, 137
Deviation, political, social, 464
Diagnosis, 28, 75
Dialogues between man and machine, *see*
  Man
Dichotomizer, *also see* Classification, 46
Diffraction pattern, 291
Diffuse reflection, 273
Digitization, *also see* Transducing, 309
Digital hardware, 95
  distance calculations, 95
  CC, 98
  NNC, 98
Dimensional
  checks, 256
  reduction, *also see* Feature design, 336
Diphthong, 378
Directional processes, 162
Discontinuity, 208
Discriminant
  analysis, stepwise, 410
  function, 26, 50
Discriminate, discrimination, *also see*
  Classification, 2, 49
Disparity, 52, 56
Dispersion indices, 345
Displacement, 256
Distance, *also see* Likeness, Similarity,
  Digital hardware, 28, 68, 419
  city block, 71
  Euclidean, 24, 71
  Hamming, 47
  Manhatten, *see* D. city block,
  Minkowski, 71
  octagonal, 71
  square, 71
Distinctive features (speech recognition), 375
Division, fixed point, 11
Domain, 313
"Don't know" decision, 68
Double-sided power spectrum, *see* Power
  spectrum,
Drilling of printed circuit boards, 231
Duty cycle, 323

Dynamic range (vehicles), 335

Economic advantages of PR, 4, 468
Edge
  detection, 33
  enhancement, 183
  tracing, 240
EEG, *see* Electroencephalography
Eigenvector, value, 419
8-connected arrays, 154
Electrical networks, 7
Electrical 'world', 3
Electroencephalography (EEG), 399
  cardiac arrest, *also see* Cardiac arrest, 406,
    417
  clinical EEG, automation of, 420
  dyslexia, 418
  electrocorticography, 406
  epilepsy, 403
  evoked potentials, 401, 410
  psychopharmacology, 406
  sensory pathway, integrity of, 402
  sleep, 406, 413
  tumor, 404
Electronic warfare, 465
Electronics, 4
Energy
  bursts, 349
  signature, 281
Engine, noise, 333
Engine, firing rate, 339, 360
Engineering drawing, 179, 259, 268
Enhancement, edges, 183
Entropy, of decision tables, 128
Envelope, 268
Environment noise, 333
Equivalence of Boolean expressions, 129
Ergonomics, 419
Erosion, 316
Error rate, 461
Estimation, PDF, 82, 423
Euclidean distance, 24, 71
Evoked response analysis, 402, 410
Existential inspection, 270, 275
EXPAND, 148
Exemplars, *also see* Archetypes, Para-
  digms, Reference vectors, 1
Exhaust system, noise, 333
Expansion, *also see* Defocus operation, 313
Experience, 12
Experimentation, 12

F-ratio test, 410
Fabry–Perot interferometer, 29
Faces, 34
False negatives, positives, *see* Error rate,
    Biasing a classifier
Familiar patterns, *also see* Limits of varia-
    tion, Boundary, 103
Feature, 213, 375, 380, 433
  design, selection, 407
  extraction, 309, 316, 337, 407
  geometric, 191
  global, 191
  lists, 7
  topological, *see* Topological features
  space, *also see* Space, 345, 416, 441
  space-mapping, 418
Feedback
  also see Learning, Adaptation, 10, 44
  tactile, 193
  visual, 193, 234
FFT, (Fast Fourier Transform), *also see*
    Transform, 29, 408
Filter, *also see* Transform
  analogue, 415
  anti-aliasing, 335
  cepstrum, 356
  homomorphic, 353
  images, 179, 183
  local area, 183
  matched, 411
  multiple band-pass, 408
  numerical, 183
  optical, *see* Optical
  rotational image, 292
  spatial, 18, 20
Finding
  cells, 325
  lines, *see* Line
  regions, 224
Finger-prints, 20, 179, 190, 273
Finite-state grammar, machine, 9, 383
Firing rate, 339, 360
Flying spot scanner, *also see* Television,
    Photodetector array, 19, 23, 153
Font, 68
Formant, 370, 377
4-connected arrays, 154
Fourier
  analysis, 296
  transform, *also see* Transform, 32, 335, 349
    inverse, 339

Fourier (*cont'd*)
  transform (*cont'd*)
    space, 296
    spatial, 19, 292, 296
Fraunhöfer diffraction, 19
Frequency
  crossing, 309
  gradient weighted, 328
  optical densities, 315
  'period', 340
  range, vehicles, 335
  rules, 123
Fricatives, 368, 372

Gaussian distribution, *also see* Class, Clus-
    ter, 410
Generalize, 46, 71
Glide, 371, 375
Global information, 29
Global properties, visual patterns, 52
Glottis, 367
Gradient, mean, as a function of level, 315
Gradient weighted frequency, 328
Grading, *also see* Automatic inspection,
    254
Grammar, 383
Graph, *also see* Tree, Lattice, 7
Graphics, *also see* Visual display unit, 10,
    178, 195, 470
Grass, 207, 226
Gray-level, scale, 146, 158, 166, 257, 273
Growth of decision rules, 137
Growing rules, CC, 107

Haar transform, *see* Transform,
Hadamard transform, *see* Transform
Hammer, 225
Hamming distance, *see* Distance
Hand–eye machines, 234
Hand-print, *also see* OCR, 28, 190, 253
Handwriting, *also see* Signature, Automatic
    inspection, 122
Hardwired arithmetic, 145, 163
Heterarchy, Heterarchical, 211, 391
Heuristic, 12, 211, 383
Hexagonal array, *see* Array
Hierarchy, 209, 224, 384, 433
Hill-climbing, 322
Histogram, 163
Hologram, 19, 27
Homomorphic filtering, 353

Human
  faces, 207, 452
  perception, 8
Hybrid system, 166
Hypercube, 127
Hyperplane, 76, 417
Hyperspace, 68, 419
Hypersphere, 78
Hypersurface, 119

IC, see Integrated circuit
Icosahedral symmetry, 295
Identification, vehicles, 333, 346
Image
  analysis, also see Analysis, 145, 189, 264,
    289
  digitization, 146
  deblurring, 195
  dissector, 309
  enhancement, 152, 178
  filtering, see Filter
  improvement, 181
  interpretation, 183
  preprocessing, 179
  processing, 145, 162, 177
  recognition, 188, 191
  reconstruction, 296
  sharpening, 183
  simulation, 295, 296
  trimming, manual, 328
Implementation
  classifiers, 26, 92
  cost, 11
Improving performance, 4, 470
Incomplete vectors, 102
Incremental simulation, 390
Indices of dispersion, see Dispersion
Industrial
  automation, 178
  revolution, 463
  sensory components, 255
  uses, 231, 253
Inequality, 6
Information, 5, 10, 419
Inspection, also see Automatic inspection
  axially symmetric objects, 259, 265
  calibration, 283
  existential, 270
  functional, 283
  integrity, 283
  magnetic ink character recognition, see
    Magnetic

Inspection (cont'd)
  metrological, see Metrological
  plane surfaces, 273, 275, 283
  roundness, 259
  shape conformity, 283
  silhouettes, 257
  size, 259
  spindular objects, 259
  straightness, 259, 268
  surface finishes, 273, 283
  texture, 276
  turned components, 259
  violent processes, 279, 283
Instrumentation, 13, 254
Integer descriptors, 6
Integrated circuits, 22, 153, 192, 283
Intelligence
  artificial, see Artificial intelligence
  human, 418
Intensity map, 207
Interaction, 10, 34, 195, 323
Interior, 211
Interparse, 389
Interrupted rule mask technique, 124
Intervals, image scan, 311
Intonation, 381
Intuition, 5, 12
Invariance
  shear, 30
  shift, position, 23, 30
  size, see Size
Invariants, world model, 240
Inverse Fourier transform, 339
Inversion, 262
Irrelevent information, 5
Isolated word recognizers, 374
Isomorphism, graph, 8
Isotropic processes, 160
Iterative
  design of discriminator, also see Learn-
    ing, Feedback, 415
  normalization, 321
  processes, 160
Iverson Language, see APL, 104

Junctions, 190

k-Nearest neighbor PDF estimator, 82
Karhunen–Loève transformation, see
    Transform
Karnaugh map, 127
Karyotyping, see Chromosome

Knowledge, 225
  sources, speech, 337, 382
Kurtosis, 345

Labelled
  nodes, 9
  sets, 102
Labelling
  chromosomes, 316
  lines, *see* Line
Language
  CLIP, *see* CLIP
  for P.R., *also see* APL, 11
Laplacian, 455
Large-store classifier, 80
Larynx, 367, 372
Laser, 21
Lathe, 259, 265
Lattice, *also see* Graph, Tree, 119, 122,
  390
LC, *see* Linear classifier, 75
  implementation, 92
Learn(ing), *also see* Adapt, 2, 11, 103, 461
  automata, 53
  CC, 106
  decision rules, 137
  LC, 106
  NNC, 111
Leaves, 120
Length of signal, *also see* Segmentation,
  335
Letters, printed, *also see* OCR, 68
Lexicon, 385
Likeness, *also see* Similar, Distance, 435
Limits of variation (of a class), *also see*
  Boundary, 420
Line
  ends, 190
  finding, 208, 212
  labelling, 218
Linear
  classifier, 75
    implementation, 92
  correlation, *see* Correlation
  decision surface, *see* Decision
  discontinuity detector, 208
  discriminant analysis, *also see* Linear
    classifier, 26
  separator, *also see* Linear classifier, 44,
    49
Linguistic cues, 382
Linguistics, 34

LINK instruction, 152
Lists of features, *see* Feature
LOAD instruction, 159, 173
Local
  area filter, 183
  operators, 190
Locates, *also see* Archetypes, Paradigms,
  Reference vectors, 77
Location
  objects, 256, 257, 262
  peaks, 341
Logic design, 9
Logically equivalent subtables, 134
"Look ahead," decision table conversion, 127
Loops, 190

Machine
  intelligence, *also see* Artificial, 276, 383
  perception of continuous speech, 381
  print, 190
Magnetic
  anomaly, 466
  ink character recognition, 253, 467
Making decisions, *see* Decision
Maltese cross detection, *see* Detector
Man
  -aids, 273
  -machine dialogue, 389
  replacement, 273
Management information systems, 10
Manhatten distance, *see* Distance, city
    block
Manual methods of spectral analysis, *see*
    Analysis
Map-making, 187
Mapping, 418
Masking, 190, 257, 264
Mass
  -screening, 4, 464, 468
  -spectrometer, spectra, 28, 465
Matched filter, *see* Filter
Matching topological features, *see* Topo-
    logical features
Matchstick figure, *see* Chromosome
Mathematical statistics, *see* Statistics
Matrix
  maximum likelihood, 321
  variance−covariance, 419
Maximindist, 90, 113
Maximum-likelihood matrix, *see* Matrix
MCPS, *see* Machine perception of con-
    tinuous speech

ME, *see* Memory element, 44
Mean, 345
Measurement space, 67, 188, 441
Medical decision-making, 27, 468
Meiotic cell division, *see* Cell
Memory element, 44
Meta-P.R., 3
Metrological inspection, 270, 275, 283
Meyer-Eppler technique, cross correlation, 18
MI, *see* Machine intelligence
Microprocessors, 119, 148, 469
Microscope, 170, 190, 289, 304, 307, 324
Minimum
    distance rule, *also see* Nearest neighbor
      classifier, 24
    -error recognition, *see* Recognition
    processing time for decision trees, 125
    storage for decision trees, 125
Minkowski r-distance, *see* Distance
Misapplication of P.R., 465
Missing measurements, 6, 101
Mobility, 346, 409
Moments, 30, 345
Monitor, 254, 282, 391
MOS/LSI, *also see* Electronics, 170
Motivation for P.R., 2, 468
Multi-dimensional vector, space, *also see*
    Hyperspace, 9, 13
Multimodal, 345
Multiplication, 152
Multivariate analysis, *see* Analysis

n-dimensional space, *also see* Hyperspace,
    24
n-tuples, 43
Navigation, 183
Nasal, 368
Nearest neighbor classifier, 24, 77
Neighboring cell accumulator, 152
Neighbors, cellular array, *see* Array
Networks, 7
    simulated neural, 152
New abilities, 468
NNC, *see* Nearest neighbor classifier
NO
    count, 128
    subtable, *see* Subtables
Nodes, 120
Noise
    acoustic, 333

Noise (*cont'd*)
    road, *see* Road
    spatial, 183
Nonholographic optical techniques, 18
Nonnumeric descriptors, 6
Nonpictorial data, image to, 183
Nonsinusoidal transforms, *see* Transform
Nonstationary signals, 335, 338
'Normal' class, *see* Class
Normal distribution, *also see* Gaussian
    distribution, 29, 410
Normalization, 5, 379
    by recognition, 31
    chromosomes, 320
    orientation, 28
    position, 29, 262
    size, *see* Size
    speech, 379
Normalized
    moments, 345
    cross correlation, 23
Numeral recognition, 121
Numerical taxonomy, *see* Taxonomy

Object location, *see* Location
OCR (Optical character recognition), 5,
    152, 178, 188, 253, 467
Octagonal distance, *see* Distance
Omissions, vectors, 6
One-class classification, 422
Operator, blobs, 239
Optical, 3, 5, 195
    character recognition, *see* OCR
    correlation, *see* Correlation
    density, 146, 309
    diffraction, *see* Diffraction pattern
    fibers, 22
    filtering, 291, 292
    inputs, 173
    processor, 148
Optimal solutions, 12
Opto-electronics, *see* Transduction
Oral, 368
OR-ed feedback, 58
Orientation, 2, 28, 268
Orthogonal function expansion, PDF, 83
Orthographic, 382
Overlap between classes, 81

Paradigms, *also see* Archetype, Reference
    vectors, Locates, Template, 1, 435

Paradox, connectivity, *see* Connectivity paradox
Parallel
  languages, 471
  measurements, 6
  processing, 52, 145
Parsing, 386, 393
Partial tree, *see* Tree
Partition, 9, 73
Parzen PDF estimate, 83
Pattern, 5, 24, 432, 460
  binary, 433
  classification, 205
  description, 1, 5, 102
  diffraction, 291
  recognition, 429
PAX, 152
PCB (Printed circuit board), 170, 234, 273, 283
PDF (Probability density function), 82
  estimator, *also see* Parzen, 423
Peakiness, 345
Peaks, location, *see* Location
Perceptrons, 44, 47, 51
Perimeter, 271
Period analysis, 409
Periodic structures, 290
Periodicity, 276
Perspective, 422
Phonation, 367
Phoneme, 360, 366, 375, 379, 382, 384, 390
Phonetic, 365, 382
  typewriter, 365
Phonological, 383, 390
Photochromic plate, 27
Photodetector arrays, *see* Array
Photographs, aerial, satellite, 23, 29, 34, 179, 190, 273, 466, 469
Pictorial representation, hyperspace, 87
Picture processing, 146
Pixels, 183, 197, 236, 262
Piece-wise linear decision surface, classifier, 77, 93
Plosive consonants, 369
Pollack's decision table algorithms, 125
Polish string representation of tree, 131
Polyhedra, 32, 191, 207
Polynomial function, 28
Position
  invariance, *see* Invariance
Position normalization, *see* Normalization

Post darkening, 337
Power
  cepstrum analysis, 339
Power spectrum, 293
  banded, 408, 416
  double-sided, 338
  homomorphed, 359
  short-term, 338, 345
  single-sided, 338
PR, *see* Pattern recognition
Practical techniques, 12
Pragmatics, 383, 388, 393, 467
Predicates, 52
Pre-processing, 5, 179, 189, 334, 376, 454
Pre-whitening, 336
Printed circuit boards, *see* PCB
Probability
  conditional, 12
  density function, *see* PDF
  joint, 12
  theory, 11
PROCESS instruction, 159
Processor
  array, 95, 146, 150, 163
  Boolean, 155
  built-in, *see* Computer
  parallel, 197
  special purpose, 195
Programs
  APL, *see* APL
  CLIP, *see* CLIP
Projected graticule, 241
Projection, hyperspace, 419
Propagation signal, 161
Properties, 433
Prosodics, 383, 390
Pseudocolor, 183
Pseudo gray-level, 186
Psychological problems, tree design, 121
Psychology, 8
Psycho-visual coding, 183

Q-factor, 408
Quality inspection, 275, 466
Quantizing effects, 260, 272
Quasiperiodic forcing function, 345
Quefrency, 339
Questionnaire, 6, 102
Queue of subtables, 136

Radar, 187, 254, 465

Radio, 5
Radiography, 469
Rahmonics, 340
Rain, noise, 341, 347
Random-access memory (RAM), 43, 170
Range map, 207
Rare, 470
Read-only memory, 43, 72, 95, 284
Reading, 13
Recognition, 431
  class, 17
  minimum-error, 27
  speech, *see* Speech
  vehicles, 333, 350
Reconnaissance, 183, 191
Recurrence relations, 11
Redundancy of decision trees, 122
Redundant variables, 5
Reference vectors, 419
Refinement, decision rules, 137
Regions, finding, *see* Finding
Regular periodicities, 291
Regularity, 264
Reject
  class, *see* Class
  patterns, category, 125
Relation, structured, 430
Reliable decisions, *see* Decision
Representation of patterns, 5
Representatives, *also see* Archetypes,
    Paradigms, 1
Resistor mask, 26
Resonating optic fibers, *see* Optical
Retrieval, fingerprints, 273
Road, 207, 226
  networks, *see* Networks
  noise, 341
Robotics, 255, 429, 444
Robust pattern description, 102
Robustness, 121
Rockets, 466
ROM, *see* Read-only memory
Root, 120
Rotation, 262
Rotational
  image filtering, *see* Filter
  power spectrum, 293
  symmetry, *see* Symmetry
Row weight, decision tables, 128
Rule mask technique, 124
Run-length coding, 185

Satellite photographs, *see* Photographs
Scanner, flying spot, *see* Flying spot scanner
Scanning, 2-phase, 309
Scattergrams, 419
Scene analysis, *see* Analysis
Seed rules, 137
SEF, *see* Single equivalent formant
Segment lattice, 390
Segmentation, 31, 189, 309, 316, 345,
    378, 382
Seismic detectors, 466
Self-setting inspection machines, 269, 284
Semantics, 382
Sensory modalities, *also see* 'World', 256
Separating function, 26, 50
Sequency, 351
Sequential P.R., 8, 33
Set theoretic languages, 471
Sex, 42, 73
Shadow, 222
  mask, 192
Shannon's channel capacity, 337
Shape factor, 271
Shear, invariance, *see* Invariance
Shift
  invariance, *see* Invariance
  operations, 148, 162
Ships, 465
Short-term power spectrum, *see* Power
    spectrum
SHRINK, 152, 161
Shwayder's algorithm for decision tables, 128
Signals, segmentation, 343
Signature
  energy, *see* Energy
  vehicles, 338
Silhouette, 256, 277
Similar, Similarity, *also see* Likeness, Dis-
    tance, 47, 68, 390
Simplification, Boolean expression, 129
Simply-connected objects, 149
Simulated neural nets, *see* Networks
Simulation of images, *see* Image
Single equivalent formant, 377
Single-sided power spectrum, *see* Power
    spectrum
6-connected arrays, 154
Size, 5, 265
  invariance, 30
  normalization, 28, 320
  parameters, 78

Skeletonizing, 152
Skewness, 345
Sky, 207, 226
Slope descriptor analysis, 422
Small-store classifier, 80
Social
  advantages, 468
  aspects of PR, 463, 468
Software, *also see* Language, APL, 13
Sonagram, 370
Sonar, 254, 465
Sorting, *also see* Inspection, 254
Source knowledge, *see* Knowledge
Space
  also see Hyperspace, 68, 188, 441
  Fourier transform, *see* Fourier transform
  research, *also see* Photographs, 179, 191
Spanning
  chord, 318
  tree, 419
Spatial
  distributions, 419
  filtering, *see* Filter
  Fourier transform, *see* Fourier transform
  frequency, 29
  noise, *see* Noise
  patterns, 146
Special-purpose computer, 146
Specificity, 52
Spectacles, 207, 225
Spectral
  analysis, *see* Analysis
  arrays, 421
  peaks, bandwidth, 408
Spectrogram, 190, 370
Spectrograph, *see* Sonagram
Speech, 3, 10, 21, 347, 365
  production, 366
  recognition, 21, 365, 375
  spectrograms, 190
  spectrum, 375, 378
  synthesis, 374
  -understanding systems, 365, 381, 387
Spheres, 223
Splitting, 316
Spoken language detector, 465
Square
  array, *see* Array
  distance, *see* Distance
Stack of decision subtables, *see* Decision
State vector, 240

Statistical
  coding, *see* Coding
  PR, 8
Statistics, 8, 11
Stepwise discriminant analysis, *see* Discriminant
Stochastic computer, 27
Stop consonant, 369
Store, *also see* Small-store classifier, Large-store classifier, ME, Random-access memory, Read-only memory, 80, 470
Straight lines, 190
Strain, 280
Strays, 86
Stress, 381
Stroke analysis, 190
Strong response, decisions, 44, 48
Structure
  data, *see* Data
  pattern, 430, 433, 460
Submaps, 127
Subpatterns, 433
Subtables, 124
Subtrees, 122
Sum of products, Boolean expressions, 135
Supermarket check-out, 466
Surveillance, 183
Symmetry, 290
  index, 319
Symptoms, 189, 469
Syntactic PR, 8, 466
Syntax, 383

Table, 119, 123
  splitting, 124
Tactile, *see* 'World',
Tanks, 465
Taxonomy, *also see* Cluster, 9
Teach terminal, *see* ME
Teacher, 72
Television (TV), *also see* Flying spot scanner, Array, Photodetectors, Transduction, 191, 207, 236, 257, 270, 324, 466
Temperature, 256
Template, *also see* Archetype, Paradigms, Reference vectors, Locates, 18
  matching, *also see* Likeness, Similarity, Distance, 28

Tesselation, cellular arrays, 154
Test, condition, *see* Condition
Test set, *also see* Training set, 105
Texture, 29, 276, 283
Thematic, 388
Thermal detectors, 466
Thinning, 152, 160, 190
Three-dimensional
  density distribution, *see* Density
  image reconstruction, *see* Image
  structures, 295
Threshold, 380, 385
Thyroid disease, 469
Time
  interval, 376
  normalization, speech, *see* Normalization
Tolerance
  also see Limits of variation, Boundary,
    268
  position, 23
Tomography, 188
Top-down, 385, 389
Topological features, 190, 213
Tori, 225
Training, 1
  set, *also see* Test set, 27, 43, 105, 321
Transducing, 4, 470
Transduction, opto-electronic, *also see*
    Television, Flying spot scanner,
    Array, Photodetector, 25
Transform, *also see* Filter, 292
  amplitude, 181
  Fourier, 9, 29, 32, 148, 183, 195, 292,
    335, 349, 369
  Haar, 9, 337, 349
  Hadamard, 183, 197
  invariant, 270
  Karhunen-Loève, 29
  Laplace, 148
  multi-dimensional, *also see* 3-dimensional,
    292
  non-sinusoidal, 349
  spatial, 181
  spectral, 183
  2-dimensional, 292, 296
  2-dimensional Fourier, 148
  Walsh, 29, 349, 369
Transients, EEG, 410
Translation, 262
Translational symmetry, *see* Symmetry
Transparencies, 148

Tree, *also see* Decision tree, Graph, Lat-
    tice, 119, 208, 226
  minimal-spanning, 419
  partial, 122
  reduction, 131
  traversal program, 125
Trial and error, 34
Triangular array, *see* Array
TTL/MSI, *also see* Electronics, Implemen-
    tation, 170
Tube, 225
TV, *see* Television
2-dimensional
  arrays, 152
  Hadamard transform, *see* Transform
  image analysis, *see* Analysis
  image filtering, *also see* Filter, 291
  transform, *see* Transform
  2-phase scanning, *see* Scanning, 2-phase
Typical, atypical patterns, 1, 81, 86

Undecidable patterns, 125
Uniqueness, 52
Unknown values, descriptors, 6
Useful information, 5

Values, 433, 452, 460
Van der Lugt's method, *see* Correlation
Variance, 345
Variance−covariance matrix, *see* Matrix
VDU, *see* Visual display unit
Vectors, 6, 12, 67
  reference, *see* Reference vectors
Vehicle sounds, 333, 338
Velum, 368
Vertex, 213, 218
Viddisector, 207
Video, 236
  phones, 179
Visual, 192, 277
  control, 232
  display unit (VDU), *also see* Graphics, In-
    teraction, 323, 336
  feedback, *see* Feedback
  images, 3
  perception, human, *see* Human
  scenes, *also see* Analysis, Photographs, 5
  sensors, 231
  'world', *see* World
Vocabulary, 382
Vocal cords, 367

Voice
  chess, 393
  controlled data management, 387
  -prints, *also see* Sonagram, 369
Voiced
  fricatives, 369
  stops, 369
Voiceless
  fricatives, 369
  stops, 369
Voids, *also see* Barren wastes, 103
von Neumann architecture, 145
Vowel, 368, 375, 387

Walsh transform, *see* Transform
Wavelength, 256
Weather prediction, 187

Wedge, 225
Weight, 49, 75
Weighting networks, 336
Window, adaptive, *see* Adaptive
Word lattice, 390
Word verification, 390
'World', 3
'World' model, 233, 238
Write clock terminal, ME, 45
Wrong turnings, 122, 131
XAP, 152
X-ray, *also see* Tomography, 29, 183, 289, 295
YES count, 128
YES subtable, 124

Zero crossings, *also see* Base crossings, 351,
  374